VOLUME SEVENTY FOUR

ADVANCES IN BOTANICAL RESEARCH

Land Plants - Trees

ADVANCES IN BOTANICAL RESEARCH

Series Editors

VOLUME SEVENTY FOUR

ADVANCES IN BOTANICAL RESEARCH

Land Plants - Trees

Volume Editors

CHRISTOPHE PLOMION

INRA, UMR BIOGECO, Cestas, France

ANNE-FRANÇOISE ADAM-BLONDON

INRA, URGI, Versailles, France

ELSEVIER

AMSTERDAM • BOSTON • HEIDELBERG • LONDON
NEW YORK • OXFORD • PARIS • SAN DIEGO
SAN FRANCISCO • SINGAPORE • SYDNEY • TOKYO

Academic Press is an imprint of Elsevier

Academic Press is an imprint of Elsevier
125 London Wall, London EC2Y 5AS, UK
The Boulevard, Langford Lane, Kidlington, Oxford OX5 1GB, UK
225 Wyman Street, Waltham, MA 02451, USA
525 B Street, Suite 1800, San Diego, CA 92101-4495, USA

First edition 2015

ISBN: 978-0-12-398548-4
ISSN: 0065-2296

For information on all Academic Press publications
visit our website at http://store.elsevier.com

CONTENTS

CONTRIBUTORS

Albert G. Abbott
Department of Forestry, University of Kentucky, Lexington, KY, USA

Filippos A. Aravanopoulos
Laboratory of Forest Genetics and Tree Breeding, Faculty of Forestry and Natural Environment, Aristotle University of Thessaloniki, Thessaloniki, Greece; Centre for Research and Technology Hellas (CERTH), Institute of Applied Biosciences, Thessaloniki, Greece

Abdelali Barakat
Department of Biology, University of South Dakota, Vermillion, SD, USA

Katharina Bräutigam
Department of Cell and Systems Biology, University of Toronto, Toronto, ON, Canada; Department of Biological Sciences, University of Toronto Scarborough, Toronto, ON, Canada

David Chagné
Palmerston North Research Centre, The New Zealand Institute for Plant & Food Research Limited (Plant & Food Research), Palmerston North, New Zealand

Evelyne Costes
INRA, UMR AGAP, CIRAD-INRA-SupAgro, AFEF Team (Architecture et Fonctionnement des Espèces Fruitières), Montpellier, France

Ingo Ensminger
Department of Biology, University of Toronto Mississauga, Mississauga, ON, Canada; Department of Cell and Systems Biology, University of Toronto, Toronto, ON, Canada; Department of Ecology and Evolutionary Biology, Toronto, ON, Canada

Ioannis Ganopoulos
Laboratory of Forest Genetics and Tree Breeding, Faculty of Forestry and Natural Environment, Aristotle University of Thessaloniki, Thessaloniki, Greece; Centre for Research and Technology Hellas (CERTH), Institute of Applied Biosciences, Thessaloniki, Greece

Jean-Marc Gion
CIRAD, UMR AGAP, UMR BIOGECO, Cestas, France

Fikret Isik
Department of Forestry and Environmental Resources, North Carolina State University, Raleigh, NC, USA

Hiroyoshi Iwata
Department of Agricultural and Environmental Biology, Graduate School of Agricultural and Life Sciences, The University of Tokyo, Bunkyo, Tokyo, Japan

Juan P. Jaramillo-Correa
Department of Evolutionary Ecology, Institute of Ecology, Universidad Nacional Autónoma de México, México City, Mexico

Stephen R. Keller
Department of Plant Biology, University of Vermont, Burlington, VT, USA

Satish Kumar
The New Zealand Institute for Plant & Food Research Limited, Havelock North, New Zealand

Zongrang Liu
Appalachian Fruit Research Station, Kearneysville, WV, USA

John J. MacKay*
Centre for Forest Research and Institute for Systems and Integrative Biology, Université Laval, Quebec, QC, Canada

Pedro J. Martínez-García
Department of Plant Sciences University of California, Davis, CA, USA

Alejandra Moreno-Letelier
Department of Evolutionary Ecology, Institute of Ecology, Universidad Nacional Autónoma de México, México City, Mexico; Botanical Garden, Institute of Biology, Universidad Nacional Autónoma de México, México City, Mexico

Geneviève J. Parent
Centre for Forest Research and Institute for Systems and Integrative Biology, Université Laval, Quebec, QC, Canada

Julien Prunier
Department of Forestry, Centre for Forest Research and Institute for Systems and Integrative Biology, Université Laval, Québec, QC, Canada

Elie Raherison
Centre for Forest Research and Institute for Systems and Integrative Biology, Université Laval, Quebec, QC, Canada

Juliana Sena
Centre for Forest Research and Institute for Systems and Integrative Biology, Université Laval, Quebec, QC, Canada

Athanasios Tsaftaris
Centre for Research and Technology Hellas (CERTH), Institute of Applied Biosciences, Thessaloniki, Greece; Department of Genetics and Plant Breeding, Aristotle University of Thessaloniki, Thessaloniki, Greece

Alejandra Vázquez-Lobo
Department of Evolutionary Ecology, Institute of Ecology, Universidad Nacional Autónoma de México, México City, Mexico; Research Centre for Biodiversity and Conservation, Universidad Autónoma del Estado de Morelos, Cuernavaca, Mexico

*Present address: Department of Plant Sciences, University of Oxford, Oxford, UK

Toshiya Yamamoto
NARO Institute of Fruit Tree Science, Tsukuba, Ibaraki, Japan

Christine Yao-Yun Chang
Department of Biology, University of Toronto Mississauga, Mississauga, ON, Canada;
Department of Cell and Systems Biology, University of Toronto, Toronto, ON, Canada

Tatyana Zhebentyayeva
Genomics and Computational Biology Laboratory, Clemson University, Clemson, SC, USA

PREFACE

Trees form a polyphyletic group that includes a variety of plant species that have independently evolved specific features such as great size, longevity and long generation time and woodiness. Trees are of major importance to human society. They provide a wide array of ecosystem services that confer benefits to people, including, for example, the production of goods such as fruits, timber, biomass fuels and precursors to many industrial and pharmaceutical products, the sequestration of carbon (mitigating climate change) and the prevention of soil erosion. Natural forests, as well as intensively managed forest tree plantations and fruit tree orchards, support the economy of many countries worldwide, providing employment especially for rural populations. Last but not least, forests also offer habitats for flora and fauna that are an essential part of the functioning of terrestrial ecosystems.

Owing to the rapidly evolving tools and methods in genomics and bioinformatics, our understanding of the structure, functioning and evolution of tree genomes is progressing at an accelerated pace. This thematic volume, 'Genomes of trees', illustrates areas where progress in forest and fruit tree genomics has been made over the past 10 years, since the first tree genome sequence (poplar) was published. The knowledge gained from research performed in this area is expected to lead to important improvements in terms of genetic resource characterisation, management and valorisation in breeding programmes, deliveringhigh quality products, with enhanced robustness to a changing environment.

In the first two chapters, David Chagné (Whole genome sequencing of fruit tree species) and Geneviève Parent et al. (Forest tree genomics: review of progress) highlight the acceleration of delivery of reference genome sequences for several forest and fruit tree species, thanks to new sequencing technologies and despite the challenge of coping with sometimes very large genomes. They underline that these new resources have already allowed a better understanding of genome evolution in plants and given access to new and powerful tools to study their biology, their adaptation to biotic and abiotic cues and to facilitate their breeding. The third and fourth chapters review how genomics could meet the breeding and conservation challenges of the twenty-first century, especially considering that breeders need to be more efficient in improving the traits of interest to increase productivity, develop genetic resources that can adapt to a changing environment and be resistant to disease and pests. In Chapter 3, (Acceleration of forest and fruit tree domestication by genomic selection, by Fikret Isik et al.) the authors consider how genomics could be used to design novel tree-breeding schemes, while in Chapter 4 (Population and conservation genomics in forest and fruit trees, by Filippos Aravanopoulos et al.) they explore how genomics could improve the characterisation of genetic resources diversity and conservation strategies and increase our ability to select and prioritise conservation targets. The three following chapters illustrate functions that are unique to these long-lived and large-size organisms: their stature, their ability to synchronise phenological events with climatic conditions, and their plasticity in response to environmental changes, thus allowing them to subsist in a spatially and temporally highly heterogeneous environments. In Chapter 5 (The genetics and genomics of tree architecture) Evelyne Costes and Jean-Marc Gion report that genomic resources and progress in phenotyping technologies, have lead to significant advances in the understanding of the genetic, physiological and molecular determinants of tree architecture plasticity in response to environmental changes. In Chapter 6, (The genetic control of bud-break in trees) Albert Abbott et al. highlight our current state of knowledge about the molecular mechanisms underlying the timing of floral and vegetative bud-break, helping to improve our understanding of how trees coordinate their annual growth, development and reproduction with the environmental cycle. In Chapter 7, (Tree responses to environmental cues) Ingo Ensminger et al. give a very complete review of the current knowledge on the physiological responses and associated molecular regulation pathways of trees to drought, temperature and photoperiod, thus completing the emerging picture from the two former chapters. In the last chapter (Chapter 8, Molecular signatures of adaptation and selection in forest trees) Juan-Pablo Jaramillo-Correa et al. elegantly summarise forward-genetic approaches (i.e. from phenotype to gene function) currently used in forest trees to identify genes/alleles that are important for adaptation.

With the continuing decrease of cost and development of higher throughput sequencing methods, the next decade will continue to see a growing number of species (including wild relatives of highly domesticated fruit tree species) being sequenced and the discovery of their variants. Major challenges are then to: (1) exploit this new wealth of data in order to detect naturally occurring variations that matter for productivity, quality and adaptation and (2) design innovative conservation and breeding strategies to help these long-lived and sessile organisms cope with the rapidly increasing biotic and abiotic threats during their lifetime, while still providing the provisioning services for which they have been planted. Genomics opens the doors to the development of these applications in forestry and fruit tree production, and thus should contribute to sustain economic growth in rural areas.

Christophe Plomion
Anne-Françoise Adam-Blondon

CHAPTER ONE

Whole Genome Sequencing of Fruit Tree Species

David Chagné
Palmerston North Research Centre, The New Zealand Institute for Plant & Food Research Limited (Plant & Food Research), Palmerston North, New Zealand
E-mail: David.Chagne@plantandfood.co.nz

Contents

Advances in Botanical Research, Volume 74
ISSN 0065-2296
http://dx.doi.org/10.1016/bs.abr.2015.04.004

Abstract

Horticultural fruit tree species have contributed significantly to the recent revolution in genomics, with many fruit and nut crops now sequenced. This review covers current progress in fruit tree whole genome sequencing, describing the state of genome assembly and annotation in grape, apple, pear, date palm, papaya, peach, kiwifruit, banana, Clementine, jujube, mulberry, sweet orange, Chinese plum, macadamia, cocoa, coffee, pineapple, blueberry, hazelnut, raspberry and mango. The availability of these genome assemblies provides fruit biologists with valuable new tools to develop new varieties more efficiently, enabling the implementation of marker-assisted selection and genomic selection, as well as contributing to the understanding of the molecular determinants of key traits unique to horticultural species and giving important clues concerning the evolution of their complex genomes.

1. INTRODUCTION

Plant genetics has experienced a genuine revolution during the last 15 years. This revolution started with the complete sequencing of the *Arabidopsis* (Kaul et al., 2000) and rice (Goff et al., 2002; Yu et al., 2002) genomes; however, since then a large number of other plant species have had their genomes entirely decoded. Fortunately, horticultural fruit tree species have been part of this revolution, with many fruit crops being sequenced in various dicotyledonous families, such as the Rosaceae, Rutaceae and even monocotyledons, such as Musaceae and Arecaceae. The key factor for the fruit tree genomics revolution was the introduction of next generation sequencing (NGS) technologies (Metzker, 2010). NGS enables the rapid generation of a massive amount of sequencing data for a fraction of the cost of the rice sequencing project, which was undertaken using earlier technologies. This means that single laboratories or small consortia now have the ability to afford to sequence the entire genome of their favourite tree crop organism at very high sequence depth within a few months. Indeed, all fruit-bearing species genomes sequenced in the last 5 years have involved utilization of some NGS data. Fortunately, the amount of NGS data generated also coincided with progress in assembly and gene annotation algorithms, as well as database and genome querying tools. Fruit tree genome assemblies have often involved a combination of NGS technologies with old-fashioned (but accurate and consistently yielding reads over 800 bp) Sanger sequencing technology. Methods for assembling large genomes de novo using NGS were reviewed by Henson and colleagues (Henson, Tischler, & Ning, 2012) and some information is available at http://en.wikibooks.org/wiki/Next_Generation_Sequencing_(NGS)/De_novo_assembly. Briefly, the initial step of the genome assembly process involves building small contigs of overlapping NGS and/or Sanger sequences. However, it is impossible to build contigs spanning entire chromosomes because sequencing technologies and assembly software are not efficient in the presence of repetitive sequences. Hence, scaffolding is necessary to connect these contigs, using sequences obtained from large insert libraries such as long-insert mate paired-end libraries or bacterial artificial chromosome (BAC)-end sequencing. Nevertheless, even with the use of such large inserts, scaffolding often reaches a point at which an increase in the amount of short read sequencing does not further improve assembly quality. A metric that is widely used to measure the quality of a genome assembly is the N50 statistic. The N50 value corresponds to a median and indicates the length for which the assembly of all scaffolds of that length or longer contains at least half the sum of the lengths of the entire scaffold assembly. Although low-throughput methods such as fluorescence in situ hybridization can help to stitch large scaffolds together, high-density genetic maps are widely used to anchor scaffolds. Sequence-based genetic markers can be retrieved from the scaffold sequences so that the scaffolds can be placed and oriented on the linkage groups, until chromosome-length scaffolds (often called pseudo-chromosomes or super-scaffolds) are developed. Following the genome assembly, the considerable task of annotating the genome remains. This includes predicting key features such as polymorphisms, GC content, repeated sequences and genes. A suite of bioinformatics tools is available for predicting protein-coding genes and repeated sequences, based on sequence homology with other sequenced genomes and alignment of RNA sequences onto the assembly contigs.

This review encompasses the current status of genome sequencing in temperate and tropical fruit and nut tree species, the list of which includes woody perennials such as classic horticultural tree crops (e.g. apple, citrus, papaya), berry-bearing vines (e.g. grape and kiwifruit), bush-type fruits (e.g. raspberry and blueberry) and nut crops (e.g. hazelnut).

2. FRUIT TREE GENOMES SEQUENCED TO DATE

2.1 The First Fruit-Bearing Genome to be Sequenced: Grapevine (*Vitis vinifera*)

Grapevine was the first fruit-bearing species to have its genome completely sequenced (Figure 1). Grapevine (*Vitis vinifera*) is diploid (2n = 38) with high heterozygosity maintained because of dioecy. Two independent projects focused on sequencing the genome of clones derived from Pinot noir. The French–Italian Public Consortium for Grapevine Genome Characterization focused on an inbred genotype derived from Pinot noir (PN40024) with an estimated 93% homozygosity, in order to avoid issues with heterozygosity during contig assembly (Jaillon et al., 2007). The first

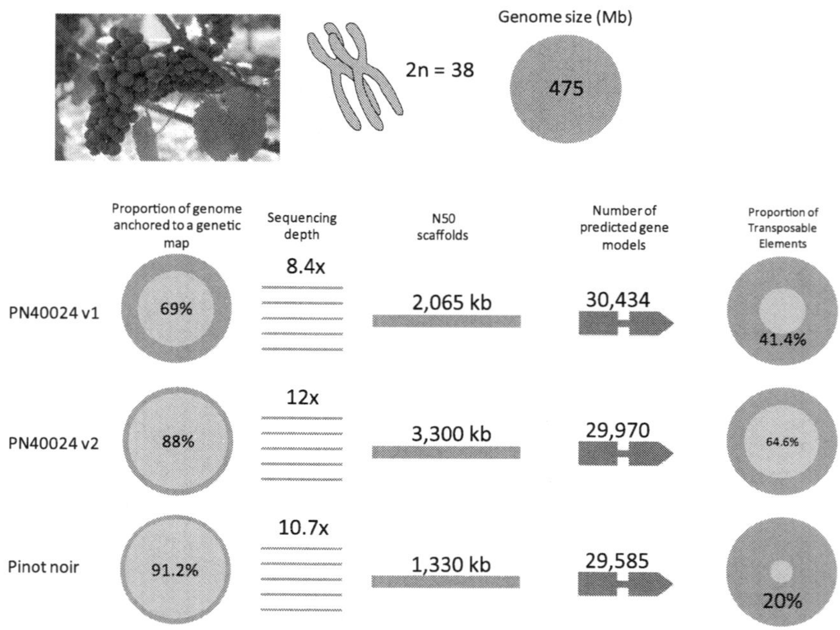

Figure 1 The *Vitis vinifera* genome at a glance. (See colour plate)

assembly of PN40024 was obtained using shotgun sequencing with 8.4x coverage. Half of the 19,577 contigs (N50) were larger than 65.9 kb and the assembly covered a total of 487 Mb, of which 69% was anchored to a genetic map. In total, 30,434 protein-coding genes were predicted. A simultaneous project led by the Istituto Agrario San Michele all'Adige (IASMA) focused on the heterozygous Pinot noir clone ENTAV115 (Velasco et al., 2007). A combination of Sanger (6.5x) and Roche/454 (4.2x) sequencing technologies were used to develop sequences and unlike the assembly of PN40024, the procedure for Pinot noir took into account the high degree of heterozygosity in assembling 2093 metacontigs totalling 477 Mb, 91.2% of which were anchored to a single nucleotide polymorphism (SNP)-based genetic map. The number of gene-coding sequences was similar to that in the PN40024 genome, with 29,585 predicted protein-coding genes. A new improved version of the PN40024 genome is now available on the EnsemblPlant Website (http://plants.ensembl.org/Vitis_vinifera/Info/Index), including new gene annotations and a genome browser. This new assembly has a depth of ~12x and covers 481 Mb in total, of which 426 Mb are anchored to 19 pseudo-chromosomes (Adam-Blondon, pers. comm.). The total number of predicted gene models in this new assembly is 29,970 (Vitulo et al., 2014).

High-density saturated SNP genetic maps have been developed for grapevine in the last few years and the draft genome of Pinot noir ENTAV115 was anchored to a such a map (Troggio et al., 2007) developed from SNPs screened using the SNPlex technique (Pindo et al., 2008). Furthermore, an Illumina Infinium® SNP array of 9000 attempted bead types has been developed for *Vitis* and named the Vitis9k array (Myles et al., 2010). The Vitis9k array has been used for multiple applications, including screening of a large germplasm set consisting of domesticated and wild *Vitis* accessions (*V. vinifera* subspecies *sylvestris*), which supported a Near East origin of domesticated grape and some introgression from subspecies *sylvestris* (Myles et al., 2011). Recently restriction site-associated DNA sequencing has been used as an alternative to SNP arrays for generating large number of markers for grape (Wang, Fang, Xin, Wang, & Li, 2012).

2.2 An Insight into Recent Whole Genome Duplication: Apple (*Malus* × *domestica*)

Apple is a diploid species with 17 pairs of chromosomes and an estimated genome size of 750 Mb. A high degree of heterozygosity is found within the apple because of the self-incompatibility locus. The draft genome of the 'Golden Delicious' cultivar was completed in 2010 using a combination of Sanger (26%) and Roche/454 (74%) sequencing technologies (Velasco et al., 2010) (Figure 2). An assembly of 1629 metacontigs covered 598 Mb of sequences anchored to an SNP genetic map, with the remaining sequences thought to be repetitive elements. The number of predicted gene models was 57,386, which was a greater number than for other plant species, reflecting the recent duplication of the apple genome. Indeed, most of the other Rosaceae species have seven, eight or nine pairs of chromosomes, except for the members of the Pyreae subfamily, which have 17 pairs of chromosomes, again suggesting an ancient duplication event.

A new version of the 'Golden Delicious' genome has been recently developed using two double haploid (DH) lines of 'Golden Delicious', thus enabling the construction of true haplotypic metacontigs. The first version and DH assemblies of 'Golden Delicious' are available through the Genome Database for Rosaceae (GDR, Jung & Main, 2014; www.rosaceae.org). High-density genetic marker screen technology has been developed for apple, including SNP arrays (Bianco et al., 2014; Chagné et al., 2012) and genotyping-by-sequencing (GBS) (Gardner et al., 2014). These technologies have been successfully used for developing high-density genetic maps (Antanaviciute et al., 2012; Clark, Schmitz, Rosyara, Luby, & Bradeen, 2014), fine mapping of major loci (Souleyre et al., 2014), genome-wide association studies (GWAS) (Kumar et al., 2013), and genomic selection (Kumar, Chagné, et al., 2012).

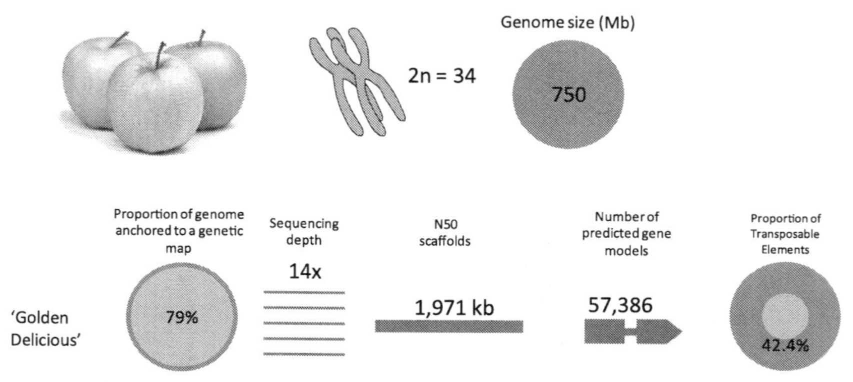

Figure 2 The *Malus × domestica* 'Golden Delicious' genome at a glance. (See colour plate)

2.3 Pear (*Pyrus* spp.) Has Fewer Repeated Elements than Apple

Pear is a diploid (2n = 34) with an estimated genome size of approximately 600 Mb. It is closely related to apple and a high degree of synteny is found between these genera (Celton et al., 2009; Montanari et al., 2013). Two groups have independently sequenced the *Pyrus* genome (Figure 3). The first group focused on sequencing the Chinese pear genome (*Pyrus* × *bretschneideri*) (Wu et al., 2013) and the second targeted European pear (*Pyrus communis*) (Chagné et al., 2014). The Chinese pear genome project sequenced 'Dangshangsuli' using a combination of NGS and BAC-by-BAC sequencing. The 'Dangshangsuli' assembly was achieved at an estimated 194x coverage and spanned 512 Mb in total, which represents 97% of the estimated pear genome. A set of 2005 SNP markers was used to anchor 75% of the genome assembly. In total, 42,812 protein-encoding genes were predicted. The European pear genome project sequenced 'Bartlett', employing Roche/454 NGS technology with single- and paired-end reads. The Bartlett v1.0 draft assembly covers 577 Mb, of which 171 Mb was anchored to a genetic map using 2279 SNP markers. A total of 43,419 protein-coding gene models were predicted. Both the 'Dangshangsuli' and 'Bartlett' genome assemblies contained more gene models than do Rosaceae species outside the Maloideae subfamily. A high number of predicted genes were also observed in apple, because of the recent whole genome

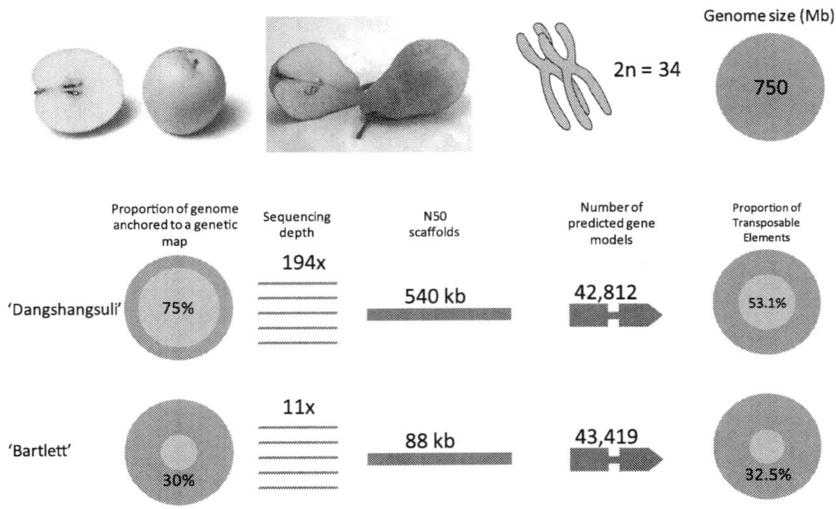

Figure 3 The Asian and European *Pyrus* genomes at a glance. (See colour plate)

duplication in the Maloideae, to which both apple and pear belong. The Chinese and European pear genomes are available at peargenome.njau.edu.cn and www.rosaceae.org/species/pyrus/pyrus_communis/genome_v1.0, respectively.

While they share the same number of chromosomes and have highly collinear genomes, the Chinese and European pear genomes are smaller in size than that of apple (~600 Mb for pear versus ~750 Mb for apple). The analyses of the pear genomes indicated that this difference in genome size is due to the presence of a greater number of repeated elements being present in the apple genome than in pear.

2.4 The Sequence of a Giant Monocotyledon: Date Palm (*Phoenix dactylifera*)

Date palm is a dioecious monocotyledon with 18 pairs of chromosomes. Two groups have sequenced the genome of the 'Khalas', which is a female cultivar (Figure 4). A first draft assembly covering gene-rich regions was obtained using Illumina Genome Analyzer IIx sequencing at an estimated depth of 53.4x (Al-Dous et al., 2011). In total, 57,277 scaffolds were obtained with an N50 size of 30 kb. The assembly covered 381 Mb in total, which is only half the estimated ~660 Mb date palm genome. Re-sequencing of date palm cultivars enabled the detection of genome-wide

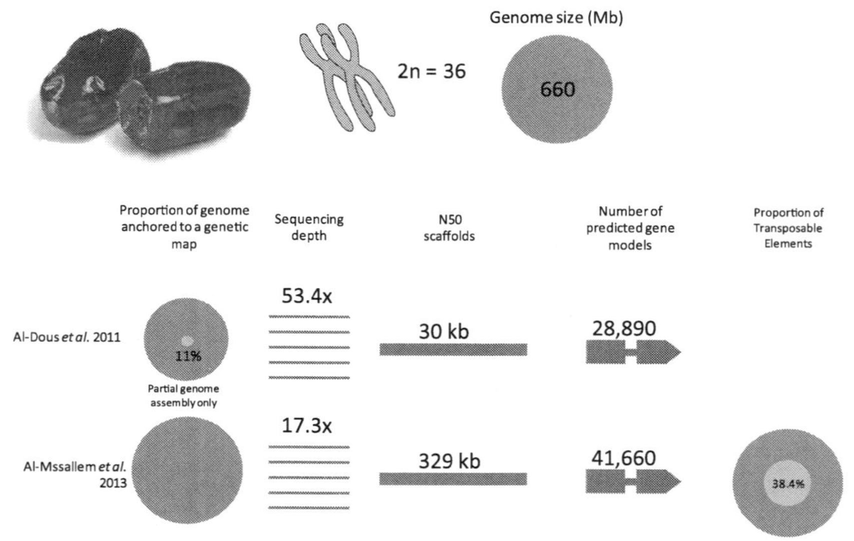

Figure 4 The *Phoenix dactylifera* 'Khalas' genome at a glance. (See colour plate)

polymorphisms including SNPs and copy number variations (CNVs). A genetic map of 'Khalas' developed using GBS included 1199 markers (Mathew et al., 2014). The map covers a total of 1293 cM and anchors ~20% of the 381 Mb genome assembly (equivalent to ~11% of the total genome). A second more advanced genome assembly was obtained concurrently by a consortium using Roche/454 shotgun sequencing of libraries of various fragment lengths, as well as SOLiD (Life Technologies) reads of large insert libraries and BAC-end sequences that were used to increase scaffold length (Al-Mssallem et al., 2013). The final assembly consisted of 82,354 scaffolds of 558 Mb in total, which is equivalent to 90% of the genome. This second more complete genome assembly is not anchored to a genetic map. In total, 41,660 protein-coding genes were predicted using the second 'Khalas' assembly.

2.5 The First Genome Sequence of a Transgenic Tree Crop: Papaya (*Carica papaya*)

Carica papaya is a tropical dioecious species of the order Brassicales with a compact genome (372 Mb) organized into nine pairs of chromosomes. A draft genome sequence was developed for papaya using the highly inbred cultivar 'SunUp' (Ming et al., 2008) (Figure 5). In total, 271 Mb contigs spanning 370 Mb of scaffolds containing gaps were assembled from NGS shotgun and BAC-end sequencing. A third of the scaffolds (235 Mb) were anchored to the 12 linkage groups of the current papaya genetic maps. A total of 24,746 protein-coding gene models were predicted in the genome of 'SunUp', which is smaller to the number of genes found in *Arabidopsis*

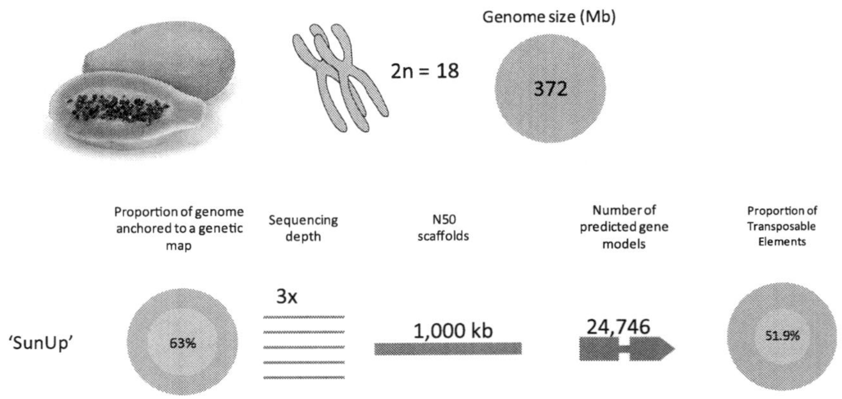

Figure 5 The *Carica papaya* 'SunUp' genome at a glance. (See colour plate)

(31,114), with which papaya shared a common ancestor ~72 Mya. The originality of the draft papaya genome sequence assembly lies in the fact that it was developed from a transgenic cultivar, making it the first example of transgenic crop to be fully sequenced and providing the opportunity to identify the transgene insertion sites. 'SunUp' contains a coat protein transgene inducing resistance to ring-spot virus (Gonsalves, 1998). Three insertion sites were found in the papaya genome; however, only one site displayed a functional transgene.

2.6 A Very High Quality Fruit Genome Sequence: Peach (*Prunus persica*)

Peach has a compact diploid (2n = 16) genome of 265 Mb. A high quality assembly based on a DH genotype of 'Lovell' was developed using Sanger sequencing at a final 8.4x coverage (Verde et al., 2013) (Figure 6). The assembly covered 226 Mb, of which 96% was aligned to a high-density genetic map based on SNPs (Verde et al., 2012), which enabled the assembly of eight pseudo-molecules corresponding to the eight haploid chromosomes. A total of 27,852 protein-coding genes were predicted. Unlike the case for the apple and pear genomes, there was no evidence that peach had experienced any recent whole genome duplication. A second version of the peach genome has been released (peach genome v2.0) and is available at services.appliedgenomics.org/gbrowse/prunus_persica_v2/ and at the Genome Database for Rosaceae (Jung & Main, 2014). The new version has improved annotation of repeated and gene sequences, correction of regions that were previously mapped erroneously and improved map

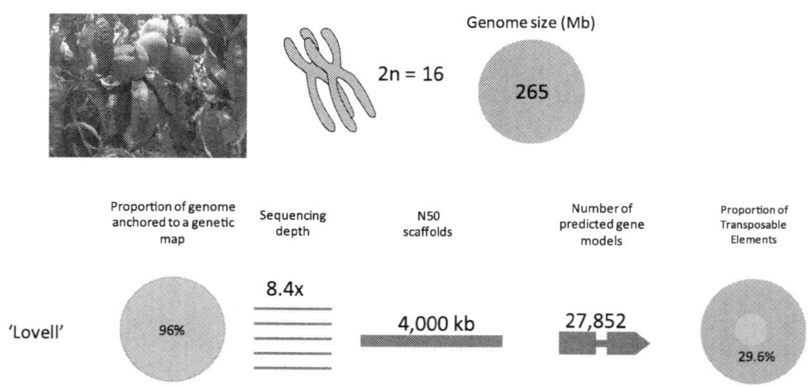

Figure 6 The *Prunus persica* 'Lovell' genome at a glance. (See colour plate)

coverage, achieving 99.2% of mapped sequences in total, with half the assembly contained in 250 contigs of more than 255 kb in length.

Resequencing of 56 peach accessions using Illumina and Roche/454 technologies enabled the detection of 1 million genome-wide SNPs, which were used to design a 9k SNP array (Verde et al., 2012). This SNP array has been used for high-density genetic map construction, quantitative trait locus (QTL) mapping (Eduardo et al., 2013; Pirona et al., 2013) and for faster and more efficient new cultivar breeding (www.rosbreed.org).

Because of the high synteny and sequence conservation within the genus *Prunus*, the peach genome was used to map NGS reads developed from a range of stone fruit species such as sweet and sour cherry (*Prunus avium* and *Prunus cerasus*, respectively), apricot (*Prunus armeniaca*) and plum (*Prunus domestica*). For example, the resequencing data of 16 sweet and 8 sour cherry cultivars using Illumina paired-end sequencing were mapped to the reference peach genome of 'Lovell'. While only 14% of the reads mapped to the peach genome assembly, a total of 1 million SNPs could be detected and were used to develop a Illumina Infinium® 6k cherry SNP array (Peace et al., 2012).

2.7 Genome Sequencing of a Species Rich in Vitamin C: Kiwifruit (*Actinidia chinensis*)

Kiwifruit is a dioecious vine belonging to the genus *Actinidia*. A range of ploidy is found in the *Actinidia* genus, in multiples of the 29 haploid chromosomes. For example, the popular green kiwifruit, *Actinidia deliciosa* 'Hayward' is hexaploid. A draft genome assembly has been developed based on the diploid (2n = 58) *Actinidia chinensis* 'Hongyang' using the Illumina HiSeq 2000 technology (Huang et al., 2013) (Figure 7). A draft genome

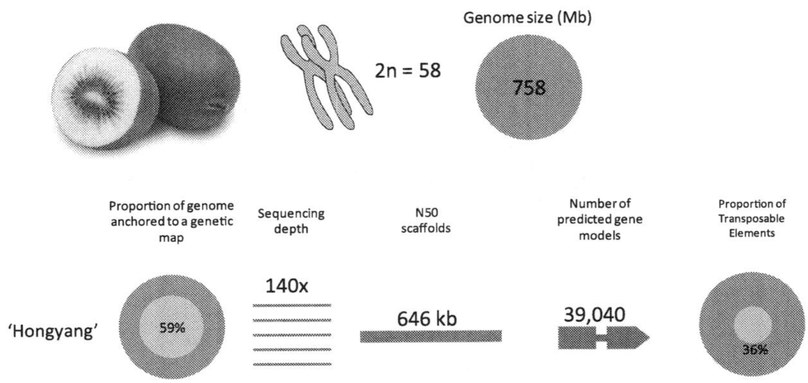

Figure 7 The *Actinidia chinensis* 'Hongyang' genome at a glance. (See colour plate)

of 616 Mb obtained from an estimated 140x coverage was assembled and represented 81% of the total kiwifruit genome. Half the genome was captured within contigs greater than 58 kb in length and an SNP-based genetic map of 3378 markers was used to assemble 73% of the assembly into 29 pseudo-molecules. In total, 39,040 protein-coding genes were predicted.

Kiwifruit has an unusually high concentration of vitamin C (ascorbic acid) in the fruit. The analysis of the kiwifruit predicted gene set and comparison with other plant species indicated that no expansion had occurred for the genes in the L-galactose pathway, which is the main route for ascorbate biosynthesis. However, other genes encoding enzymes involved in its biosynthesis (*Aldononolactonase* and *L-ascorbate peroxidase*), as well as regeneration from its oxidized form (*monohydroascorbate reductase*) have expanded in kiwifruit, compared with those in tomato, sweet orange and *Arabidopsis*.

2.8 Decoding the Complex Subgenomes of a Triploid Crop: Banana (*Musa acuminata*)

Commercial banana cultivars such as 'Cavendish' are triploid, which complicates both breeding and genome assembly. A DH cultivar of the species *Musa acuminata* (A genome; 2n = 22) originating from Malaysia ('Pahang'), which contributes one of the three subgenomes of 'Cavendish', was selected for genome sequencing by the Global Musa Genomics Consortium (D′Hont et al., 2012) (Figure 8). A combination of Roche/454 and Sanger sequencing produced a draft assembly of 472 Mb of the estimated 523 Mb at 20x coverage, with Illumina short reads used to correct errors. Half the genome assembly was contained in scaffolds greater than 1.3 Mb and 70% of the scaffolds were anchored to the 11 linkage groups of a *Musa* genetic

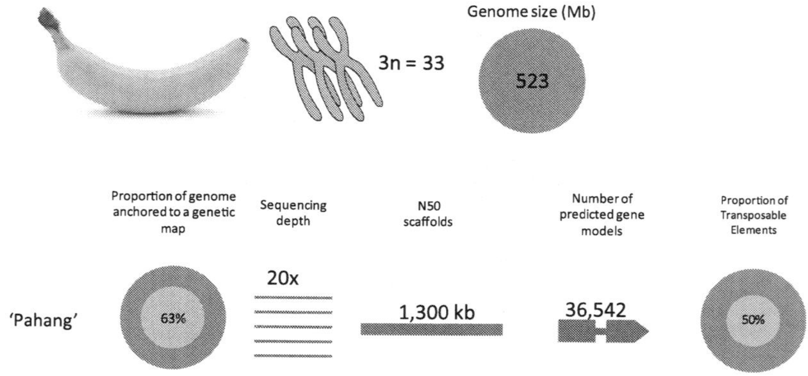

Figure 8 The *Musa acuminata* 'Pahang' genome at a glance. (See colour plate)

map obtained using simple sequence repeats (SSR) and diversity array technology markers. In total, 36,542 protein-coding gene models were predicted. The latest version of the banana genome is available at http://banana-genome.cirad.fr/ and community annotation system and querying tools were developed at the Banana Genome Hub (Droc et al., 2013).

Following the sequencing of the A genome, an attempt was made to sequence the genome of the B genome from *Musa balbisiana*, another contributor to the triploid genome of commercial cultivars. The wild diploid *M. balbisiana* 'Pisang Klutuk Wulung' (PKW) was sequenced using Illumina HiSeq short paired-end sequencing technology (Davey et al., 2013). The obtained sequencing reads were mapped to the A genome assembly of 'Pahang'. In total, 86.9% of the PKW sequencing data mapped to the 'Pahang' assembly and the consensus sequence of PKW covered 78.9% of the 'Pahang' genome. The remaining reads that did not map were de novo assembled; however, this assembly was fragmented, with a large number of contigs (63,245) of small size (N50 = 467 bp). The divergence between the A and B genome was one SNP every 23 bp on average, while one heterozygous SNP occurred on average every 55 bp in the *M. balbisiana* genome. Strikingly, the number of predicted protein-coding genes from *M. balbisiana* (36,638) was very similar to that in *M. acuminata*.

2.9 On the Origin of *Citrus* Fruit: Clementine Mandarin (*Citrus* × *Clementina*)

A high-quality draft genome sequence of Clementine was developed from haploid material derived from single ovule obtained after induced gynogenesis of diploid Clemenules (Figure 9). This strategy avoided the technical

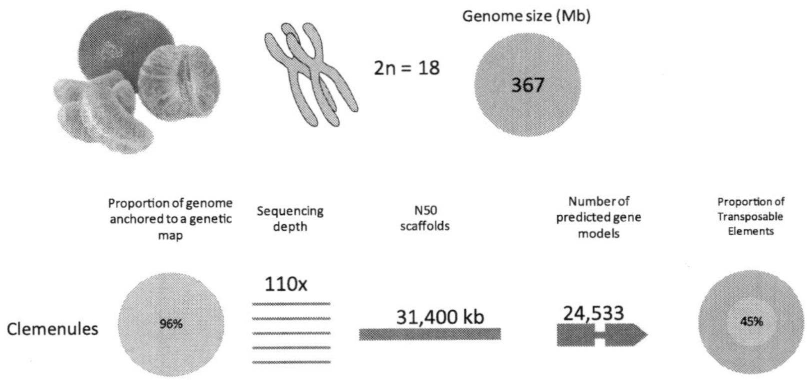

Figure 9 The *Citrus* × *clementina* 'Clemenules' genome at a glance. (See colour plate)

difficulties that are associated with assembling heterozygous diploid genomes. The genome assembly of Clemenules covers 301 Mb in total, with 96% of the scaffolds anchored to the nine citrus chromosomes, which enabled the assembly of nine high-quality pseudo-chromosomes. The *Citrus* × *clementina* genome is accessible at http://www.citrusgenomedb.org/species/clementina.

The Clemenules genome assembly was useful as a reference to map high-depth resequencing of a number of well-known citrus fruit, including sweet (*Citrus sinensis*) and sour orange (*Citrus aurantium*), pummelo (*Citrus maxima*) and other mandarins (*Citrus reticulata* and *Citrus* × *deliciosa*). The comparative genome analysis enabled the clarification of some controversial theories about the origin of these cultivated species. Surprisingly, mandarin-type citrus is admixed with pummelo and none of the cultivated mandarin accessions resequenced is derived purely from the wild mandarin *C. reticulata*. For example, Ponkan and Willowleaf mandarins result from the introgression of pummelo (*C. maxima*) into *C. reticulata* (wild mandarin).

2.10 Jujube (*Ziziphus jujuba*)

Jujube has a diploid genome with 12 pairs of chromosomes (2n = 24) and an estimated size of 444 Mb. A draft genome sequence of 'Dongzao' was developed using NGS and a combination of BAC-by-BAC and mate-paired libraries of various sizes (Liu et al., 2014) (Figure 10). The assembly spanned a total of 437 Mb (98% of the jujube genome) with a scaffold N50 size of 301 kb. A large proportion of the 'Dongzao' assembly (73.5%) was anchored to a genetic map of 2419 SNP markers obtained using restriction site-associated DNA sequencing (RAD-seq) and covering all 12 chromosomes.

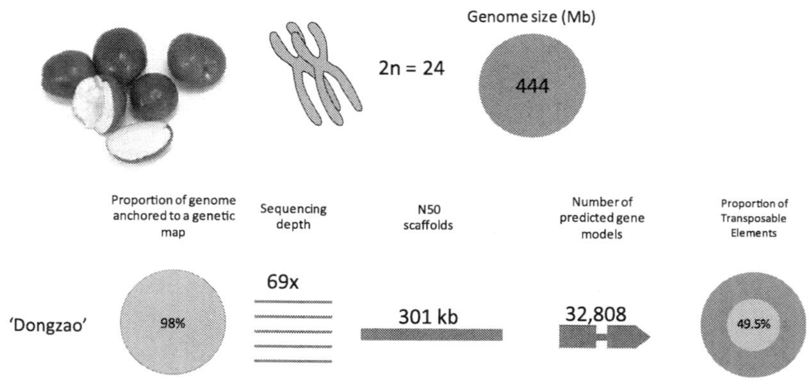

Figure 10 The *Ziziphus jujuba* 'Dongzao' genome at a glance. (See colour plate)

In total, 32,808 gene models were predicted from the 'Dongzao' genome assembly.

2.11 Mulberry (*Morus notabilis*)

Mulberry belongs to the Moraceae family and possesses seven pairs of chromosomes, with an estimated genome size of 357 Mb. A draft genome sequence was obtained with a total coverage of 236x covered 330 Mb of the *Morus* genome (He et al., 2013) (Figure 11). The draft genome assembly was not anchored to a genetic map. In total, 29,338 protein-coding genes were predicted.

2.12 Sweet Orange (*Citrus sinensis*)

Sweet orange is a diploid species with nine base chromosomes and an estimated genome size of 367 Mb. One of the unique botanical features of sweet orange and other members of the *Citrus* genus is the development of nucellar embryos, which are apomictic embryos that are genetically identical to the mother plant. A draft genome sequence of the Valencia sweet orange has been developed using a combination of shotgun sequencing, mate paired-end NGS sequencing with various insert size and BAC-end sequencing (Xu et al., 2013) (Figure 12). In total, 320.5 Mb were assembled into a draft genome (87.3% of the *Citrus* genome) with half the genome assembled in scaffolds greater than 1.69 Mb in length. In total, 239 Mb (79% of the assembly or 65% of the estimated genome size) of the sweet orange genome was anchored to a genetic map, employing a combination of Sanger and 454 shotgun sequencing (Wu et al., 2014). This assembly covers 319 Mb in total and half the genome is contained within scaffolds greater than 250 kb. The

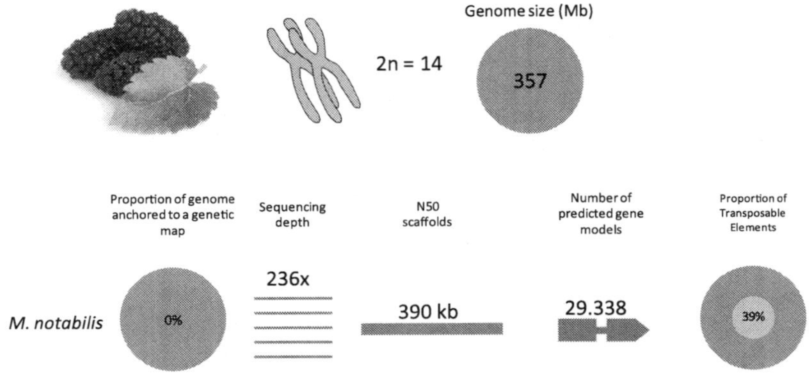

Figure 11 The *Morus notabilis* genome at a glance. (See colour plate)

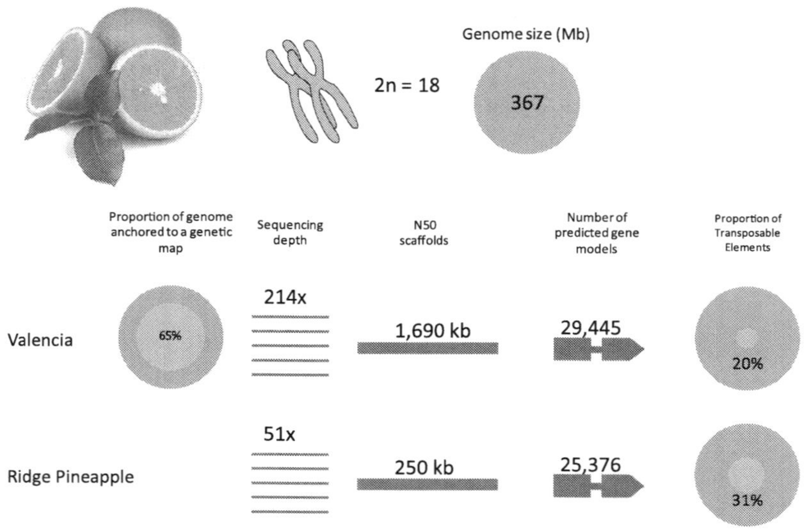

Figure 12 The *Citrus sinensis* Valencia and Ridge Pineapple genomes at a glance. (See colour plate)

Ridge Pineapple genome assembly and annotations are available at http://www.citrusgenomedb.org/species/sinensis/genome1.0. The Valencia and Ridge Pineapple genomes contain 29,445 and 25,376 predicted protein-coding genes, respectively.

2.13 Chinese Plum (*Prunus mume*)

Prunus mume or Chinese plum is an ornamental species that has a genome size similar to other members of the *Prunus* genus, such as peach. A wild accession of *P. mume* collected in Tongmai town, Bomi county, Tibet, China, was sequenced using the Illumina GAII technology and an assembly covering 237 Mb of the estimated 280 Mb (84%) *Prunus* genome was obtained (Zhang et al., 2012) (Figure 13). Optical mapping data and a genetic map obtained by RAD-sequencing were used to anchor 83.9% the assembly scaffolds to eight pseudo-chromosomes. The *P. mume* genome is available at the *P. mume* database (PMDB; http://prunusmumegenome.bjfu.edu.cn/).

2.14 Macadamia Nut (*Macadamia integrifolia*)

Macadamia was first domesticated around 100 years ago. Commercial cultivars, a few generations removed from wild relatives in Australian subtropical rain forests, are highly heterozygous (Nock, Elphinstone, et al., 2014). The

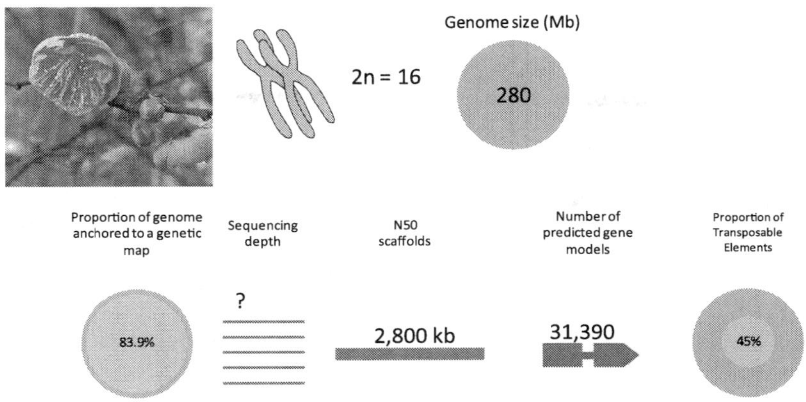

Figure 13 The *Prunus mume* genome at a glance. (See colour plate)

Macadamia Genome Consortium is working on the genome of the widely grown *Macadamia integrifolia* HAES 741 (2n = 28) (Figure 14). A draft genome assembled from Illumina paired-end and mate-pair reads is estimated to cover 80% of the 780 Mb genome at >40x coverage, with 32% repetitive content. De novo prediction methods identified 29,086 protein-coding genes. The chloroplast genome was used to determine the phylogenetic position of *Macadamia* among flowering plants (Nock, Baten, & King, 2014). Gene similarity of predicted macadamia genes to those of the closest available complete genome, the sacred lotus *Nelumbo nucifera*, was 83%. Transcriptome data have been generated and will be used to confirm gene predictions.

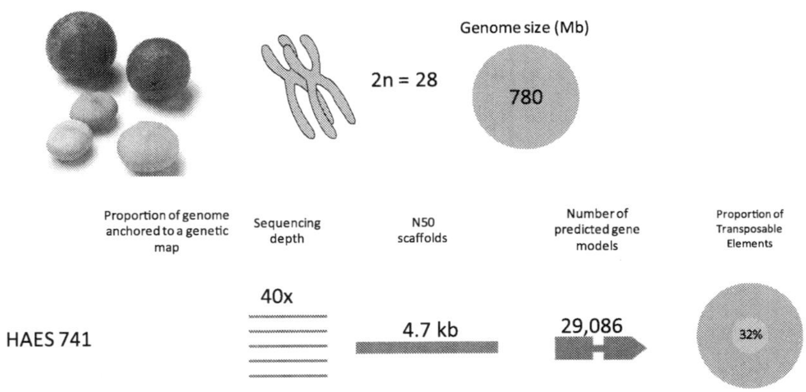

Figure 14 The *Macadamia integrifolia* genome at a glance. (See colour plate)

2.15 Pod Colour Fine Mapping in Cocoa (*Theobroma cacao*)

Theobroma cacao is a diploid species with 10 pairs of chromosomes and an estimated genome size of 430 Mb. Two consortia have worked on the complete sequencing of the cocoa genome. The first genome assembly of cocoa was developed from the highly homozygous Belizian Criollo cultivar B97-61/B2 (Argout et al., 2011) (Figure 15). The sequencing was performed using a combination of Roche/454 and Sanger BAC end reads at a read depth of 16.7x. The assembly spanned a total of 327 Mb, corresponding to 76% of the estimated genome of cocoa and half the genome assembly was contained in scaffolds greater than 473 kb in length. A genetic map was used to anchor 67% of the B97-61/B2 assembly (50% of the total genome size). In total, 28,798 protein-coding gene models were predicted from this assembly. A second consortium focused on sequencing the genome of the 'Matina' clone 'Matina 1-6', which belongs to the Amelonado germplasm group (Motamayor et al., 2013). The sequencing consisted of Roche/454 and Sanger sequencing and an iterative process was used to develop a chromosome-scale assembly of 346 Mb, with scaffold N50 of 34.4 Mb. The genome sequence of 'Matina 1-6' is available at the Cacao Genome Database (http://www.cacaogenomedb.org/). The usefulness of the 'Matina 1-6' genome assembly was demonstrated by making use of it to develop a genetic marker linked to pod colour. Fine mapping of the major locus

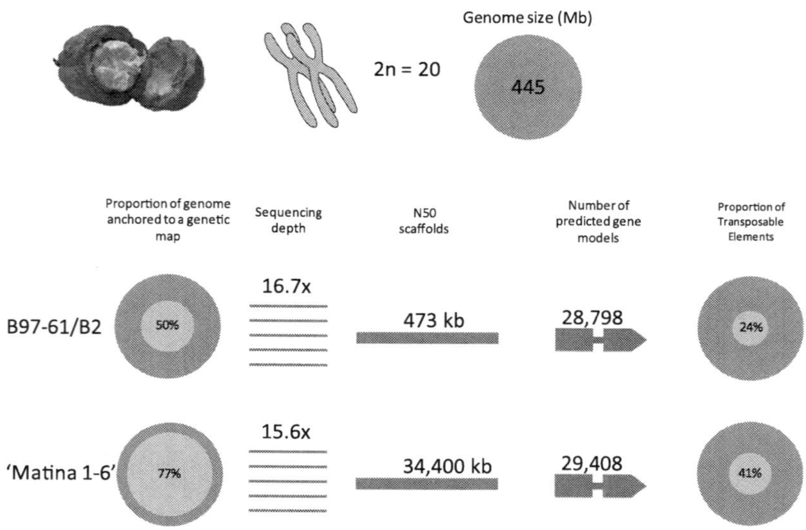

Figure 15 The *Theobroma cacao* genomes at a glance. (See colour plate)

controlling pod colour in segregating populations was performed and an interval of ~1.2 Mb containing the pod colour candidate gene identified on chromosome 4. A candidate gene homologous to *Arabidopsis MYB113* was detected, and resequencing of red and green pod parents indicated that *TcMYB113* is associated with pod colour. A polymorphism in a TAS4-si81 target site was proposed as potentially causing the variation in phenotype. Although further functional analysis is required for full confirmation of function, this polymorphism is now available for application in marker-assisted selection (MAS) in cocoa.

2.16 Coffee (*Coffea canephora*) and the Evolution of Genes Controlling Caffeine Biosynthesis

Coffee beans are harvested from the tetraploid *Coffea arabica*, a hybrid of *Coffea canephora* and *Coffea eugenioides*. A draft genome assembly was developed for the double haploid *C. canephora* accession DH200-94 using Roche/454 and Sanger sequencing (Denoeud et al., 2014) (Figure 16). The assembly covered a total of 568 Mb (80% of the estimated 710 Mb genome), of which 64% was anchored to a genetic map. Half the assembly was contained in scaffolds greater than 1.26 Mb in length. In total, 25,574 protein-coding gene models were predicted from the DH200-94 assembly. The coffee genome is available at the Coffee Genome Hub (http://coffee-genome.org/). The coffee genome assembly indicated that while coffee has undergone fewer genome duplications than other sequenced plant species, the genes encoding enzymes involved in the biosynthesis of caffeine, such as *xanthosine methyltransferase* (*XMT*) and *methylxanthine methyltransferase* (*MXMT*), have expanded because of tandem duplications. As caffeine

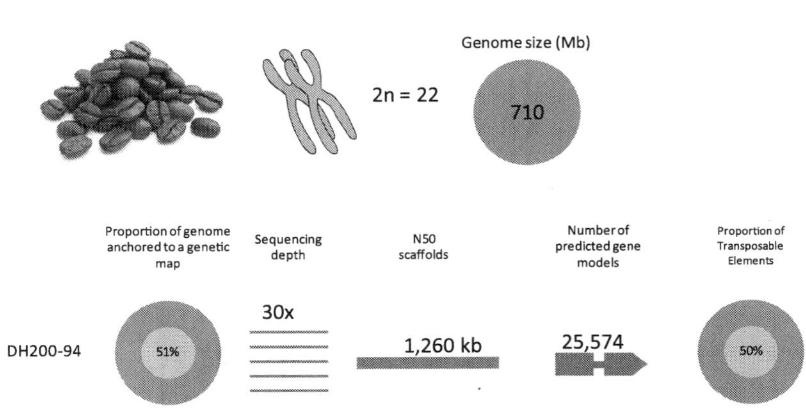

Figure 16 The *Coffea canephora* genome at a glance. (See colour plate)

has insecticidal properties, it is likely that the expansion of caffeine biosynthesis genes may have played the role during the evolution of the *Coffea* genus.

2.17 The Genome Sequence of a Crassulacean Monocotyledon: Pineapple (*Ananas comosus*)

Pineapple is a diploid monocotyledon with 25 pairs of chromosomes and an estimated genome size of 526 Mb. A draft genome assembly was developed for the variety F153 using a combination of Roche/454 and Illumina shotgun sequences (Zhang, Liu, & Ming, 2014) (Figure 17). The draft assembly covers 375 Mb and half of it is contained in scaffolds greater than 408 kb. No high-density genetic map has been developed to anchor this genome assembly so far. In total, 25,862 protein-coding genes were predicted. This set of genes will be instrumental to develop an understanding of the molecular control of Crassulacean acid metabolism photosynthesis.

2.18 Blueberry (*Vaccinium corymbosum*)

Blueberry species and modern commercial cultivars have complex genomes ranging from diploid to hexaploid. A draft genome of blueberry has been developed (Bian et al., 2014; Gupta et al., 2014) for a diploid accession (W8520) of the species *Vaccinium corymbosum*, which has 12 pairs of chromosomes and an estimated genome size of 500 Mb (Figure 18). The W8520 genome was sequenced using Roche/454 sequencing of varying fragment size libraries. The draft genome assembly obtained had 13,757 scaffolds with half the genome contained in scaffolds greater than 145 kb in size.

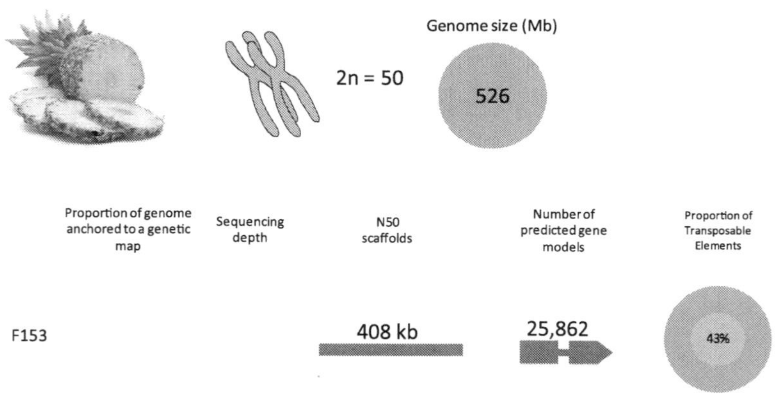

Figure 17 The *Ananas comosus* F153 genome at a glance. (See colour plate)

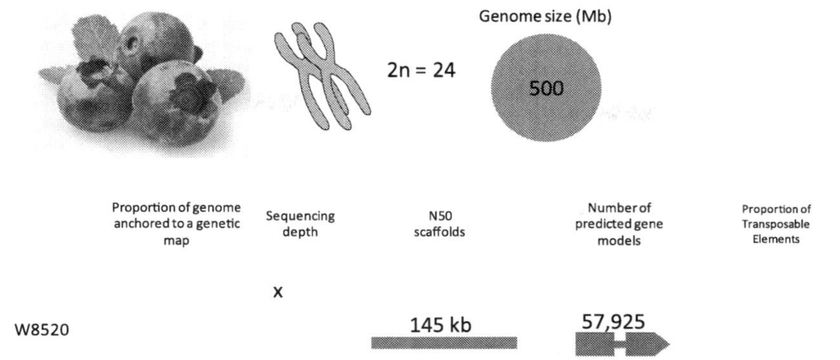

Figure 18 The *Vaccinium corymbosum* W8520 genome at a glance. (See colour plate)

RNA sequencing combined with ab initio analysis enabled the detection of 57,925 protein-coding gene models (Gupta et al., 2014). A genetic map was not used to anchor the genome assembly and no estimation of the proportion of repeated elements is given in the original publication. The draft genome of W8520 has been used to mine new SSR markers (Bian et al., 2014). Information about the blueberry genome and genetic markers developed from it is available at the Genome Database for *Vaccinium* (http://www.vaccinium.org/).

2.19 Hazelnut (*Corylus avellana*)

European hazelnut (*Corylus avellana* L.) is an agriculturally important tree crop in the family Betulaceae, with considerable genetic diversity among hundreds of accessions. Breeding is focused on maximizing nut yield and quality, while concomitantly reducing susceptibility to diseases. The diploid (2n = 22) 'Jefferson' was selected for the reference hazelnut genome. Illumina sequence was employed to generate a de novo draft assembly comprising 345 Mb (91%) of the flow cytometry-determined genome size of 379 Mb, with ~90x coverage (Figure 19). Half the assembly is contained in scaffolds greater than 21.5 kb, with the largest scaffold comprising 274.5 kb. Homology-based annotation identified 34,754 putative protein-coding gene models having high homology to the best annotated and most closely related plant species. Resequencing of seven additional accessions has enabled the identification and characterization of ~2 million polymorphisms. GBS analysis has now added thousands of additional markers to the genetic linkage map of European hazelnut. The current version of the hazelnut genome is available for homology-based queries at hazelnut.

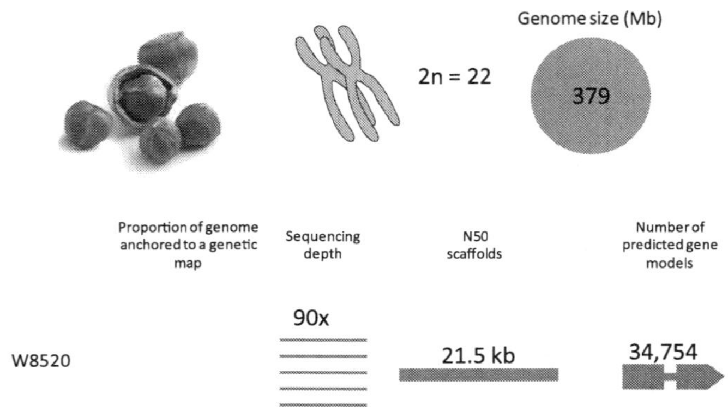

Figure 19 The *Corylus avellana* 'Jefferson' genome at a glance. (See colour plate)

mocklerlab.org, including JBrowse tracks that allow visualization of gene models, functional annotations and polymorphism effects.

2.20 Raspberry (*Rubus* spp.)

Raspberry is a member of the Rosaceae family, with seven chromosomes and an estimated 250 Mb genome size in total. The genome sequencing of two species that can cross-hybridize are currently being completed: red raspberry (*Rubus idaeus*) and black raspberry (*Rubus occidentalis*). Both species are highly syntenic with the strawberry genome (Bushakra et al., 2012), and chromosome numbering follows that of woodland strawberry (*Fragaria vesca*) (Shulaev et al., 2011). The black raspberry genome assembly of the highly homozygous accession ORUS-4115-3 (estimated 0.06 heterozygosity) has been developed using an estimated 940x coverage of Illumina NGS data (Figure 20; Mockler, pers. comm.). The assembly has 2226 scaffolds and 82% of the assembly is anchored to a genetic map, which covers 200 Mb in total. A total of 28,005 protein-coding gene models are predicted.

2.21 Mango (*Mangifera indica*)

Mango is an allotetraploid ($4n = 40$) with a genome size of 450 Mb. A combination of Roche/454 and Illumina shotgun sequencing was used at a depth of 160x to develop a preliminary draft assembly of 'Amrapali'. While the genome sequence has not yet been published, a preliminary draft assembly was reported at the Plant & Animal Genome Conference in 2013, with 432 Mb assembled and 63,130 predicted protein-coding genes (Figure 21). Further refinement of the genome assembly is underway to take account of the high heterozygosity found in mango.

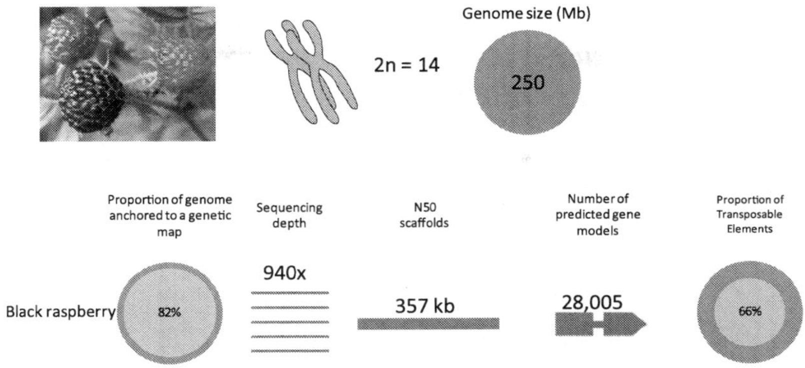

Figure 20 The *Rubus occidentalis* genomes at a glance. (See colour plate)

2.22 Summary of Genome Assemblies in Fruit Trees

A summary of the fruit tree genome assemblies is presented in Figure 22. A striking feature of these genome assemblies is the great heterogeneity of quality of assembled sequences. While NGS has enabled the decoding of most of these genomes at a lower cost than for previous genomes, such as the human or rice genomes, the quality of NGS-based genomes is not optimal. In general, the assembly of genomes for highly heterozygous species, using solely NGS short reads technology, is hazardous and the assembly can end up very fragmented. Furthermore, uncertainties remain as to how well the haplotypes of heterozygous species are determined during the contig and scaffold assemblies. However, high-quality genome assemblies have been developed for species such as banana, peach, raspberry, cocoa, papaya,

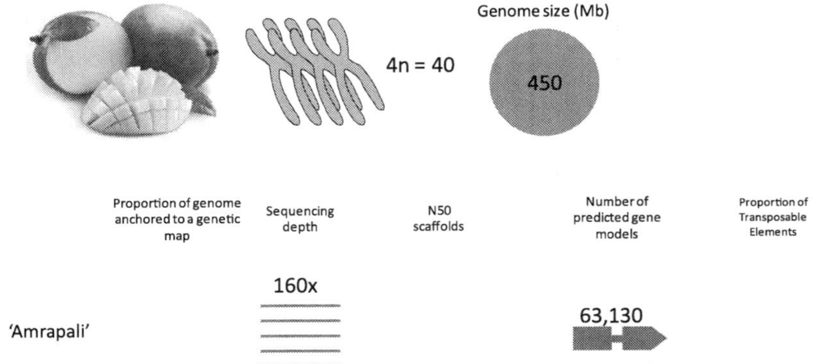

Figure 21 The *Mangifera indica* genome at a glance. (See colour plate)

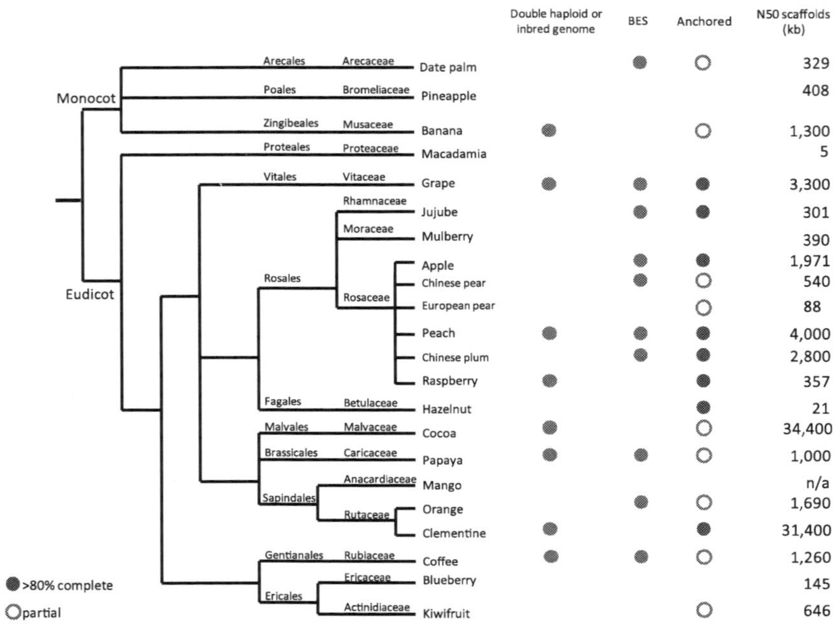

Figure 22 Quality metrics of fruit tree genomes sequenced to date. The species are presented using a simplified phylogeny. The use of double haploid or highly inbred accessions for sequencing, as well as whether BAC-end sequencing (BES) and genetic maps (anchored) were used, or not. The N50 for the scaffolds is indicated in kilobases. (See colour plate)

Clementine mandarin, coffee and grape, by employing highly inbred accessions. BAC-end sequencing (including Sanger sequencing) also improves assembly metrics greatly. For example, the N50 metrics are much higher in Chinese pear than in European pear because of the use of BAC-end sequencing in Chinese pear, while the European pear genome was produced using only NGS data (Chagné et al., 2012; Wu et al., 2013). High-density genetic maps are useful for anchoring of scaffolds, and high-quality draft genome such as peach and Clementine mandarin have almost complete scaffolds anchored to pseudo-chromosomes (Verde et al., 2013; Wu et al., 2014).

3. FRUIT TREES GENE EXPANSIONS AND GENOME EVOLUTION

3.1 Gene Families Expansions in Fruit Trees

A rapid glance at the annotation of fruit tree genomes indicates that the number of genes varies greatly among species. Gene losses and duplications are major driving forces in the adaptive evolution of genomes. Duplicated genes are considered to be the primary mechanism for generating novel gene functions, which may lead to the creation of new species. Duplicated genes normally evolve through one of three scenarios: one copy may lose its function by pseudogenisation (or nonfunctionalization); or one copy evolves a new beneficial function by neo-functionalization and the other copy retains the original function or partitioning of ancestral gene function among duplicated copies may occur by subfunctionalization. The comparative analysis of gene families among species is often carried out in assembled genomes, with the hypothesis that expanded gene families may have evolved new functions that are novel to a particular species.

Bioinformatics tools such as OrthoMCL (http://www.orthomcl.org/orthomcl/) are useful for the comparison of the entire gene sets of sequenced genomes across multiple species. For example, the analysis of the European pear genome demonstrated that a total of 5350 protein-coding clusters identified in European pear were conserved across the proteomes of 13 species. Interestingly, 556 clusters were unique to European pear (Chagné et al., 2014).

Genome analyses of fruit tree species published to date have focused on particular gene families that might explain some of the unique biological features of the species. The grape genome has highlighted the expansion of the *terpene synthase* (*TPS*) gene family, in comparison to *Arabidopsis*, with the hypothesis that the expansion in the number of grape *TPS* genes may be related to terpenoid compound composition and abundance in wine (Jaillon et al., 2007). Sweet orange, jujube and kiwifruit are well-known for being rich in vitamin C and their genome sequence analysis has focused on the genes involved in ascorbate biosynthesis. An expansion of the *D-galacturonic acid reductase* (*GalUR*) family was observed in citrus (Xu et al., 2013), while *aldononolactonase*, *L-ascorbate peroxidase* and *myo-inositol oxygenase* families are expanded in kiwifruit (Huang et al., 2013) and the *monodehydroascorbate reductase* gene family has expanded in jujube (Liu et al., 2014). In coffee, an expansion in the number of N-*methyltransferase* genes that are involved in caffeine biosynthesis was observed (Denoeud et al., 2014). In date palm, the late embryogenesis abundant gene family has expanded, which may relate to their role in conferring resistance to arid conditions (Al-Mssallem et al., 2013).

3.2 Whole Genome Replications

Polyploidy is a widespread feature in plant genomes, including fruit tree species. For example, commercial varieties of kiwifruit, blueberry, mango and banana are polyploid. It is noteworthy that the analysis of the grapevine and apple genomes has contributed greatly to the understanding of genome evolution in the eudicotyledons and such research has illustrated the importance of whole genome duplication (WGD) as a prevailing driving force for creating new gene functions. The conclusions from these analyses support the hypothesis that even diploid species have experienced past polyploidy events (paleoploidy) during their evolution. Grape experienced a paleoploidy event approximately 100—150 million years ago (Mya) and this triplication event (often called the γ event) is shared among eudicotyledon species (Jaillon et al., 2007; Jiao et al., 2011). Lineage-specific events (α and β events) have been demonstrated in other fruit tree species. For example, analysis of the *Malus* genome demonstrated a paleotetraploidisation event (Velasco et al., 2010).

Comparative analysis of genomes often highlights conservation of gene order over long stretches of chromosomal regions among species (synteny). The analysis of the number of paralogous copies of syntenic segments within, between and among species is useful for the detection of WGD events. For example, alignment of homologous genes among chromosomes within the grapevine PN40042 genome and between chromosomes of grape, poplar, rice and *Arabidopsis* was used to reveal the ancestral polyploidisation event that is common to all eudicots (Jaillon et al., 2007). Triplicated paralogous regions within the grape genome emphasize the γ event and comparison of three paralogous grape chromosomes originating from the γ event with the poplar genome indicates that the triplication γ event predated the divergence between poplar and grape. Furthermore, as each of these three paralogous grape chromosomes displayed two regions paralogous with poplar, it is likely that a WGD event has occurred since these two species diverged. However, comparison of the three paralogous grape chromosomes with the rice genome showed that only one orthologous region exists in rice, supporting the hypothesis that the γ event occurred after the divergence between monocotyledons and dicotyledons. For apple, comparative analysis of the assembled chromosomes indicated that homologies exist between pairs of entire chromosomes (e.g. chromosomes 5 and 10) or parts of chromosomes (e.g. distal portions of chromosomes 7 and 2), indicating interchromosomal rearrangements (Velasco et al., 2007).

Another common method of finding evidence for WGD is to analyze the number of synonymous substitutions per synonymous site (Ks value or synonymous distance) calculated for pairs of homologous gene copies (Gaut & Doebley, 1997). Ks estimates neutral evolution and provides an indication of the relative age of the duplication. High and low Ks values indicate highly and less diverged pairs of duplicated genes, which have evolved from ancient or more recent duplication events, respectively. Examples of application of Ks analysis include the detection of recent WGD events in apple between 30 and 45 Mya (Velasco et al., 2010), date palm ~90 Mya (Al-Mssallem et al., 2013) and kiwifruit ~26.7 Mya (Huang et al., 2013).

4. APPLICATIONS OF WHOLE GENOME SEQUENCE AND GENOME-WIDE GENOTYPING TECHNOLOGIES FOR BREEDING

Draft genome sequence assemblies are available for a number of fruit tree species as listed in the previous sections. The challenge is now to translate these decoded genomes into new tools that can be implemented by tree breeders for plant variety improvement. High-quality genome assemblies greatly facilitate this task by enabling the complete inventory of DNA variations in a species, including point mutations, CNVs, indels and epigenetic variations, such as DNA methylation. Whole genome resequencing and DNA variant calling of cultivated varieties and wild species are the ultimate technologies for linking genotype to phenotype for a wide range of traits. Large sets of reproducible and genetic markers anchored to the genome have been developed for most of the 22 species listed above. Some examples of commonly used marker technologies are described in the following sections. The construction of dense genetic maps, which was in the past a matter of months or years of hard work, now takes a few weeks or less, thanks to these genome-wide marker systems.

4.1 Commonly Used Marker Technologies

4.1.1 Simple Sequence Repeats

SSR (Tautz, 1989) are still attractive markers, as they are simple to design and their genotyping requires only basic equipment that is available in most molecular biology laboratories, including PCR machines and gel and capillary electrophoresis apparatus. Genome-wide SSR markers can be efficiently developed by searching for tandemly repeated motifs in genome sequences using bioinformatics tools (for example the software SPUTNIK developed by Chris Abajian at the University of Washington is popular for SSR searches) (Robinson, Love, Batley, Barker, & Edwards, 2004) and is a must less laborious alternative to methods that were used in the past, such as constructing enriched libraries (Billotte, Lagoda, Risterucci, & Baurens, 1999). Numerous successful examples of SSR development from in silico search of cDNA and whole genome sequences have been published in the last 15 years for a wide range of organisms, including fruit tree species. For example, in grape a set of high-quality SSR markers was developed by searching long stretches of di-, tri- and tetranucleotide motifs within the PN40024 v1.0 genome assembly (Cipriani et al., 2008). This set of evenly distributed highly polymorphic SSRs was used for uncovering parentage, genetic diversity and synonymy in a large grape germplasm collection (Cipriani et al., 2010) as well as for genetic map construction in a related genus (Riaz, Hu, & Walker, 2012). One attractive feature of SSRs is that they often cross-amplify between related species because of conservation of the PCR primer sequences (Wunsch & Hormaza, 2002). However, to be truly deemed transferrable between species, SSR markers must have conserved polymorphic repeated motifs.

Despite the advantages of SSR markers, they are laborious to use and not cost-effective compared with SNP-based technologies when a high density of markers is required. Consequently, SNP-based high-throughput assays have replaced SSR analyses for most fruit trees with a genome sequence completed. Many techniques exist for SNP genotyping; however, two types of techniques are particularly noteworthy because of the high number of markers that can be screened simultaneously at a reasonable cost: SNP arrays, and reduced representation GBS.

4.1.2 Single Nucleotide Polymorphism Arrays

SNP arrays such as the Illumina Infinium® II assays (Chagné, Bianco, Lawley, Micheletti, & Jacobs, 2015) are based on the single base extension technique. The principle of the single base extension method is as follows: a probe is designed to target a sequence immediately flanking an SNP and in the case of the Illumina Infinium® II assay, a flanking sequence of 60 bp. Next, the 3′-terminus of the oligonucleotide is extended by a DNA polymerase using ddNTPs labelled with a fluorescent dye. Hence, each terminating fluorescent dye corresponds to an individual base/allele. The Illumina Infinium® II array uses two fluorescent dyes and this makes it possible to detect up to two allelic variants for each variable SNP site. However, because of the limitation imposed by using two colours only, the SNP types cannot be A:T or G:C. To enable high-throughput genotyping of several thousands to millions of SNP simultaneously, the Illumina Infinium® II probes are attached to beads and deployed in a fixed glass slide format.

SNP arrays of the Illumina Infinium® II type have been developed for a range of fruit tree species (Bianco et al., 2014; Chagné et al., 2012; Fujii et al., 2013; Montanari et al., 2013; Myles et al., 2010; Peace et al., 2012; Verde et al., 2012). Myles et al. (2011) characterized genome-wide patterns of genetic variation in several hundred cultivars of *V. vinifera* and its wild relative *V. sylvestris* using the grape 9000 SNP Infinium® array (Myles et al., 2010). They demonstrated that *V. vinifera* was domesticated from *V. sylvestris* in the Near East and have identified parent—offspring and sibling connections, most of them first-degree relationships, among some well-known varieties. To date, three SNP arrays have developed for apple, with 8000 (Chagné et al., 2012), 20,000 (Bianco et al., 2014) and 487,000 (www.fruitbreedomics.com) markers. The apple IRSC SNP array of 8000 SNPs (Chagné et al., 2012) has been used to assess the efficiency of genomic selection for improving fruit quality in an apple breeding programme (Kumar, Chagné, et al., 2012) and to develop a dense SNP-based linkage map of an apple rootstock progeny (Antanaviciute et al., 2012).

4.1.3 Genotyping-by-Sequencing

A recent alternative to SNP arrays is GBS. GBS harnesses the high-throughput power of NGS to enable the detection of large number of polymorphisms (mostly SNPs) throughout the genome. The advantage of GBS over SNP arrays is that this technique does not rely on a fixed and hence biased set of SNPs for genotyping. Instead, de novo SNP detection is performed simultaneously with SNP, calling in sets of 100s or 1000s of individuals.

GBS can take multiple forms and a range of techniques have been described: GBS (Elshire et al., 2011; Poland, Brown, Sorrells, & Jannink, 2012), restriction site-associated DNA sequencing (RAD-seq) (Baird et al., 2008), specific-locus amplified fragment sequencing (SLAF-seq) (Sun et al., 2013) and exome capture (Choi et al., 2009), to name a few. However, all methods cited have the same principle, as they rely on resequencing a selected fraction of the genome that can be captured with high reproducibility across samples. The technique uses DNA barcodes specific to each sample to enable the multiplexing of samples within a single sequencing run. In brief, the GBS protocol is as follows: genomic DNA is digested with restriction enzymes and the digested fragments are then ligated to barcoded adapters, with each DNA sample having its own barcode. The DNA libraries obtained are then amplified using PCR, pooled and sequenced using NGS. Multiple bioinformatics pipelines have been developed to process raw sequencing data and transform them into genotype calls. For example, the TASSEL pipeline can analyze GBS data (Glaubitz et al., 2014).

GBS has been used for developing large sets of markers in raspberry (Ward et al., 2013), apple (Gardner et al., 2014) and pummelo (Guo et al., 2015). The red raspberry GBS analysis was carried out using a segregating population from a *R. idaeus* 'Heritage' × 'Tulameen' cross. DNA samples from 71 progeny and both parents were digested using the *Ape*KI restriction enzyme, following the protocol of Elshire and colleagues (Elshire et al., 2011; Ward et al., 2013). In total, 9143 SNPs segregated in the 'Heritage' × 'Tulameen' population, of which 4521 were used to construct a high-density genetic map.

4.2 Utilization of Genome-Wide Markers for Fruit Tree Breeding

High-density and genome-wide markers systems enable fruit tree scientists to develop an understanding of the genetic control of key horticultural traits. These technologies are used for genotyping large sets of individuals, such as germplasm collections and segregating populations. High-density genetic mapping and GWAS enable the researcher to hone into the genetic loci linked to trait variability, for example by delimitating a genomic region associated with the trait by QTL interval mapping. Once a region linked to a trait is located (typically specified by centiMorgan coordinates for flanking markers on the genetic map surrounding the QTL), it is then possible to scan the genome assemblies within the region for positional candidate genes. As genetic markers are sequence-based and can be retrieved from the genome assemblies, i.e. centiMorgan positions (genetic location) are translated into base pairs (physical location). Further studies are then require to validate the relationship between candidate genes and trait variability, for example by elucidating their putative function on the basis of sequence homology with gene sequences from model systems such as *Arabidopsis*, examining the differential gene expression between individuals carrying the allele or not and by knocking out or complementing the allele in accessions that carry or do not carry the desired allele, respectively. Some examples of the use of genomics to study key horticultural traits are given in other chapters of this book. One outstanding example is the use of the *Prunus* genome to locate a mutation coding for the pillar trait (*br* locus) in peach (Dardick et al., 2013). Two DNA pools consisting of pillar and standard phenotype individuals from a F_2 segregating population were sequenced using Illumina NGS. The resulting sequences were mapped to the reference genome assembly of *P. persica* 'Lovell' and SNPs linked to the *br* locus were identified at the distal end of chromosome 2. Further fine mapping using high-resolution melting SNP markers and a search for structural variants pointed towards Ppa10082 as a potential gene model that might be mutated in the pillar phenotype. Gene expression analysis and transformation in *Arabidopsis* indicated that Ppa10082, which encodes for a homologue of rice *OsTAC1*, is responsible for control of the pillar trait by the *br* locus.

Resequencing of germplasm accessions that contribute an allele linked to a unique phenotype adds to the development of markers useful for improving the efficiency of breeding by application of MAS. Briefly, genetic markers associated with the trait of interest are used by fruit tree breeders to select parents and seedlings carrying the desired allele. A remarkable example was recently published describing the development of SNP markers useful for selecting disease-resistant seedlings in apple breeding populations (Jaensch et al., 2015). In brief, markers flanking eight major loci controlling apple scab, fire blight and powdery mildew resistances were searched in the reference *Malus × domestica* 'Golden Delicious' genome and then resequenced in breeding founders carrying the resistance alleles as well as in susceptible cultivars possessing high fruit quality that are widely used as breeding parents. This enabled the authors to choose a set of markers that will be highly efficient for MAS across a range of breeding germplasm.

In addition to MAS, the availability of high-density genome-wide marker screening systems has enabled the implementation of a further new method for improving the efficiency of fruit tree breeding: genomic (or genome-wide) selection (Kumar, Bink, Volz, Bus, & Chagné, 2012; Kumar, Chagné, et al., 2012). A complete review of the principles and potential of genomic selection is in chapter "Acceleration of Forest and Fruit Tree Domestication by Genomic Selection" by Fikret.

5. THE FUTURE OF FRUIT TREE GENOMICS

An impressive number of fruit tree genomes have been sequenced in the last 5 years. The genome sequences of other popular species such as olive, European plum and avocado, as well as tropical orphan crops, are currently being deciphered (http://africanorphancrops.org/). While the development of draft genome assemblies are key milestones for fruit biologists, the quality of the current draft genomes has room for improvement, for example by improving the size of the scaffolds, reducing the gaps, increasing the proportion of the assembly anchored to genetic maps and haplotyping of the contigs. New molecular and bioinformatics methods that are capable of sequencing large fragments of several kilobases, assembling larger haplotyped scaffolds and visualizing the nanometric structure of DNA molecules, will improve the current genome assemblies in the near future. In the meantime, the working draft genome sequences developed so far are extremely useful, as access to full genome sequence was the biggest challenge for fruit tree species geneticists and breeders in the past. In the near future, newly developed genome assemblies and annotations will translate into a remarkable increase in the number of loci for which the candidate genes and the causative genome variants are identified for key traits such as fruit quality, adaptation to biotic and abiotic factors and improved yield. When coupled to new technologies to reduce the length and number of breeding cycles (Nocker & Gardiner, 2014), the future for fruit breeders will indeed be bright.

ACKNOWLEDGEMENTS

I thank Dr Susan E. Gardiner for critical reading of the manuscript draft, Drs Catherine Nock (Southern Cross University, Australia), Shawn Mehlenbacher (Oregon State University, USA), Erik Rowley and Todd Mockler (Danforth Plant Sciences Center, USA) and Nagendra Singh (National Research Center on Plant Biotechnology, India) for sharing preliminary information on the macadamia nut, hazelnut, black raspberry and mango genome assemblies prior to publication.

REFERENCES

Al-Dous, E. K., George, B., Al-Mahmoud, M. E., Al-Jaber, M. Y., Wang, H., Salameh, Y. M., et al. (2011). De novo genome sequencing and comparative genomics of date palm (*Phoenix dactylifera*). *Nature Biotechnology, 29*(6), 521–U584.

Al-Mssallem, I. S., Hu, S., Zhang, X., Lin, Q., Liu, W., Tan, J., et al. (2013). Genome sequence of the date palm *Phoenix dactylifera* L. *Nature Communications, 4.*

Antanaviciute, L., Fernandez-Fernandez, F., Jansen, J., Banchi, E., Evans, K. M., Viola, R., et al. (2012). Development of a dense SNP-based linkage map of an apple rootstock progeny using the Malus Infinium whole genome genotyping array. *BMC Genomics, 13.*

Argout, X., Salse, J., Aury, J.-M., Guiltinan, M. J., Droc, G., Gouzy, J., et al. (2011). The genome of *Theobroma cacao*. *Nature Genetics, 43*(2), 101–108. http://dx.doi.org/10.1038/ng.736.

Baird, N. A., Etter, P. D., Atwood, T. S., Currey, M. C., Shiver, A. L., Lewis, Z. A., et al. (2008). Rapid SNP discovery and genetic mapping using sequenced RAD markers [Article]. *PLoS One, 3*(10), e3376.

Bian, Y., Ballington, J., Raja, A., Brouwer, C., Reid, R., Burke, M., et al. (2014). Patterns of simple sequence repeats in cultivated blueberries (*Vaccinium* section *Cyanococcus* spp.) and their use in revealing genetic diversity and population structure. *Molecular Breeding, 34*(2), 675–689.

Bianco, L., Cestaro, A., Sargent, D. J., Banchi, E., Derdak, S., Di Guardo, M., et al. (2014). Development and validation of a 20K single nucleotide polymorphism (SNP) whole genome genotyping array for apple (*Malus* × *domestica* Borkh.). *PLoS One, 9*(10).

Billotte, N., Lagoda, P. J. L., Risterucci, A. M., & Baurens, F. C. (1999). Microsatellite-enriched libraries: applied methodology for the development of SSR markers in tropical crops [Article]. *Fruits (Paris), 54*(4), 277–288.

Bushakra, J. M., Stephens, M. J., Atmadjaja, A. N., Lewers, K. S., Symonds, V. V., Udall, J. A., et al. (2012). Construction of black (*Rubus occidentalis*) and red (*R. idaeus*) raspberry linkage maps and their comparison to the genomes of strawberry, apple, and peach [Article]. *Theoretical and Applied Genetics, 125*(2), 311–327.

Celton, J.-M., Chagné, D., Tustin, S. D., Terakami, S., Nishitani, C., Yamamoto, T., et al. (2009). Update on comparative genome mapping between *Malus* and *Pyrus*. *BMC Research Notes, 2*, 182.

Chagné, D., Bianco, L., Lawley, C., Micheletti, D., & Jacobs, J. M. (2015). Methods for the design, implementation, and analysis of illumina infinium SNP assays in plants. *Methods in Molecular Biology, 1245*, 281–298.

Chagné, D., Crowhurst, R. N., Pindo, M., Thrimawithana, A., Deng, C., Ireland, H., et al. (2014). The draft genome sequence of European pear (*Pyrus communis* L. 'Bartlett') [Article]. *Plos One, 9*(4), 12.

Chagné, D., Crowhurst, R. N., Troggio, M., Davey, M. W., Gilmore, B., Lawley, C., et al. (2012). Genome-wide SNP detection, validation, and development of an 8K SNP array for apple. *PLoS One*, 7(2), e31745.

Choi, M., Scholl, U. I., Ji, W. Z., Liu, T. W., Tikhonova, I. R., Zumbo, P., et al. (2009). Genetic diagnosis by whole exome capture and massively parallel DNA sequencing [Article]. *Proceedings of the National Academy of Sciences of the United States of America, 106*(45), 19096–19101.

Cipriani, G., Marrazzo, M. T., Di Gaspero, G., Pfeiffer, A., Morgante, M., & Testolin, R. (2008). A set of microsatellite markers with long core repeat optimized for grape (*Vitis* spp.) genotyping [Article]. *BMC Plant Biology, 8*, 13.

Cipriani, G., Spadotto, A., Jurman, I., Di Gaspero, G., Crespan, M., Meneghetti, S., et al. (2010). The SSR-based molecular profile of 1005 grapevine (*Vitis vinifera* L.) accessions uncovers new synonymy and parentages, and reveals a large admixture amongst varieties of different geographic origin [Article]. *Theoretical and Applied Genetics, 121*(8), 1569–1585.

Clark, M. D., Schmitz, C. A., Rosyara, U. R., Luby, J. J., & Bradeen, J. M. (2014). A consensus 'Honeycrisp' apple (*Malus × domestica*) genetic linkage map from three full-sib progeny populations. *Tree Genetics & Genomes, 10*(3), 627–639.

D'Hont, A., Denoeud, F., Aury, J.-M., Baurens, F.-C., Carreel, F., Garsmeur, O., et al. (2012). The banana (*Musa acuminata*) genome and the evolution of monocotyledonous plants. *Nature, 488*(7410), 213–217. http://dx.doi.org/10.1038/nature11241.

Dardick, C., Callahan, A., Horn, R., Ruiz, K. B., Zhebentyayeva, T., Hollender, C., et al. (2013). *PpeTAC1* promotes the horizontal growth of branches in peach trees and is a member of a functionally conserved gene family found in diverse plants species [Article]. *Plant Journal, 75*(4), 618–630.

Davey, M., Gudimella, R., Harikrishna, J. A., Sin, L. W., Khalid, N., & Keulemans, J. (2013). A draft *Musa balbisiana* genome sequence for molecular genetics in polyploid, inter- and intra-specific *Musa* hybrids. *BMC Genomics, 14*(1), 683.

Denoeud, F., Carretero-Paulet, L., Dereeper, A., Droc, G., Guyot, R., Pietrella, M., et al. (2014). The coffee genome provides insight into the convergent evolution of caffeine biosynthesis. *Science, 345*(6201), 1181–1184.

Droc, G., Lariviere, D., Guignon, V., Yahiaoui, N., This, D., Garsmeur, O., et al. (2013). The banana genome hub. *Database-The Journal of Biological Databases and Curation, 2013*, bat035.

Eduardo, I., Chietera, G., Pirona, R., Pacheco, I., Troggio, M., Banchi, E., et al. (2013). Genetic dissection of aroma volatile compounds from the essential oil of peach fruit: QTL analysis and identification of candidate genes using dense SNP maps. *Tree Genetics & Genomes, 9*(1), 189–204.

Elshire, R. J., Glaubitz, J. C., Sun, Q., Poland, J. A., Kawamoto, K., Buckler, E. S., et al. (2011). A robust, simple genotyping-by-sequencing (GBS) approach for high diversity species. *PLoS One, 6*(5), e19379.

Fujii, H., Shimada, T., Nonaka, K., Kita, M., Kuniga, T., Endo, T., et al. (2013). High-throughput genotyping in citrus accessions using an SNP genotyping array [Article]. *Tree Genetics & Genomes, 9*(1), 145–153.

Gardner, K. M., Brown, P., Cooke, T. F., Cann, S., Costa, F., Bustamante, C., et al. (2014). Fast and cost-effective genetic mapping in apple using next-generation sequencing. *G3 (Bethesda), 4*(9), 1681–1687.

Gaut, B. S., & Doebley, J. F. (1997). DNA sequence evidence for the segmental allotetraploid origin of maize. *Proceedings of the National Academy of Sciences of the United States of America, 94*(13), 6809–6814.

Glaubitz, J. C., Casstevens, T. M., Lu, F., Harriman, J., Elshire, R. J., Sun, Q., et al. (2014). TASSEL-GBS: a high capacity genotyping by sequencing analysis pipeline [Article]. *PLoS One, 9*(2), e90346.

Goff, S. A., Ricke, D., Lan, T. H., Presting, G., Wang, R. L., Dunn, M., et al. (2002). A draft sequence of the rice genome (*Oryza sativa* L. ssp. *japonica*). *Science, 296*(5565), 92–100.

Gonsalves, D. (1998). Control of papaya ringspot virus in papaya: a case study. [Review]. *Annual Review of Phytopathology, 36*, 415–437.

Guo, F., Yu, H., Tang, Z., Jiang, X., Wang, L., Wang, X., et al. (2015). Construction of a SNP-based high-density genetic map for pummelo using RAD sequencing. *Tree Genetics & Genomes, 11*(1).

Gupta, V., Estrada, A. D., Blakley, I. C., Reid, R., Patel, K., Meyer, M. D., et al. (2014). *RNA-Seq analysis and annotation of a draft blueberry genome assembly identifies candidate genes involved in fruit ripening, biosynthesis of bioactive compounds, and stage-specific alternative splicing.* Retrieved from http://biorxiv.org/biorxiv/early/2014/10/08/010116.full.pdf.

He, N., Zhang, C., Qi, X., Zhao, S., Tao, Y., Yang, G., et al. (2013). Draft genome sequence of the mulberry tree *Morus notabilis* [Article]. *Nature Communications, 4*.

Henson, J., Tischler, G., & Ning, Z. (2012). Next-generation sequencing and large genome assemblies. *Pharmacogenomics, 13*(8), 901–915.

Huang, S. X., Ding, J., Deng, D. J., Tang, W., Sun, H. H., Liu, D. Y., et al. (2013). Draft genome of the kiwifruit *Actinidia chinensis* [Article]. *Nature Communications, 4*, 9.

Jaensch, M., Broggini, G. A. L., Weger, J., Bus, V. G. M., Gardiner, S. E., Bassett, H., et al. (2015). Identification of SNPs linked to eight apple disease resistance loci. *Molecular Breeding, 35*(1).

Jaillon, O., Aury, J.-M., Noel, B., Policriti, A., Clepet, C., Casagrande, A., et al. (2007). The grapevine genome sequence suggests ancestral hexaploidization in major angiosperm phyla. *Nature, 449*(7161), 463–U465.

Jiao, Y. N., Wickett, N. J., Ayyampalayam, S., Chanderbali, A. S., Landherr, L., Ralph, P. E., et al. (2011). Ancestral polyploidy in seed plants and angiosperms [Article]. *Nature, 473*(7345), 97–U113.

Jung, S., & Main, D. (2014). Genomics and bioinformatics resources for translational science in Rosaceae. *Plant Biotechnology Reports, 8*(2), 49–64.

Kaul, S., Koo, H. L., Jenkins, J., Rizzo, M., Rooney, T., Tallon, L. J., et al. (2000). Analysis of the genome sequence of the flowering plant *Arabidopsis thaliana*. *Nature, 408*(6814), 796–815.

Kumar, S., Bink, M. C. A. M., Volz, R. K., Bus, V. G. M., & Chagné, D. (2012). Towards genomic selection in apple (*Malus* × *domestica* Borkh.) breeding programmes: prospects, challenges and strategies. *Tree Genetics & Genomes, 8*(1), 1–14.

Kumar, S., Chagné, D., Bink, M. C., Volz, R. K., Whitworth, C., & Carlisle, C. (2012). Genomic selection for fruit quality traits in apple (*Malus* × *domestica* Borkh.) [Article]. *PLoS One,* 7(5), e36674.

Kumar, S., Garrick, D. J., Bink, M. C., Whitworth, C., Chagné, D., & Volz, R. K. (2013). Novel genomic approaches unravel genetic architecture of complex traits in apple. *BMC Genomics, 14*, 393.

Liu, M.-J., Zhao, J., Cai, Q.-L., Liu, G.-C., Wang, J.-R., Zhao, Z.-H., et al. (2014). The complex jujube genome provides insights into fruit tree biology [Article]. *Nature Communications, 5.*

Mathew, L. S., Spannagl, M., Al-Malki, A., George, B., Torres, M. F., Al-Dous, E. K., et al. (2014). A first genetic map of date palm (*Phoenix dactylifera*) reveals long-range genome structure conservation in the palms. *BMC Genomics, 15.*

Metzker, M. L. (2010). Applications of next-generation sequencing technologies — the next generation. [Review]. *Nature Reviews Genetics, 11*(1), 31—46.

Ming, R., Hou, S., Feng, Y., Yu, Q., Dionne-Laporte, A., Saw, J. H., et al. (2008). The draft genome of the transgenic tropical fruit tree papaya (*Carica papaya* Linnaeus). *Nature, 452*(7190), 991—996. http://dx.doi.org/10.1038/nature06856.

Montanari, S., Saeed, M., Knaebel, M., Kim, Y., Troggio, M., Malnoy, M., et al. (2013). Identification of pyrus single nucleotide polymorphisms (SNPs) and evaluation for genetic mapping in european pear and interspecific pyrus hybrids. *PloS One, 8*(10).

Motamayor, J., Mockaitis, K., Schmutz, J., Haiminen, N., Livingstone, D., III, Cornejo, O., et al. (2013). The genome sequence of the most widely cultivated cacao type and its use to identify candidate genes regulating pod color. *Genome Biology, 14*(6), r53.

Myles, S., Boyko, A. R., Owens, C. L., Brown, P. J., Grassi, F., Aradhya, M. K., et al. (2011). Genetic structure and domestication history of the grape. *Proceedings of the National Academy of Sciences of the United States of America, 108*(9), 3530—3535.

Myles, S., Chia, J.-M., Hurwitz, B., Simon, C., Zhong, G. Y., Buckler, E., et al. (2010). Rapid genomic characterization of the genus *Vitis*. *PloS One, 5*(1).

Nock, C. J., Baten, A., & King, G. J. (2014). Complete chloroplast genome of *Macadamia integrifolia* confirms the position of the Gondwanan early-diverging eudicot family Proteaceae. *BMC Genomics, 15.*

Nock, C. J., Elphinstone, M. S., Ablett, G., Kawamata, A., Hancock, W., Hardner, C. M., et al. (2014). Whole genome shotgun sequences for microsatellite discovery and application in cultivated and wild Macadamia (Proteaceae). *Applications in Plant Sciences, 2*(4).

Nocker, S.v., & Gardiner, S. E. (2014). Breeding better cultivars, faster: applications of new technologies for the rapid deployment of superior horticultural tree crops. *Horticulture Research, 1*(14022).

Peace, C., Bassil, N., Main, D., Ficklin, S., Rosyara, U. R., Stegmeir, T., et al. (2012). Development and evaluation of a genome-wide 6K SNP array for diploid sweet cherry and tetraploid sour cherry. *PLoS One,* 7(12), e48305.

Pindo, M., Vezzulli, S., Coppola, G., Cartwright, D. A., Zharkikh, A., Velasco, R., et al. (2008). SNP high-throughput screening in grapevine using the SNPlex (TM) genotyping system. *BMC Plant Biology, 8.*

Pirona, R., Eduardo, I., Pacheco, I., Linge, C. D., Miculan, M., Verde, I., et al. (2013). Fine mapping and identification of a candidate gene for a major locus controlling maturity date in peach. *BMC Plant Biology, 13.*

Poland, J. A., Brown, P. J., Sorrells, M. E., & Jannink, J. L. (2012). Development of high-density genetic maps for barley and wheat using a novel two-enzyme genotyping-by-sequencing approach [Article]. *PLoS One,* 7(2), e32253.

Riaz, S., Hu, R., & Walker, M. A. (2012). A framework genetic map of *Muscadinia rotundifolia* [Article]. *Theoretical and Applied Genetics, 125*(6), 1195—1210.

Robinson, A. J., Love, C. G., Batley, J., Barker, G., & Edwards, D. (2004). Simple sequence repeat marker loci discovery using SSR primer [Article]. *Bioinformatics, 20*(9), 1475—1476.

Shulaev, V., Sargent, D. J., Crowhurst, R. N., Mockler, T. C., Folkerts, O., Delcher, A. L., et al. (2011). The genome of woodland strawberry (*Fragaria vesca*) [Article]. *Nature Genetics, 43*(2), 109–116.

Souleyre, E. J. F., Chagne, D., Chen, X., Tomes, S., Turner, R. M., Wang, M. Y., et al. (2014). The *AAT1* locus is critical for the biosynthesis of esters contributing to 'ripe apple' flavour in 'Royal Gala' and 'Granny Smith' apples. *Plant Journal, 78*(6), 903–915.

Sun, X. W., Liu, D. Y., Zhang, X. F., Li, W. B., Liu, H., Hong, W. G., et al. (2013). SLAF-seq: an efficient method of large-scale de novo SNP discovery and genotyping using high-throughput sequencing [Article]. *PloS One, 8*(3), 9.

Tautz, D. (1989). Hypervariability of simple sequences as a general source for polymorphic dna markers [Article]. *Nucleic Acids Research, 17*(16), 6463–6471.

Troggio, M., Malacarne, G., Coppola, G., Segala, C., Cartwright, D. A., Pindo, M., et al. (2007). A dense single-nucleotide polymorphism-based genetic linkage map of grapevine (*Vitis vinifera* L.) anchoring pinot noir bacterial artificial chromosome contigs. *Genetics, 176*(4), 2637–2650.

Velasco, R., Zharkikh, A., Affourtit, J., Dhingra, A., Cestaro, A., Kalyanaraman, A., et al. (2010). The genome of the domesticated apple (*Malus* × *domestica* Borkh.) [Article]. *Nature Genetics, 42*(10), 833–839.

Velasco, R., Zharkikh, A., Troggio, M., Cartwright, D. A., Cestaro, A., Pruss, D., et al. (2007). A high quality draft consensus sequence of the genome of a heterozygous grapevine variety. *PLoS One, 2*(12), e1326.

Verde, I., Abbott, A. G., Scalabrin, S., Jung, S., Shu, S., Marroni, F., et al. (2013). The high-quality draft genome of peach (*Prunus persica*) identifies unique patterns of genetic diversity, domestication and genome evolution [Article]. *Nature Genetics, 45*(5), 487–494.

Verde, I., Bassil, N., Scalabrin, S., Gilmore, B., Lawley, C. T., Gasic, K., et al. (2012). Development and evaluation of a 9K SNP array for peach by internationally coordinated SNP detection and validation in breeding germplasm [Article]. *PLoS One*, 7(4), e35668.

Vitulo, N., Forcato, C., Carpinelli, E. C., Telatin, A., Campagna, D., D'Angelo, M., et al. (2014). A deep survey of alternative splicing in grape reveals changes in the splicing machinery related to tissue, stress condition and genotype. *BMC Plant Biology, 14.*

Wang, N., Fang, L., Xin, H., Wang, L., & Li, S. (2012). Construction of a high-density genetic map for grape using next generation restriction-site associated DNA sequencing. *BMC Plant Biology, 12.*

Ward, J. A., Bhangoo, J., Fernandez-Fernandez, F., Moore, P., Swanson, J. D., et al. (2013). Saturated linkage map construction in *Rubus idaeus* using genotyping by sequencing and genome-independent imputation. *BMC Genomics, 14*(2).

Wu, G. A., Prochnik, S., Jenkins, J., Salse, J., Hellsten, U., Murat, F., et al. (2014). Sequencing of diverse mandarin, pummelo and orange genomes reveals complex history of admixture during citrus domestication [Research]. *Nature Biotechnology, 32*(7), 656–662.

Wu, J., Wang, Z. W., Shi, Z. B., Zhang, S., Ming, R., Zhu, S. L., et al. (2013). The genome of the pear (*Pyrus bretschneideri* Rehd.) [Article]. *Genome Research, 23*(2), 396–408.

Wunsch, A., & Hormaza, J. I. (2002). Molecular characterisation of sweet cherry (*Prunus avium* L.) genotypes using peach *Prunus persica* (L.) Batsch SSR sequences. *Heredity, 89*, 56–63.

Xu, Q., Chen, L.-L., Ruan, X., Chen, D., Zhu, A., Chen, C., et al. (2013). The draft genome of sweet orange (*Citrus sinensis*). *Nature Genetics, 45*(1), 59–66. http://dx.doi.org/10.1038/ng.2472.

Yu, J., Hu, S. N., Wang, J., Wong, G. K. S., Li, S. G., Liu, B., et al. (2002). A draft sequence of the rice genome (*Oryza sativa* L. ssp. *indica*). *Science, 296*(5565), 79–92.

Zhang, Q., Chen, W., Sun, L., Zhao, F., Huang, B., Yang, W., et al. (2012). The genome of *Prunus mume*. *Nature Communications, 3*, 1318.
Zhang, J., Liu, J., & Ming, R. (2014). Genomic analyses of the CAM plant pineapple. *Journal of Experimental Botany, 65*(13), 3395–3404.

CHAPTER TWO

Forest Tree Genomics: Review of Progress

Geneviève J. Parent*, Elie Raherison*, Juliana Sena*, John J. MacKay*,†,1

*Centre for Forest Research and Institute for Systems and Integrative Biology, Université Laval, Quebec, QC, Canada

†Present address: Department of Plant Sciences, University of Oxford, Oxford, UK

[1]Corresponding author: E-mail: John.Mackay@plants.ox.ac.uk

Contents

Advances in Botanical Research, Volume 74
ISSN 0065-2296
http://dx.doi.org/10.1016/bs.abr.2015.05.004

Abstract

Forest tree genomics is progressing at an accelerated pace owing to recent developments in next-generation sequencing (NGS) technologies. With NGS, genomics research has simultaneously gained in speed, magnitude and scope. In the last few years, the first conifer genomes at a staggering size of 20–24 gigabases and the genomes of several hardwood trees have been sequenced and analyzed. Biological insights have resulted from these sequencing initiatives as well as from genetic mapping, gene expression profiling and gene discovery research over nearly two decades. This review emphasizes major areas of progress in forest tree genomics, including insights into genome evolution, genome function arising from large-scale gene expression profiling, the genomic architecture of quantitative traits and the population genomics of adaptation. We discuss future directions in these areas with potential inputs from NGS technologies and propose avenues for developing a more integrated understanding of genetic diversity and its impacts on genome function. These directions promise to sustain research aimed at addressing emerging challenges in forestry and produce applied outputs to preserve, enhance and responsibly use world forests.

1. INTRODUCTION

With the development of next-generation sequencing (NGS) technologies, genomics research has simultaneously gained in speed, magnitude and scope, resulting in unprecedented research outputs. The potential to analyze whole genomes of thousands of individuals in model plants and animals and to rapidly apply these approaches to nonmodel systems such as forest trees is nothing less than revolutionary. In just a few years, NGS has enabled the sequencing of several conifer genomes estimated at 20–24 Gb in size (Birol et al., 2013; Neale et al., 2014; Nystedt et al., 2013) and genome resequencing in poplar (Evans et al., 2014; Porth et al., 2013). Projects such as sequencing conifer genomes still represent a major feat but the methods and capacity are being developed to overcome the inherent challenges.

Insights into forest tree genomes and their evolution arise from recent genome-sequencing initiatives, as well as developments in large-scale gene discovery, genetic mapping, gene expression profiling and association mapping over nearly two decades. We review the knowledge gained from these advances, discuss emerging questions and outline knowledge gaps with a view to potential inputs from other systems and NGS. Given the breadth and scope of the research, we have not attempted to cover all of the recent progress to equal depth but have focussed on areas of major activity and attempted to identify potentially fruitful areas for future investigation. These research directions promise to sustain and enhance research outputs and applied outcomes such as those recently developed from genomic selection.

2. WHY RESEARCH FOREST TREE GENOMICS?

Forest trees are present in many taxonomic groups among the angiosperms and the gymnosperms. Because of the major ecological and economic importance of trees and forests in many parts of the world, forestlands are facing increasing pressure from industrial uses, deforestation for agricultural production, and urban expansion. Their management and conservation is further challenged by the acceleration of environmental changes, the emergence of new diseases and the upsurge of insect pests.

2.1 Species Diversity, Ecological and Economic Importance

Forest trees are nearly as taxonomically diverse as the extant seed plants themselves. Trees species are found among the gymnosperms and the angiosperms (Magnoliophyta); however, extant angiosperm trees are overwhelmingly represented within the eudicots and largely absent from the monocots (Groover, 2005).

All but two of the 35 orders of eudicots contain tree species along with species with various degrees of woody growth such as herbs, bushes or shrubs (Stevens, 2012), clearly indicating that they do not form a monophyletic group (Groover, 2005). The evolutionary and molecular implications have been discussed by Groover (2005), among others. Angiosperm tree species number in the tens of thousands. The Amazon alone was estimated to harbour 16,000 different tree species, although it dominated by 227 species which account for 50% of the individuals (ter Steege et al., 2013). Some genera have diversified to form a large number of tree and shrub species and occupy many different habitats and regions, for example, eucalypts

(*Eucalyptus* spp., 800 species), oaks (*Quercus* spp., 400 species), willows and poplars (*Salix* and *Populus* spp., 400+ species), maples *(Acer* spp., 126 species), nothofagus (southern beeches, *Nothofagus* spp., 35 species) (Mabberley, 1987) and acacia (*Acacia* spp., a nonmonophyletic group of 1030 species; Miller, Seigler, & Mishler, 2014) (the Angiosperm Phylogeny Website v13, http://www.mobot.org/MOBOT/research/APweb/).

Gymnosperm trees on the other hand are largely represented by a single order, i.e. the conifers (Coniferales), which is the largest and most studied of the gymnosperm lineages. Conifers represent 635 recognized species out of the fewer than 1000 species of extant gymnosperms while cycads and ginko representing only a handful of species (www.catalogueoflife.org/; Farjon & Page, 1999; Gernandt, Willyard, Syring, & Liston, 2011).

Note that in this chapter, we refer to angiosperm trees as hardwoods or hardwood trees and for simplicity we will discuss conifer trees as the main representatives of the gymnosperm trees, are often referred to as softwoods or softwood trees.

Both angiosperm and gymnosperm are found in a variety of habitats across the different forested biomes (FAO, 2010). Hardwood trees including many nondeciduous species represent the dominant tree form across tropical forests (ter Steege et al., 2013) and subtropical forests around the world. Deciduous hardwood forests dominated by a variety of oaks, maples, beech and many other species are found in Eastern North America and Europe (Archibold, 1995) as well as Eastern Asia (Wen, 1999). Deciduous hardwoods also grow together with conifers, most often in temperate and boreal regions; aspen and birch also extend far into boreal regions. The conifers are often associated with boreal forests in the Northern hemisphere (e.g. *Picea mariana* in Canada, Farrar, 1995) and high mountainous locations (e.g. *Picea mexicana* in Mexico, Ledig, Jacob-Cervantes, Hodgskiss, & Eguiluz-Piedra, 1997) but they are also distributed in a variety of habitats including evergreen subtropical forests (e.g. species in Vietnam; Wang, Abbott, Ingvarsson, & Liu, 2014) and from the sea-level ranges (e.g. *Pinus pinaster* in Western Europe, Burban & Petit, 2003).

Because forest trees dominate many of the world's ecosystems, they play an important role in global carbon, nutrient and atmospheric cycles, and are essential for the provision of many ecosystem services. Trees are also widely used in reforestation programmes in tropical, temperate and boreal regions. They play a significant role in local and global economies because of their amenability to large-scale plantations to produce wood, their role in landscape management, their rapid growth potential with low input requirements, relative ease of processing to make both paper and solid wood products and wide use as source of renewal energy.

Over the last several decades, genetic selection and breeding programmes have been implemented to a wide variety of trees species as a basis to establish productive plantations and for restoration purposes in both the Northern and Southern hemispheres (e.g. see White, Adams, & Neale, 2007; Zobel & Talbert, 1984). For hardwoods, targeted genera include eucalypts (*Eucalyptus* spp.), poplars (*Populus* spp.), oaks (*Quercus* spp.) and willows (*Salix* spp.), among others. For conifers, major genera targeted by breeding include pines (*Pinus* spp.), spruces (*Picea* spp.), Douglas-fir (*Pseudotsuga menziesii*), larches (*Larix* spp.) and Japanese cypress (*Cryptomeria japonica*), among others. However, forest tree breeding on a large scale is relatively recent and the vast majority of forests and forest tree plantations are made up of largely undomesticated tree species. Furthermore, most of the world's forests are derived from natural regeneration (FAO, 2010).

2.2 Unique Features of Forest Trees

Forest trees bring together a unique combination of genetic and biological features which condition their evolution and adaptability. Forest trees are the longest lived organisms on earth which means that several generations may overlap and interbreed and, that considerable phenotypic plasticity is needed to withstand changing conditions. In terms of their genetic makeup, many tree species are highly outbreeding and heterozygous (White et al., 2007), have high levels of gene flow owing to wind pollination (Kremer et al., 2012) and tend carry a high genetic load, all of which influence population levels of differentiation and local adaptation. Forest trees encompass a wide range of genome sizes from the very large as seen in conifers, to the compact as seen in poplars and eucalypts.

2.3 Contemporary Issues and Emerging Challenges

There is growing evidence that the health and adaptation of forest trees populations is becoming increasingly challenged by ongoing environmental changes, whether is associated with the effects of globalization, climate warming or others factors. Decimation of the American chestnut by an introduced blight-causing bacteria which occurred in the first half of the twentieth century represents one of the earliest and most striking examples of the impacts of globalization on forests (Anagnostakis, 1987). The first decade of the twenty-first century has provided us with striking examples of shifts in insect pests and the emergence of new pathogens with devastating effects. For example, plant pathogens such as *Phythophthora* spp. have moved around the world with globalization and in some cases have jumped to new hosts. In 2009, *Phythophthora ramorum* (W. De Cock and Man in't Veld) an oomycete that causes sudden oak death in America, was reported to infect larch plantations causing an epidemic in the United Kingdom (Brasier & Webber, 2010). Meanwhile, the mountain pine beetle (*Dendroctonus ponderosae* Hopkins) has decimated tens of thousands of hectares of pine forest in Western North America (Kurz et al., 2008) because of temperature-driven range expansion (Raffa, Powell, & Townsend, 2013). Genomics is rapidly becoming part of the toolkit to develop an improved understanding of tree defences and the evolution of diseases and pests that represent threats to tree health.

Further climate changes expected before the end of the twenty-first century are likely to intensify adaptation challenges. Simulations indicate that up to 60% of tree species in boreal and temperate regions will have a hard time adjusting to warmer climates predicted for 2085 (Hamann & Wang, 2006). Aitken, Yeaman, Holliday, Wang and Curtis-McLane (2008) outlined the three possible outcomes for forest tree populations under present climate warming scenarios adaptation, migration or extirpation. The migration potential of most forest trees is very unlikely to track forecasted rates of climate changes (Aitken et al., 2008). In the warmest parts of existing ranges, extirpation is expected to occur as a result of maladaptation. Extirpation of even a single species may have short- or long-term consequences depending on the species abundance, the scale of the change and the fragmentation of the population, among others. Adaptation potential is more complex to ascertain and is likely to vary significantly depending on several interacting factors (Aitken et al., 2008). For example, adaptation will depend upon phenotypic variation and standing genetic variation (Siol, Wright, & Barrett, 2010), strength of selection, fecundity and biotic interactions. Understanding which part of standing genetic variation is adaptive as opposed to neutral is a central research theme in evolutionary biology and was identified as a major challenge to address for forest tree genomics (Neale & Kremer, 2011).

The development of forest tree genomics has been largely driven by the opportunity to accelerate tree breeding and domestication as reviewed by Harfouche et al., (2012). Recent developments have also brought into focus opportunities to address emerging issues and challenges facing trees and forests. For example, assisted migration as solution to mitigate impacts of climate change may benefit from insights from genetics and genomics research (Aitken et al., 2008; Alberto et al., 2013). This review covers the major areas of progress in forest tree genomics including genome evolution, insights being derived from gene expression profiling, the genomic bases of adaptation and explore some future directions for integrating our understanding of major types of genetic diversity in relation to genome function. This synthesis aims to set the stage for future developments and for addressing the emerging challenges in the twenty-first century.

3. GENE DISCOVERY AND DERIVED GENOMIC RESOURCES

Gene discovery based on large-scale expressed sequence tags (EST) and complimentary DNA (cDNA) sequencing has played a large role in forest tree genomics research owing to the lack of references genomes and large size of conifer genomes (Mackay et al., 2012; Neale & Kremer, 2011). A survey of public gene data repositories shows that the species with the most available sequence data belong to the Pinaceae (cryptomeria, pines, spruces and others), the Salicaceae (mainly poplars), the Fagaceae (oak, chestnut, beech) and Myrtaceae (eucalyptus) (Table 1). The outcomes have enabled the development of gene databases (Sjödin et al., 2009; Wegrzyn, Lee, Tearse, & Neale, 2008), transcriptome characterization (Rigault et al., 2011) and profiling (see below) and efficient genotyping platforms (e.g. Eckert et al., 2009), among others.

Coding sequence conservation within the plant kingdom has meant that the majority of sequences from forest trees are similar to known plant sequences and may be assigned a predicted gene function (Kirst et al., 2003; Noveas et al., 2008; Sterky et al., 1998). This clearly facilitates comparative studies; however, 30—40% of genes typically do not match proteins of known function (Kirst et al., 2003; Rigault et al., 2011). In recent years, gene sequence discovery and analysis has moved to higher throughput pyrosequencing (Parchman, Geist, Grahnen, Benkman, & Buerkle, 2010) and RNA sequencing (RNA-seq) (see Table 1, short read archive) which also has the advantage of facilitating simultaneously identification of sequence variations (single nucleotide polymorphisms, SNPs) and gene expression levels (Camargo et al., 2014; Chen, Uebbing, et al., 2012; Padovan, Lanfear, Keszei, Foley, & Kulheim, 2013; Yeaman et al., 2014). The reduced cost per unit of sequence has also led to the analysis of species not previously studied such as Chinese fir (Wang et al., 2013) and haloxylon (a desert tree) (Long et al., 2014).

Table 1 Genome characteristics and development of genomics resources in major angiosperm and gymnosperm trees

Genus	Species	Genome size 2C (pg)[a]	Chromosome number[b]	Reference genome	ESTs[c]	SNP[c]	Genetic map[d]	Short read archive[c] RNA	DNA
Angiosperms									
Acacia	*mangium*	1.3	13	no	9110	928	Butcher and Moran (2000)	3	0
Castanea	*dentata*	800 MB[e]	12	no	34,800	11,924	Sisco et al. (2005)	5	0
	mollissima	1.6[f]	12	Fang et al. (2013)	9480	1392	Sisco et al. (2005)	5	8
	sativa	2.0	12	no	613		Casasoli et al. (2006)	1	0
Eucalyptus	*camaldulensis*	1.3	11	Hirakawa et al. (2011)	58,584		Brondani, Williams, Brondani, & Grattapaglia, (2006)	12	2
	globulus	1.1	11	Ref. in Myburg et al. (2014)	28,893		Thamarus, Groom, Murrell, Byrne, & Moran, (2002)	1	5
	grandis	1.2[g]	11	Myburg et al. (2014)	42,576		Arumugasundaram et al. (2011)	14	64
	urophylla	1.3	11	no	7440	152	Grattapaglia & Sederoff, (1994)		
Fagus	*grandifolia*	1.1	12	no	23,668	1231	No	2	0
	sylvatica	1.0	12	no	31,309		Scalfi et al. (2004)	5	0
Fraxinus	*excelsior*	2.0	23	www.ashgenome.org	12,083		no	0	22

Populus	*alba*	1.0	19	no	162	470	Paolucci et al. (2010)	0	80
	deltoides	1.1[h]	19	no	14,661		Yin, DiFazio, Gunter, Riemenschneider, & Tuskan, (2004)	71	9
	nigra	1.1	19	no	51,361		Cervera et al. (2001)	0	0
	tremula	0.9	19	no	37,313		Pakull, Groppe, Meyer, Markussen, & Fladung, (2009)	17	122
	trichocarpa	1.0	19	Tuskan et al. (2006)	89,943	1154	Cervera et al. (2001)	99	1063
Quercus	*petraea*	1.6	12	no	58,230	254	Bodenes et al. (2012)	9	0
	robur	1.9	12	Plomion et al. (2015)	81,671	12,784	Bodenes et al. (2012)	68	2
	suber	1.9	12	no	6698		no	36	2
Gymnosperms									
Abies	*alba*	33.1	12	no	2806	258	no	2	0
Araucaria	*angustifolia*	44.7	13	no	10		no	24	0
Cryptomeria	*japonica*	22.1	11	no	61,500		Tani et al. (2003)	3	0
Picea	*abies*	40.0	12	Nystedt et al. (2013)	14,345	674	Lind et al. (2014)	113	15
	glauca	32.3	12	Birol et al. (2013)	313,353	219,402	Pelgas et al. (2006)	21	57
	mariana	34.9	12	no	4598	773	Kang, Mann, Major, & Rajora, (2010)	0	8
Pinus	*banksiana*	45.5	12	no	36,379		no	3	0
	contorta	44.2	12	no	40,483		Li & Yeh, (2001)	54	0
	densiflora	50.1	12	no	3316		Kim, Choi, & Kang, (2005)	0	0
	echinata	45.5	12	no	107		No	0	0
	elliottii	46.6	12	no	150		Nelson, Nance, & Doudrick, (1993)	24	0

(Continued)

Table 1 Genome characteristics and development of genomics resources in major angiosperm and gymnosperm trees—cont'd

Genus	Species	Genome size 2C (pg)[a]	Chromosome number[b]	Reference genome	ESTs[c]	SNP[c]	Genetic map[d]	Short read archive[c] RNA	DNA
	massoniana	51.4	12	no	124		Li, Chen, et al. (2010)	1	0
	patula	43.8	12	no	23		no	0	0
	pinaster	57.8	12	no	34,753	5739	de Miguel et al. (2012)	25	0
	pinea	60.8	12	no	326		no	2	0
	radiata	48.5	12	no	8717	1652	Moraga-Suazo et al. (2014)	0	1
	sylvestris	46.0	12	no	19,610	1455	Komulainen et al. (2003)	9	2
	taeda	44.2	12	Neale et al. (2014)	328,662	15,005	Echt et al. (2011)	48	115
	thunbergii	44.0	12	no	3299		Kondo et al. (2000)	6	3
Pseudotsuga	*menziesii*	38.1	13	no	18,142	470	Eckert et al. (2009)	105	72

EST, expressed sequence tags; SNP, single nucleotide polymorphism.
[a]http://data.kew.org/except for those annotated.
[b]Chromosome counts database.
[c]NCBI.
[d]One map presented.
[e]http://www.hardwoodgenomics.org/.
[f]Barow & Meister, 2003.
[g]Grattapaglia & Bradshaw, 1994.
[h]Ahuja & Neale, 2005.

One of the most significant genomic resources derived from EST and cDNA sequencing are genotyping platforms, which have led to the construction of genetic maps of higher density (Eckert et al., 2009; Geraldes et al., 2013; Neves, Davis, Barbazuk, & Kirst, 2014) and several others (see Table 1). These in turn have enabled structural analyses (Pavy et al., 2012) and comparative genomics studies (Bartholome et al., 2014; Komulainen et al., 2003; Pavy et al., 2012).

4. GENOME ANALYSIS AND EVOLUTION

Forest tree genome sequencing has accelerated significantly very recently. With the development of NGS technologies, most forest tree genomes have been reported in 2013 and 2014. To date, published forest tree genomes span both hardwood and softwood trees distributed among several genera including *Populus* (Tuskan et al., 2006), *Salix* (Dai et al., 2014), *Eucalyptus* (Myburg et al., 2014), *Betula* (Wang et al., 2013), *Fraxinus* (http://www.ashgenome.org), *Castanea* (http://www.hardwoodgenomics.org/chinese-chestnut-genome), *Quercus* (Plomion et al., 2015), *Picea* (Birol et al., 2013; Nystedt et al., 2013) and *Pinus* (Neale et al., 2014) (see Table 1). In this section, we focus on the most fully characterized hardwood genomes; *Populus* and *Eucalyptus* and on recently available conifer genomes.

4.1 Genome Sequencing and Assembly

4.1.1 Populus

The first forest tree genome sequenced was that of a *Populus trichocarpa* female tree (Nisqually-1). It was obtained by using a hybrid strategy that combined whole-genome shotgun sequencing, construction of a physical map based on bacterial artificial chromosome (BAC) restriction fragment fingerprints, BAC-end sequencing and extensive genetic mapping based on simple sequence repeat length polymorphisms that allowed chromosome reconstruction with the assembled genome (Tuskan et al., 2006). An improved version (V3.0) of the *Populus* genome assembly includes 81 Mb of finished clone sequences combined with a new high-density physical map. The genome assembly is approximately 422.9 Mb arranged in 1446 scaffolds with 181 scaffolds greater than 50 kb in size, representing approximately 97.3% of the genome. Key descriptive statistics are the N50 (number of contigs that collectively cover at least 50% of the assembly) and the L50 (length of the shortest contig among those that collectively cover 50% of the assembly); they were assessed for contigs and scaffolds. For contigs, the N50 is 206 and the L50 is 552.8 Kb; for scaffolds, the N50 is 8 and the L50 is 19.5 Mb. This assembly can be accessed in the JGI comparative plant genomics portal at: http://phytozome.jgi.doe.gov.

4.1.2 Eucalyptus

A first nonredundant chromosome-scale reference (V1.0) sequence for BRASUZ1 (an inbred *Eucalyptus grandis* tree) was assembled based on whole-genome Sanger shotgun sequencing, paired-end BAC sequencing and a high-density genetic linkage mapping (Myburg et al., 2014). A recent comparison between new high-resolution genetic maps for *E. grandis* and *Eucalyptus urophylla* (Bartholome et al., 2014) with the reference genome highlighted 85% of collinear regions and 43% noncollinear regions and 13% nonsyntenic regions. These regions were corrected in the latest version (V2.0) which is available on Phytozome 10 (http://phytozome.jgi.doe.gov/pz/portal.html#!info?alias=Org_Egrandis). The *E. grandis* assembly (V2.0) is approximately 691 Mb arranged in 4943 scaffolds with 288 scaffolds greater than 50 kb in size, representing approximately 94.2% of the genome. Approximately 641 Mb is arranged in 32,835 contigs (~7.4% gap). For the scaffolds, the N50 is 5 and the L50 is 57.5 Mb; for the contigs, the N50 is 2267 and the L50 is 67.2 kb.

4.1.3 Conifers

Genome sequences were recently reported for *Picea abies* (Nystedt et al., 2013), *Picea glauca* (Birol et al., 2013) and *Pinus taeda* (Neale et al., 2014). In addition, assemblies were released for *Pinus lambertiana* and *Pseudotsuga menziesii* (http://pinegenome.org/pinerefseq/), and reduced depth sequencing was reported for six other species (Nystedt et al., 2013). These developments are driven by progress in shotgun genome sequencing and associated bioinformatics methods (Nystedt et al., 2013; Simpson et al., 2009; Zimin et al., 2013) which have been applied to analyzing both haploid (*P. abies* and *P. taeda*) and diploid conifer DNA. Different strategies were explored to assemble the genomes into contigs and scaffolds by making use of fosmid sequences (Nystedt et al., 2013) and RNA-seq data. The sequences and assemblies are shedding new light into conifer genome evolution (De La Torre et al., 2014; Soltis & Soltis, 2013); however, assemblies reported to date remain highly fragmented, comprised of greater than 10 million unordered scaffolds and have a scaffold L50 between 6 kb and 67 kb, which is 3–4 orders of magnitude less than the *Populus* and *Eucalyptus* genomes. The very large size and the highly repetitive content of conifer genomes continue to represent a challenge for achieving more contiguous assemblies. We may also expect that the abundance of pseudogenes will complicate further analyses and finishing of assemblies.

4.2 Genome Evolution in Hardwood and Conifer Trees

It is not surprising given the very large difference in genome sizes that genome structure and evolution differ greatly between *Eucalyptus* and *Populus* on the one hand, and conifers on the other. The conifers stand out as having the largest average genome sizes among plant orders, which have been estimated between 18 to over 35 Gbp (Murray, Leitch, & Bennett, 2012). In contrast, the genomes of *Populus* (450 Mbp) and *Eucalyptus* (640 Mbp) are much more compact. For example, at 20 Gbp, the *P. glauca* genome is 31 and 44 times larger than the *Populus* and *Eucalyptus* genome, respectively (Table 1). It is well known that large genomes among angiosperms are the consequence of multiple genomes duplications and polyploidization events with intense periods of transposable elements (TEs) activity and multiplication (Bennetzen, 2002). In conifer genomes analyzed to date, there is no evidence of polyploidization or whole-genome duplications (WGD), but retrotransposons are abundant and widespread (Neale et al., 2014; Nystedt et al., 2013; Wegrzyn et al., 2014).

4.2.1 Transposable Elements

TEs are widespread in plant genomes, exceptionally abundant in species with large genomes and play a major role in their evolution.

Hardwood tree genomes comprise significant but variable TEs content. As in many plant species, retrotransposons account for a major portion of the *Eucalyptus* genome (44.5%), with LTR-RT sequences being the most abundant (21.9%) (Myburg et al., 2014). The DNA transposons (class II TEs) represent only 5.6% of the genome and Helitron elements were found to be the most abundant with an estimated 15,000 copies (3.8% of the genome) (Myburg et al., 2014). *Populus trichocarpa* has approximately 40% of repetitive elements; however, a small fraction seems to be TEs as described in RepPop (Zhou & Xu, 2009). The most abundant classes of TEs are LTR Gypsy and Copia (Douglas & DiFazio, 2010).

In conifer trees, TEs can represent a large portion of the genomes, estimated at 69% in *P. abies* (Nystedt et al., 2013) and up to 80% in *P. taeda* (Wegrzyn et al., 2014). Class I TEs, retrotransposons, are by far the most abundant and are primarily represented by long terminal repeat retrotransposons (LTR-RT). The LTR-RT sequences were estimated to represent

58% of the genome both in *P. abies* and the *P. taeda* (Neale et al., 2014; Nystedt et al., 2013; Wegrzyn et al., 2014). Only three families, the *Ty3/Gypsy*, *Ty1/Copia* and *Gymny* superfamilies make up the bulk of LTR-RTs in conifers as shown by recent genome annotations (Morse et al., 2009; Neale et al., 2014; Nystedt et al., 2013; Wegrzyn et al., 2014) and BAC sequencing (Kovach et al., 2010; Magbanua et al., 2011; Sena et al., 2014).

TEs have variable roles in the evolution of trees genomes. In *Populus*, it was suggested that very few TEs are transcriptionally active. Their estimated insertion date indicated that Gypsy and Copia elements have both been active after separation of the different poplar sections but with different time courses (Cossu, Buti, Giordani, Natali, & Cavallini, 2012). A comparison of *Eucalyptus globulus* (530 Mbp) and *E. grandis* (640 Mbp) indicated that recent TE activity only accounts for 2 Mbp of genome size difference and that a very large number of small nonactive TEs account for most of the difference. A parallel may be drawn to comparison between the congeneric *Arabidopsis thaliana* (125 Mbp) and *Arabidopsis lyrata* (~200 Mbp) genomes, but in the case of *Arabidopsis* most of the difference in genome size could be accounted for by hundreds of thousands of small deletions, mostly in noncoding DNA (Hu et al., 2011). By comparison, conifers present a completely different evolutionary history. The accumulation of TEs in conifers is very ancient and has occurred over a very long time frame spanning tens to hundreds of millions of years (Nystedt et al., 2013). The lack of removal of replicated LTR-RTs appears to be responsible for their massive accumulation rather than a higher rate of multiplication (Morgante & Poali, 2011; Nystedt et al., 2013).

4.2.2 Gene Content

Gene content, i.e. the number of predicted genes, was estimated to be in the same range for *Populus* and *Eucalyptus*, but could be slightly higher in conifers. In *Populus*, Tuskan et al. (2006) identified a first-draft reference set of 45,555 protein-coding gene loci in the nuclear genome using a variety of ab initio, homology-based and expressed sequence tag. Since then, the gene models have been improved by using RNA-seq transcript assemblies. Phytozome v10.1 (http://phytozome.jgi.doe.gov) contains 41,335 loci containing protein-coding transcripts for poplar. In *E. grandis*, 36,349 protein-coding transcripts were predicted based on EST and cDNA data. The gene models are also available in Phytozome v10.1 (http://phytozome.jgi.doe.gov).

Gene content estimates ranged from 50,174 in *P. taeda* (Wegrzyn et al., 2014) to 70,968 in *P. abies* (Nystedt et al., 2013), but only about one-third of them were reported as high confidence, i.e. supported by expressed sequences. Conifer genome annotations have revealed a surprisingly large fraction of sequences classified as genes or gene-like fragments. Gene-like sequences represented 2.4% and 2.9% of the *P. abies* and *P. taeda* genome, respectively, (Neale et al., 2014; Nystedt et al., 2013) and as high as 4% from earlier analyses (Morgante & Paoli, 2011). This is far larger than that would be expected for the number of predicted genes. This discrepancy may be explained by the abundance of pseudogenes reported in conifers (Bautista et al., 2007; Kovach et al., 2010; Magbanua et al., 2011) for which a genome-wide characterization is still lacking.

One factor that may explain the difference in gene number between poplar, eucalyptus and conifer species is their different polyploidization histories. There is no evidence of polyploidization in the Pinaceae and a well-documented history of polyploidy events in *Populus* and *Eucalyptus*. Other factors which may have an influence are tandem duplication frequency, gene evolution rates and the evolutionary forces that influence the fate of duplicated copies.

4.2.3 Retention of Tandem Duplications versus WGD in Populus *and* Eucalyptus

Single gene and WGD have played a major role in evolution of angiosperm plants. The genome sequence of *Populus* and *Eucalyptus* provided evidence of two WGD, an ancient paleohexaploidy event shared with many dicotyledonous plants, and a more recent and lineage-specific WGD. The recent WGD detected in *Populus* was specific of *Salicaceae* family and occurred 60–65 Myr ago (Tuskan et al., 2006) whereas, in *Eucalyptus*, the lineage-specific WGD occurred about 106–114 Myr ago. Interestingly, the *Eucalyptus* WGD is older than those detected in other rosids and could have played an important role in the origin of Myrtales (Myburg et al., 2014).

Over the course of evolution, duplicated gene copies resulting from WGD events may be retained as indicated by the 8000 pairs of duplicated genes in *Populus*. Duplicated genes may retain the same set of functions as the ancestral copy (Davis & Petrov, 2004), retain only a subset of the original set of functions (subfunctionalization) (Lynch & Force, 2000), acquire a new function (neofunctionalization) or degrade into a nonfunctional gene (nonfunctionalization) (Ohno, 1970). Rodgers-Melnick et al. (2012) used microarray expression analyses of a diverse set of tissues in *Populus* and functional annotation to evaluate the factors that are associated with the retention of duplicate genes. They hypothesized that duplicate gene retention from WGD in *Populus* is driven by a combination of sub-functionalization of duplicate pairs and purifying selection favouring retention of genes encoding proteins with large numbers of interactions as proposed by the gene balance hypothesis. This hypothesis posits that genes encoding components of multi-subunit complexes are more likely to evolve in concert because the dosage change in the quantities of subunits affects the interaction and function of the whole complex (Birchler & Veitia, 2007).

Gene loss in *Populus* after the salicoid genome duplication has been less extensive than following the previous WGD (c. 120 Myr), suggesting that the *Populus* genome reorganization is a dynamic process in progress. In contrast to *Populus*, most of the *Eucalyptus* duplicates have been lost after their most recent WGD. The extensive loss of duplicates in *Eucalyptus* has been shown by a pairwise comparison of syntenic segments with *Vitis*, which was selected for comparison because it is a basal rosid lineage that is a paleohexaploid and without evidences of more recent WGD events as were detected in *Populus* and *Eucalyptus* (Jaillon et al., 2007).

In contrast to genes encoding proteins with large numbers of interactions, genes with poorly connected products in a network would have an elevated probability of retention following tandem duplication (Ren et al., 2014). A study of the gene family of class III peroxidase (PRX) in *Populus* identified other mechanisms that play a role in gene retention such as protein subcellular relocalization associated with a new function. Class III PRX are involved in stress responses in plants but some PRX duplicates have been recruited to cell wall metabolism, including lignin polymerization, or to the vacuole as part of defence responses to abiotic and biotic stresses (Ren et al., 2014). Although the *E. grandis* genome has lost many paralogous genes that appeared following the recent WGD, it has retained genes in tandem duplications (34% of the total genes) at a much higher frequency than observed in the *Populus* genome (Myburg et al., 2014; Tuskan et al., 2006). Some of the expanded gene families are related to lignocellulosic biomass production, secondary metabolites and oils (e.g. phenylpropanoid biosynthesis, terpene synthase and phenylpropanoid gene families). It was proposed that tandem duplication has a significant role in shaping functional diversity in *Eucalyptus* (Myburg et al., 2014).

4.2.4 Gene Structure

Similar exons lengths have been reported when comparing homologous genes between *P. glauca* and *P. trichocarpa* (Sena et al., 2014) and *E. grandis* (Myburg et al., 2014). In contrast, introns lengths are more variable among these species. Conifers genes tend to accumulate long introns with the largest introns surpassing 60 kb in spruce (Nystedt et al., 2013) and 120 kb in pine (Wegrzyn et al., 2014). On average the *Picea* introns are 1000 bp in length, *Populus* 380 bp and *Eucalyptus* approximately 425 bp (Myburg et al., 2014; Nystedt et al., 2013; Tuskan et al., 2006). The intron average length is higher in conifer genes which typically accumulate one or a few very long introns although the majority introns are in the 100 to 200-bp range and are comparable in size to those found in angiosperms (Sena et al., 2014).

A comparative analysis of selected orthologous genes between *P. glauca* and *P. taeda* clearly showed the conservation of gene structure and the distribution of intron sizes in spite a divergence time of 100—140 MYA (Sena et al., 2014). The conservation of long introns was also observed across gymnosperm taxa, where a group of long introns in *P. abies* was identified as orthologous to long introns in *Pinus sylvestris* and *Gnetum gnemon* (Nystedt et al., 2013). These observations suggest that the long introns observed in conifers likely date back to a period predating the divergence of major conifer groups. The gene content of contemporary conifer genomes is also ancient and largely conserved between species as shown by high levels of synteny in comparative genetic mapping in the Pinaceae and the ancient origin of gene duplicates (Pavy et al., 2012).

5. GENE EXPRESSION AND TRANSCRIPTOME PROFILING

The expression of a gene is by definition the activity of its protein product. In this section, we review and discuss research on RNA transcript profiling, which has been developed as the principal — but not the only — approach for gaining insights into gene expression. Protein profiling has also been applied to investigations of forest trees but on a more limited scale of analysis and on relatively few species (Abril et al., 2011).

5.1 Large-Scale RNA Transcript Profiling Methods

Large-scale RNA transcript profiles have been mostly studied using two approaches which are hybridization-based microarrays and RNA-seq (Table 2).

Table 2 Gene expression and transcriptome profiling in forest trees

Species	Methods	Comparisons	No. differentially expressed genes (%)	Statistical significance	No. analyzed genes	References
A – Comparative analyses of tissue types						
Picea glauca	Oligo MA	Comparison of seven vegetative tissue types from aerial and below ground organs	18,052 (76)	adj$P \leq 0.05$	23,853	Raherison et al. (2015)
Pinus contorta	RNA-seq	Foliage vs root plus stem tissues	8131 (34)	adj$P \leq 0.01$	23,889	Yeaman et al. (2014)
P. glauca × *Picea engelmannii*	–	–	6695 (28.5)	–	23,519	–
Populus maximowiczii × *Populus nigra*	MA	Vegetative tissues including bark, phloem, cambial zone, secondary xylem, leaves, whole stems and different developmental stages	17,179 (28)	$P \leq 0.01$; $\lvert \text{ratio} (\log_2) \rvert \geq 1$	61,251	Ko, Kim, Hwang, and Han (2012)
Eucalyptus grandis	*RNA-seq*	Early floral bud vs roots	15,544 (43)	NA	36,376	Vining et al. (2014)
Quercus spp.	RNA-seq	Ecodormant bud, swelling bud, secondary xylem, root, leaf and differentiated callus	7574 (8.3)	adj$P \leq 0.05$	90,786	Lesur, Le Provost, et al. (2015)
B – Comparative analyses of developmental stages						
Pinus taeda	cDNA MA	Xylem at five time points within a growing season	667 (19)	adj$P \leq 0.0001$	3512	Paiva et al. (2008)
Cryptomeria japonica	Oligo MA	Early (wood formation) vs latewood (cessation of growth and dormancy)	10,380 (57)	$P \leq 0.05$; adj$P \leq 0.2$; $\lvert \text{ratio} (\log_2) \rvert \geq 1$	18,082	Mishima et al. (2014)

Cunninghamia lanceolata	RNA-seq	Cambial tissues at the active vs dormant stages	4415 (7.3)	adj$P \leq 0.001$; $\lvert$ratio $(\log_2)\rvert \geq 2$	59,669	Qiu et al. (2013)
—	—	Cambial tissues at the active vs reactivating stages	883 (1.5)	—	—	—
—	—	Cambial tissues at the reactivating vs dormant stages	4018 (6.7)	—	—	—
Pinus radiata	cDNA MA	Early vs latewood at the juvenile stage (5 yr)	687 (21)	adj$P \leq 0.05$	3320	Li, Wu, et al. (2010)
—	—	Early vs latewood at the transition stage (9 yr)	995 (30)	—	—	—
—	—	Early vs latewood at the mature stage (30 yr)	381 (12)	—	—	—
P. taeda	cDNA MA	Early vs latewood of low specific gravity	87 (4)	adj$P \leq 0.01$	2171	Yang and Loopstra (2005)
—	—	Early vs latewood of high specific gravity	110 (5)	—	—	—
—	—	Earlywood of low vs high specific gravity	51 (2.3)	—	—	—
—	—	Latewood of low vs high specific gravity	131 (6)	—	—	—
P. radiata	cDNA MA	Earlywood of high vs low stiffness	112 (3.4)	$P \leq 0.05$	3320	Li et al. (2011)
—	—	Latewood of high vs low stiffness	295 (8.9)	—	—	—
Picea sitchensis	cDNA MA	Needles at late summer (transition stage) vs early winter (dormancy stage)	2224 (10.2)	adj$P \leq 0.05$; $\lvert$ratio $(\log_2)\rvert \geq 2$	21,840	Holliday et al. (2008)
E. grandis	RNA-seq	Young vs mature leaves	474 (1.3)	NA	36,376	Vining et al. (2014)
—		Early vs late floral bud	607 (1.7)	—	—	—

(*Continued*)

Table 2 Gene expression and transcriptome profiling in forest trees—cont'd

Species	Methods	Comparisons	No. differentially expressed genes (%)	Statistical significance	No. analyzed genes	References
Quercus petraea	RNA-seq	Endodormant vs ecodormant buds	75 (1.2)	adj$P \leq 0.05$	6471	Ueno et al. (2013)
Fagus sylvatica	RNA-seq	Ecodormant vs swelling buds	205 (1.0)	adj$P \leq 0.05$	21,057	Lesur et al. (2015)
C – Defences and responses to biotic factors						
P. glauca × *P. engelmannii*[a]	cDNA MA	Bark of trees that are susceptible vs resistant to the white pine weevil (*Pissodes strobi*)	191 (1)	adj$P \leq 0.05$; $\lvert$ratio $(\log_2)\rvert \geq 0.6$	17,825	Verne et al. (2011)
P. sitchensis[a]	cDNA MA	Apical shoots with vs without removing bark	610 (0.4)	adj$P \leq 0.01$; $\lvert$ratio $(\log_2)\rvert \geq 1$	16,700	Friedmann et al. (2007)
P. glauca[a]	Oligo MA	Needles of trees that are susceptible vs resistant to the spruce budworm (*Choristoneura occidentalis*)	486 (2.1)	adj$P \leq 0.05$	23,853	Mageroy et al. (2015)
Pinus monticola[b]	RNA-seq	Needles of resistant trees; uninfected vs infected with white pine blister rust (*Cronartium ribicola*)	789 (3.4)	adj$P \leq 0.05$; $\lvert$ratio $(\log_2)\rvert \geq 0.6$	23,000	Liu et al. (2013)
–	–	Needles of susceptible trees: uninfected vs infected with *C. ribicola*	562 (2.4)	–	–	–
Larix gmelinii[b]	RNA-seq	Needles of control vs jasmonic acid-treated trees	2383 (4.7)	adj$P \leq 0.001$; $\lvert$ratio $(\log_2)\rvert \geq 1$	51,157	Men et al. (2013)
–	–	Needles of control vs methyl jasmonate-treated trees	2767 (5.4)	–	–	–

Pinus sylvestris[b]	*P. taeda* cDNA MA	Roots of control vs saprotrophic fungus (*Trichoderma aureoviride*) inoculated, 15 days postinoculation	10 (0.5)	adj$P \leq 0.01$; |ratio ($\log_2$)| ≥ 0.3	2109	Adomas et al. (2008)
–	–	Roots of control vs mutualistic fungus (*Laccaria bicolor*) inoculated trees, 15 days postinoculation	16 (0.8)	–	–	–
–	–	Roots of control vs pathogenic fungus inoculated (*Heterobasidion annosum*), 15 days postinoculation	294 (13.9)	–	–	–
P. sitchensis[b]	cDNA MA	Bark of control vs *P. strobi*-treated trees	2382 (24.5)	adj$P \leq 0.05$; |ratio ($\log_2$)| ≥ 0.6	9720	Ralph et al. (2006)
–	–	Bark of control vs mechanically wounded trees	3089 (31.8)	–	–	–
–	–	Shoot tips of control vs western spruce budworm (*C. occidentalis*)-treated trees	358 (3.7)	–	–	–
–	–	Shoot tips of control vs *C. occidentalis*-treated trees, 3 h posttreatment, 52 h posttreatment	3490 (35.9)	–	–	–
P. radiata[b]	*Pinus* oligo MA	Mucilaginous xylem of control vs ethephon-treated trees, 8 weeks posttreatment	23,084 (13)	adj$P \leq 0.01$; |ratio ($\log_2$)| ≥ 1	175,614	Dubouzet et al. (2014)
–	–	Xylem (woody fibrous tissue) of control vs ethephon-treated trees, 8 weeks posttreatment	12,718 (7.2)	–	–	–

(Continued)

Table 2 Gene expression and transcriptome profiling in forest trees—cont'd

Species	Methods	Comparisons	No. differentially expressed genes (%)	Statistical significance	No. analyzed genes	References
—	—	Bark of control vs ethephon, 8 weeks posttreatment	1761 (1)	—	—	—
D – Responses to abiotic factors						
Pinus pinaster	cDNA MA	Compression vs normal wood	496 (7.2)	adj$P \leq 0.001$; $\lvert$ratio ($\log_2$)$\rvert \geq 1.5$	6841	Villalobos et al. (2012)
P. radiata	cDNA MA	Compression vs opposite wood	970 (29)	adj$P \leq 0.05$; $\lvert$ratio ($\log_2$)$\rvert \geq 0.6$	3320	Li et al. (2013)
Chamaecyparis obtusa	RNA-seq	Compression vs normal wood	2875 (7.1)	adj$P \leq 0.05$	40,602	Sato, Yoshida, Hiraide, Ihara, and Yamamoto (2014)
P. contorta	RNA-seq	Needles of trees grown under seven treatments varying in temperature, humidity and day length	11,658 (48.8)	adj$P \leq 0.01$	23,889	Yeaman et al. (2014)
P. glauca × *P. engelmannii*	—	—	6413 (27.3)	—	23,519	—
P. sylvestris	*P. taeda* cDNA MA	Hypocotyls which were grown under continuous red vs far-red light	644 (5.1)	adj$P \leq 0.05$; $\lvert$ratio ($\log_2$)$\rvert \geq 0.95$	12,523	Ranade, Abrahamsson, Niemi, and García-Gil (2013)

Picea abies	RNA-seq	Embryonic callus generated at cold (18 °C) vs warm (30 °C) temperature	1608 (1.1)	$\lvert$ratio (log$_2$)$\rvert \geq 1$	143,723	Yakovlev et al. (2014)
Populus balsamifera	MA	Leaves of well-watered vs water-stressed trees	280 (0.4)	adj$P \leq 0.05$; $\lvert$ratio (log$_2$)$\rvert \geq 2$	61,313	Hamanishi et al. (2010)
Populus euphratica	RNA-seq	Control vs salt-stressed callus	23,512 (27)	adj$P \leq 0.001$; $\lvert$ratio (log$_2$)$\rvert \geq 1$	86,777	Qiu et al. (2011)
Populus trichocarpa	cDNA MA	Shoot apex of control vs nitrogen-treated trees	1037 (1.8)	adj$P \leq 0.05$; $\lvert$ratio (log$_2$)$\rvert \geq 1$	56,055	Euring, Bai, Janz, and Polle (2014)
P. euphratica	*RNA-seq*	Control vs salt-stressed callus	884 (2.4)	adj$P \leq 0.05$; $\lvert$ratio (log$_2$)$\rvert \geq 1$	36,144	Zhang et al. (2014)
Eucalyptus camaldulensis	*RNA-seq*	Leaves of well-watered vs water-stressed trees	4320 (28)	adj$P \leq 0.01$	15,538	Thumma et al. (2012)
Eucalyptus melliodora	RNA-seq	Leaves of trees with resistant vs susceptible phenotype to insect or vertebrate herbivores	1406 (10.7)	adj$P \leq 0.05$	13,104	Padovan et al. (2013)
Eucalyptus urophylla × *E. grandis*	RNA-seq		1469 (4.2)	adj$P \leq 0.01$; $\lvert$ratio (log$_2$)$\rvert \geq 0.6$	34,919	Camargo et al. (2014)

(*Continued*)

Table 2 Gene expression and transcriptome profiling in forest trees—cont'd

Species	Methods	Comparisons	No. differentially expressed genes (%)	Statistical significance	No. analyzed genes	References
Haloxylon ammodendron	RNA-seq	Tissues of control vs drought-treated trees	1060 (1.3)	adj$P \leq 0.1$	79,918	Long et al. (2014)
Eucalyptus spp.	RNA-seq	Leaves of irrigated vs nonirrigated trees	155 (1.1)	adj$P \leq 0.05$	14,460	Villar, Plomion, and Gion (2011)

Methods: cDNA MA and oligo MA are cDNA and oligonucleotide microarray, respectively; RNA-seq, RNA sequencing; RNA-seq in normal and in italic indicate de novo and reference-based assembly, respectively.
Comparisons: jasmonic acid, methyl jasmonate and ethephon are phytohormones that regulate growth and involve in defence signalling processes (Guo and Ecker. 2004; Schnurr, Cheng, & Boe, 1996; Wasternack, 2007).
No. differential genes, transcripts or probes (%): The number in parentheses corresponds to the percentage (%) of differential genes (transcripts or probes) relative to the total number of analyzed genes (transcripts or probes).
Statistical significance criteria: *P*, *P* value; adj*P*, adjusted *P* value; NA, not available.
[a]**Species**: Constitutive defence.
[b]**Species**: Induced defence.

Hybridization methods used microarray which contains a collection of probes spotted or printed onto a glass surface. The probes are either cDNA amplicons (generated by PCR amplification) or oligonucleotides which are selected to represent a known gene and to detect its expression levels in a sample. Microarray-based approaches involve several steps which are briefly: to convert mRNA into cDNA, to label cDNA with fluorescent dyes, to hybridize labelled cDNA samples to microarrays which are then scanned for image processing to quantify the fluorescent signal intensities. Expression levels of a gene are proportional to the signal intensities of its corresponding probes. Microarray sensitivity and specificity are partly related to the probe length. In general, cDNA probes (>500 nucleotides) are less specific than oligonucleotide probes (25–70 nucleotides) because they are more prone to nonspecific cross hybridization (Chou, Chen, Lee, & Peck, 2004). Inversely, shorter oligonucleotides (<25 nucleotides) are more sensitive to DNA sequence polymorphisms and are less well suited for heterologous analyses (Pullat et al., 2007). Transcript profiling in forest trees has used both cDNA (e.g. Li, Yang, & Wu, 2013; Ralph et al., 2006; Villalobos et al., 2012) and long-oligonucleotide microarrays (e.g. Dubouzet et al., 2014; Maganaris et al., 2011; Raherison et al., 2012).

The RNA-seq approach developed with recent advances in NGS technologies. It consists of converting mRNA to cDNA, sheering the cDNA into fragments of desirable lengths to facilitate high throughput, sequencing, processing the reads and mapping them onto a reference genome or transcriptome (reference-based assembly) or joining reads that overlap into larger fragments (de novo assembly), each representing mRNA. The expression level of a gene corresponds to the number of transcripts derived from that same gene in the sample. Reference-based methods have been used only in few studies in poplar (Zhang et al., 2014) and eucalyptus (Thumma, Sharma, & Southerton, 2012; Vining et al., 2014). In conifers, authors used de novo assembly approach based on different analysis approaches, and then generated a large variation of transcript numbers between studies. For example, Yakovlev et al. (2014) reported sixfold higher number of sequences than Liu, Sturrock and Benton (2013) and Yeaman et al. (2014) who realigned their sequences with reference genomes.

Many studies reported the high consistency of results generated from microarray and RNA-seq approaches (e.g. Kogenaru, Qing, Guo, & Wang, 2012; Zhao, Fung-Leung, Bittner, Ngo, & Liu, 2014). For example, correlation between the gene expression profiles obtained from RNA-seq and microarray is estimated at r^2 about 90% (Zhao et al., 2014). RNA-seq confirmed differential expression of 99% genes identified using microarray (Raherison et al., 2012). Overall, RNA-seq offers significant advantages over microarrays because of its higher detection capacity. One of the widely recognized shortcomings of microarrays is that they only detect transcripts that corresponding to sequences included in the array design, while RNA-seq enables investigation of both known and novel transcripts. Second, microarrays have lower and upper limits for quantification due to the background signal and probe saturation, while RNA-seq affords a wide dynamic range with the potential for very deep analysis and discovery of rare transcripts. A 70-fold range was recorded in a study of human blood (Zhao et al., 2014); and a range of 8000-fold for about 16 million *Saccharomyces cerevisiae* sequences reads (Nagalakshmi et al., 2008). Finally, RNA-seq offers another signal-to-noise advantage by eliminating cross hybridization that can be seen with microarray technology.

5.2 Insights into Biological Processes

5.2.1 Tissue Comparison and Transcriptome Organization

Tissue differentiation has generally been linked to deep transcriptome reorganizations compared to that associated with developmental stages or environmental conditions in plants including *Arabidopsis* (Ma et al., 2005) and maize (Downs et al., 2013). The survey of forest tree transcriptome studies presented in Table 2 is consistent with this observation. The proportion of differentially expressed genes was generally much higher in tissue comparisons (Table 2A, ranging from 8.3% to 76% of genes tested) than in comparisons of different developmental stages (Table 2B), and in studies of abiotic and biotic interactions (Table 2C—D).

In a recent study in *P. glauca*, we classified 22,781 genes as variable (79%, 24 co-expression groups) or invariant (21%) by profiling across several vegetative tissues, and delineated co-expression groups that are indicative of the modular organization of the transcriptome (Raherison, Giguere, Caron, Lamara, & MacKay, 2015). Our results showed that deep transcriptome reorganization is associated with tissue differentiation compared to developmental stages or environmental conditions, and that patterns are conserved between spruce species as might be expected given the ancient evolutionary origins of tissue differentiation.

5.2.2 Growth and Development

Temporal reorganization of the transcriptome across developmental stages has been investigated in different tissues in forest trees (Table 2B). Many of the studies have focussed on changes in the wood transcriptome that occur over the course of a growth season, with two major active growth phases known as earlywood (beginning of the growth season) and latewood (end of season), while other studies examined foliar tissue at different stages of maturity or during dormancy, floral buds and adventitious root development (Table 2B). Here we summarize conclusion from these studies.

- Gene expression varies between earlywood and latewood. When comparing trees with different wood physical properties, i.e. low versus high specific gravity (Yang & Loopstra, 2005) and low versus high stiffness (Li, Wu, & Southerton, 2011) the number of genes differentially expressed in latewood was at least twice greater than in earlywood.
- In an annual cycle of trees, transitional stages that are linked to dormancy were associated with large changes in transcriptome makeup. Qiu et al. (2013) carried out pairwise comparisons of different stages of cambium development and reported relatively higher numbers of differential genes in transition phases to and from dormancy and the lowest number of genes differentially expressed was recorded in comparison between reactivating and active stages.
- Seasonal transcriptome reorganization varies with cambial age. The change from earlywood to latewood formation was compared in juvenile trees, mature trees and trees in transition between the juvenile and mature status (Li, Chen, Gao, & Yin, 2010). The proportion of genes differentially expressed in transition wood was higher in 9-year-old trees (30%) than in juvenile (21%; 5-year-old trees) and mature (12%; 30-year-old trees) trees.
- Transcriptome change involved in foliage developmental stages seems to be more important between transition and dormant stages in conifer trees (Holliday, Ralph, White, Bohlmann, & Aitken, 2008) than between young and mature stages in *Eucalyptus* (Vining et al., 2014). Floral bud development involved the same order of magnitude of differential gene number as foliage development in *Eucalyptus* (Vining et al., 2014).

5.2.3 Responses to Biotic Factors

Trees are long-lived plants and have diverse strategies to cope with biotic attack. For example, they are capable of counteracting biotic attacks through pre-established physical and chemical barriers known as constitutive defences. If these barriers are breached, signalling pathways may be activated to trigger targeted or general immune responses as the next line of defence, known as induced defences. Few studies have investigated transcriptomic alterations related to constitutive defences in trees (Table 2C). Several of these studies have been carried in spruces (*Picea* spp.) and reported a relatively low proportion of differentially expressed transcripts between stems with and without physical barrier or bark (Friedmann et al., 2007) and between resistant and susceptible trees to the white pine weevil (Verne, Jaquish, White, Ritland, & Ritland, 2011) or to the spruce budworm (Mageroy et al., 2015). Analyses of the functional annotations showed indicated that many of the genes were stress responsive (e.g. Friedmann et al., 2007). Secondary metabolism and stress-related genes were overexpressed in resistant trees (Mageroy et al. (2015); Verne et al., 2011).

Many of the studies of transcript profiling in forest trees have investigated induced defences and biological response following different types of induction (insect herbivores, mechanical wounding, phytohormones and fungi) (Table 2C). By comparing tissues of treated and untreated trees, it was generally shown that a large proportion of genes were differentially expressed in response to the treatment compared to studies comparing different levels of constitutive defences. The main conclusions from these studies are as follows:

- Transcriptome changes or reorganization increase with stress exposure time. For example, Ralph et al. (2006) reported 10 times more differentially expressed genes at 52 h posttreatment than 3 h posttreatment in western spruce budworm-infected trees of *Picea sitchensis*. Dubouzet et al. (2014) found 10–30 times more differentially expressed genes in *Pinus radiata* when comparing responses in xylem and bark tissues of 1 week and 8 weeks posttreatment with ethephon. A similar pattern was reported for *P. sylvestris* trees inoculated with a pathogenic fungus (Adomas et al., 2008). The dynamic transcriptome response to different exposure durations may vary depending on the nature of the stress. For example, Adomas et al. (2008) found the number of differentially expressed genes after inoculation with a nonpathogen decreased with the exposure times.
- Resistant trees may exhibit a greater transcriptomic response (more responsive genes) to biotic stress than susceptible trees. For example (Liu et al., 2013), the proportion of differentially expressed genes between uninfected and infected trees with white pine blister rust was higher in resistant (3.4%) than in susceptible trees (2.4%).
- The proportion of responsive genes among those tested ranged very widely, i.e. from 0.5% to 36%, which is likely due to technical variation between studies. For example, the lowest and the highest proportions

were obtained from studies using different significance criteria and different number of genes (Adomas et al., 2008; Ralph et al., 2006).

Functional annotation analyses show expression patterns of genes involved in some biological processes. Genes implicated in stress response include enzymes of primary and secondary metabolisms such as

- Genes encoding for lipoxygenase (LOX), allene oxide cyclase (AOS) and allene oxide synthase (AOC) were upregulated in insect-attacked, in mechanically wounded and in jasmonate-treated trees (Men, Yan, & Liu, 2013; Ralph et al., 2006). LOX, AOS and AOC are enzymes responsible for synthesis of jasmonic acid, which is a signal molecule in defence.
- Genes involved in stress response and in secondary metabolism are strongly preferentially expressed in trees under biotic stresses (Adomas et al., 2008; Liu et al., 2013; Men et al., 2013; Ralph et al., 2006).
- Primary metabolism genes may have different expression patterns depending on the type of biotic interaction. They are downregulated in trees under stress caused by insect attack and fungal pathogen infection but upregulated in trees inoculated with symbiotic fungus (Adomas et al., 2008; Ralph et al., 2006). Expression patterns of primary metabolism genes may vary also between tree genotypes. They had higher expression levels in resistant than in susceptible trees infected with white pine blister rust (Liu et al., 2013).

5.2.4 Responses to Abiotic Factors

Abiotic factors play a major role in tree growth and development. They include temperature, light, water and nutrients among others, which play a role in normal developmental processes such as conditioning trees to changing conditions during the annual growth cycle (e.g. cold-induced dormancy). In Table 2D, we report a number of studies that have investigated transcriptomic responses associated with stresses caused when many of these same factors reach a level that is outside of the bounds that are favourable for development, such as a drought or a heat shock for example.

Transcriptome responses to drought and high-salt-induced stress have been widely investigated in *Populus* (e.g. Qiu et al., 2011), *Eucalyptus* (e.g. Thumma et al., 2012) and the desert tree *Haloxylon ammodendron* (Long et al., 2014; Table 2D). Other factors investigated included responses to abrupt changes in day length (photoperiod), temperature and nitrogen supply (see Table 2D). The transcriptome response associated with formation of reaction wood in both hardwoods and conifers caused by a mechanical stress has also been largely investigated.

Many different classes of genes were found to be transcriptionally responsive to abiotic factors and they cannot adequately be summarized here given the diversity of stresses involved. The set of genes that respond to abiotic factors overlaps to those for biotic factors. The numbers of responsive genes appear to be highly variable between studies even when the same factor was investigated. For example, water stress affected the expression of as little as 0.4% of genes in *Populus* leaves (Hamanishi et al., 2010) and as many as 28% of genes in *Eucalyptus* (Thumma et al., 2012) which suggests that meta-analyses or more standardized protocols may be needed to delineate trends as to the types of genes involved.

6. TRAIT VARIATION OF FOREST TREES

Forest trees have been studied with a new approach that combines genomics and trait variation to address issues relevant to economic production or issues relevant to ecological questions. This new approach may be used to accelerate breeding programmes and have economic impacts by achieving genetic gains more rapidly (Neale, 2007). For instance, trees may be selected based on genomic markers associated at a high frequency with traits of interest (e.g. growth, resistance; Thavamanikumar, Southerton, Bossinger, & Thumma, 2013). Theoretically, this will lead to early selection and shorten the selection steps at each generation of breeding (see Harfouche et al., 2012). Identification of correlations between traits and genes may also allow identifying the gene pathways and genetic architecture underlying functional traits. For ecologic issues, the factors affecting differentiation between populations or species may be characterized in order to assess gene flow or identify putative adaptive loci. Consequently, a better understanding can be developed of the effects of factors such as deforestation on gene flow (e.g. Lander, Boshier, & Harris, 2010) and adaptation during global warming (Aitken et al., 2008).

Most traits are under the control of multiple genes. To identify these genes, three different approaches are used in forest trees; these are quantitative trait loci (QTL) mapping, transcriptome comparison and association studies.

The traditional QTL mapping approach aims to delineate chromosomal regions that underpin phenotypic variation; it generates linkage disequilibrium between genetic markers and QTLs, by crossing individuals and creating a segregating population. The mapping precision of QTLs is determined by the number of genetic markers, the size of the progeny array and appropriateness of statistical tools (Gonzalez-Martinez et al., 2006; Paterson, 1998). Under certain conditions (marker-saturated fine-linkage map), QTL studies have also permitted the identification of candidate regions for further in-depth genomic characterization (see Gonzalez-Martinez et al., 2006 for more information on limitations of this approach). The second approach consists of comparing transcriptomes or gene expression between individuals or groups of different individuals using microarrays or RNA-seq. In this approach, gene expression comparisons are carried out between individuals who may represent different geographic regions (e.g. Holliday et al., 2008) or contrasted phenotypes (e.g. Mageroy et al., 2015) to identify genes that are differentially expressed between the groups compared. The third approach consists of using association studies in which correlations between genotype and phenotype are tested in unrelated individuals (see Gonzalez-Martinez et al., 2006 for more details); this approach is used to overcome limitations of pedigree-based on QTL mapping. Association studies necessitate large sample size (N > 500) to detect causative polymorphism of small effect (~5% of phenotypic variance explained) (Long & Langley, 1999).

All of these three approaches link genes to phenotypes but only association studies link specific genotypes to phenotypes. Association studies are used in population genomics which can be broadly defined as the simultaneous study of alleles at loci across the genome. Population genomics is a discipline that combines genomic concepts and technologies with the population genetics objective of understanding evolution (Luikart, England, Tallmon, Jordan, & Taberlet, 2003). Presently, the most used markers to characterize loci variability are SNPs. SNPs are found in coding and noncoding regions. This contrasts with markers that were previously used in most population genetics studies, such as amplified fragment length polymorphism (AFLP) and variable number tandem repeats (VNTR), for which the position was typically unknown. In association studies, specific genotypes can also be linked to variable traits or environments.

In the next two sections, we present studies of trait variation that pursue two general aims. On the one hand, investigations of the genomic architecture of traits are aimed at describing the internal factors (e.g. genes) underlying traits of interest, and on the other hand, investigations of genetic differentiation attempt to link external factors (e.g. temperature) to adaptive genes. These two general lines of investigation are not mutually exclusive but tend to be used to study trait variations from an economic and ecological perspective, respectively. These sections are not intended to provide an exhaustive review of all of the literature, but a general overview of recent progress and potential directions for future studies.

6.1 Genomic Architecture of Traits

Here, we present results from QTL mapping, transcriptome comparison and association studies as they represent a significant part of the body of literature in this field.

6.1.1 Growth and Wood Properties

In the last two decades, growth and wood properties have been traits of major focus in forest genomics, unsurprisingly. The proportion of phenotypic variation explained by QTLs or SNPs for traits such as stem volume, diameter growth, lignin and cellulose content were estimated in *Populus* (Wegrzyn et al., 2010), *Eucalyptus* (Grattapaglia, Bertolucci, Penchel, & Sederoff, 1996; Gion et al., 2011; Kirst et al., 2004; Thumma et al., 2009), *Castanea* (Casasoli et al., 2004), *Pinus* (Cumbie et al., 2011; Jaramillo-Correa et al., 2015; Pot et al., 2006) and *Picea* (Beaulieu et al., 2011; Prunier et al., 2013). Overall, the variation in quantitative traits explained by individual QTL was low and varied from 7% to 19%, and was lower with individual SNP and rarely exceeds 5% (Grattapaglia & Resende, 2011). The relatively small proportion of variance explained by QTL or SNP is consistent with multigenic control (Gonzalez-Martinez, Huber, Ersoz, Davis, & Neale, 2008; Prunier et al., 2013). In some cases, the total character variance accounted for all QTLs was much higher. For instance, proportion of phenotypic variance of height growth explained by all QTLs was 59% in *P. glauca* (Pelgas, Bousquet, Meirmans, Ritland, & Isabel, 2011). A major trend from studies in forest trees is that wood properties are generally under moderate to strong additive genetic control in contrast to growth, which is under lower genetic control (Stackpole, Vaillancourt, de Aguigar, & Potts, 2010). Some studies have also identified genes associated with growth (González-Martínez, Wheeler, Ersoz, Nelson & Neale, 2007) and wood properties such as cell structure (Gonzalez-Martinez et al., 2008), lignin production (Wong, Cannon, & Wickneswari, 2011), cellulose content (Lepoittevin, Harvengt, Plomion & Garnier-Géré, 2012) and microfibril angle (Gonzalez-Martinez et al., 2007). Studies identifying genes related to growth and wood properties are available for numerous forest tree taxa (see review Grattapaglia et al., 2012 for *Eucalyptus* sp.). It was found that several MYB and NAC genes also regulate secondary cell wall formation in xylem tissues and control lignin biosynthesis genes in transgenic functional tests in pines and spruce (Bomal et al., 2008; Craven-Bartle, Pascual, Canovas, & Avila, 2013; Duval et al., 2014; Patzlaff et al., 2003). One of these genes, *PgNAC-7* was identified as a major hub gene that is preferentially expressed during the formation of earlywood (Raherison et al., 2015). Few association studies have been able to bridge the interspecific gap and associate putative orthologs with similar traits in several species but this trend may change in the near future. These comparisons may help to identify key genes involved in the litigious parallel or convergent evolution of elongated stems in tree taxa (Groover, 2005).

6.1.2 Resistance

Phenotypic variance of resistance traits explained by a single QTL or SNP varies from low (Lind et al., 2014; Quesada et al., 2010) to high (Freeman, O'Reilly-Wapstra, Vaillancourt, Wiggins, & Potts, 2008) in forest trees. Phenotypic variation of resistance traits can be estimated as the ability to prevent the infection from establishing, lesions from expanding, fungal spread and global damage (e.g. defoliation) for pathogens or insect herbivores. In *P. abies*, each QTL explained between 4.6% and 10.1% of the phenotypic variation of resistance against the pathogen *Heterobasidion parviporum* (Lind et al., 2014). In contrast, 52% of phenotypic variance of resistance against another pathogen *Mycosphaerella cryptica* was explained by two QTLs in *E. globulus* (Freeman et al., 2008). Comparisons of gene expression between individuals that present different resistance phenotypes have also been used to identify candidate genes and pathways underlying defence mechanisms. For instance, the expression level of a gene encoding β-glucosidase is up to 1000-fold higher in resistant than nonresistant trees of *P. glauca* (Mageroy et al., 2015). The gene product was functionally and able to catalyze the release of two acetophenone compounds (Mageroy et al., 2015) that are toxic for the spruce budworm, *Choristoneura fumiferana* (Delvas, Bauce, Labbé, Ollevier, & Bélanger, 2011). Similarly, transcriptome comparison between *Thuja plicata* producing contrasted amounts of monoterpinoids allowed the identification of CYP450 catalyzing the hydroxylation of (+)-sabinene to *trans*-sabin-3-ol, associated with resistance against herbivores such as ungulates (Gesell et al., 2015). Association studies have also been conducted with resistance traits (e.g. Quesada et al., 2010). In *P. taeda*, 10 SNPs have small effects and putative roles in basal resistance, direct defence and signal transduction during infection with pitch canker, *Fusarium circinatum* (Quesada et al., 2010). A trend observed in recent studies is that comparative transcriptome profiling between genotypes with contrasting response against pathogens or herbivores is proving to be a fruitful approach for finding key genes in defensive pathways.

6.2 Genomic Differentiation in Trees

Identifying patterns of genomic diversity and differentiation at the geographic scale is a central question of evolutionary biology, and trees are well-suited species for its study for different reasons (Aitken et al., 2008; Gonzalez-Martinez et al., 2006). Various biological and geographical features are expected to increase the randomness of diversity within a species distribution, and thus, enable the detection of genes affecting key traits for local adaptation and selective sweeps (see Aitken et al., 2008 for more details). These features are large populations, high outcrossing rates, large distributions, a sessile life habit, wide dispersal (e.g. gene flow through pollen), long life span and availability of natural populations. These features are common to most forest tree species, but not to all, so that interspecific comparisons within or between genera are highly interesting to disentangle the effects of evolutionary forces. Another interesting aspect in the study of forest tree genomics is that managed populations (e.g. progeny trials) may be available to estimate heritability of traits (Neale & Ingvarsson, 2008) and thus, extrapolate the effects of selection in natural populations.

6.2.1 Intraspecific and Interspecific Gene Flow

Over the last two decades, population structure or gene flow between species of forest trees has been mostly characterized with markers other than SNPs, such as AFLPs and VNTRs. Recently, efforts to identify intraspecific and interspecific patterns of gene flow have been intensified by increasing genomic resources. Here, we describe general trends combining results from studies ranging from small (e.g. $N = 6$) to large (e.g. $N > 200$) numbers of markers.

Although most tree species have large population size and potential for wide ranging dispersal, they may present intraspecific population structure within their natural range. This includes tropical species such as *E. globulus* (Cappa et al., 2013) and *Acacia mangium* (Butcher, Moran, & Perkins, 1998) and temperate or boreal species such as *P. mariana* (Prunier, Gerardi, Laroche, Beaulieu, & Bousquet, 2012), *Pinus contorta* (Parchman et al., 2012) and *Populus tremuloides* (Callahan et al., 2013). In temperate and boreal regions, population structure is mostly associated with isolation in distinct glacial refugia during Pleistocene followed by land recolonization poleward (see Shafer, Cullingham, Cote, & Coltman, 2010 for review, Prunier et al., 2012). Within natural ranges, lineages or populations may be characterized by independent demographic histories; however, they may share similarity in their demographic disequilibrium (Excoffier, Hofer, & Foll, 2009). This means that the effect of evolutionary forces between these groups could be similar.

Interspecific gene flow also affects numerous tree species. Hybridization provides an opportunity for introgression, where genes from one parental species infiltrate the other through multiple backcrossing events. For instance, hybridization and introgression are abundant between *Eucalyptus* spp. (Arumugasundaram, Ghosh, Veerasamy, & Ramasamy, 2011), *Quercus* spp. (Burgarella et al., 2009), *Populus* spp. (Geraldes et al., 2014), *Pinus* spp. (Cullingham, Cooke, & Coltman, 2014) and *Picea* spp. (De La Torre et al., 2015). In recent years, population genomics allowed to characterize not only the extent of interspecific gene flow between species, but also the heterogeneity of gene flow across the genome. It was observed that divergent selection can reduce gene flow at sites linked to the direct targets of selection before alleles at those sites have a chance to recombine away and introgress into the other population (Feder, Egan, & Nosil, 2012). Islands of divergence may then occur throughout the genome which favours speciation.

6.2.2 Adaptation

Minimum temperatures limit the poleward expansion of forest tree species, whereas limited water availability interacting with high temperatures limits expansion in the opposite, or equatorial, direction in many regions (Allen & Breshears, 1998; Woodward & Williams, 1987). Thus, climate alters the geographic distribution of plant species from local to global scales. One major goal of population genomics in the last years has been to identify the adaptive genes underlying these geographic patterns. The association studies approach is now frequently used to target adaptive genes. The combination of at least two statistical methods (e.g. F_{st} outlier, regression, differentiation) and the union or intercept of their results are generally used to identify adaptive loci (e.g. Eckert et al., 2010; Prunier et al., 2012). However, a review of the statistical methods used to identify adaptive loci proposed to improve their detection by first using multivariate statistical models (see Sork et al., 2013 for more details).

Temperature is an important factor influencing the timing of bud flush and bud set. Bud phenology traits delineate the annual growth period in tree species most strongly in boreal and temperate regions, and vary in a manner that is tightly linked to latitudinal and altitudinal clines (Alberto et al., 2013). These geographic patterns may result in locally adapted populations (reviewed in Aitken et al., 2008). Bud phenology traits are under the control of 11–13 QTLs in *Quercus robur* (Scotti-Saintagne et al., 2004), *P. glauca* (Pelgas et al., 2011), *P. menziesii* (Eckert et al., 2009). In *Populus tremula*, two nonsynonymous SNPs in the phytochrome B2 gene were independently associated with variation in the timing of bud set and explained between 1.5% and 5% of its phenotypic variation (Ingvarsson, Garcia, Luquez, Hall, & Jansson, 2008). Besides, allele frequency at different loci correlates with latitudinal position in numerous other species (e.g. Chen, Kallman, et al., 2012; Eckert et al., 2010; Prunier, Laroche, Beaulieu, & Bousquet, 2011).

Aridity is the other important climate variable influencing species distribution. The genomics of drought tolerance has been studied extensively and has been reviewed relatively recently (Hamanishi & Campbell, 2011). QTL mapping studies have generally identified few loci and explained a relatively small proportion of drought tolerance variation (Tschaplinski et al., 2006). In a study of *P. trichocarpa* and *Populus deltoides* hybrids, seven identified QTLs explained greater than 7.5% of phenotypic variance in drought tolerance (Tschaplinski et al., 2006, see Street et al., 2006 for more details in *Populus*). In *P. taeda*, five loci were associated to the aridity gradient found across the natural range (Eckert et al., 2010). The primary functions of the five gene products encoded by these loci were related to abiotic and biotic stress responses (Eckert et al., 2010) but none of them were related directly to osmosis control pathway gathered by Hamanishi and Campbell (2011).

7. FUTURE DIRECTIONS: INTEGRATING GENETIC DIVERSITY AND GENOME FUNCTION

This chapter has provided an overview of major areas of progress in forest tree genomics, including genome evolution, genome function focussing on gene expression and the transcriptome, the genetic architecture of quantitative traits and the population genomics of adaptation. The emerging issues surrounding adaptability to changing environmental conditions may hinge on the interplay between genetic diversity and genome function, representing a major avenue for future developments. Genomics research into forest trees has developed a solid foundation upon which to study this interface and to fully exploit the power of genomics and NGS. Adaptability to changing conditions depends on phenotypic plasticity, standing genetic variations and associated phenotypic variability. The lessons learnt and experimental approaches developed in human genomics and population genomics in model systems present us with fruitful avenues to develop this new knowledge in forest trees. Research on the human genome has developed a broad understanding of different types of genetic or genomic variations and their functional consequences associated with heritable disorders, cancer, development and ageing, among others. Structural variations such as gene copy-number variations (CNVs), epigenetic changes such as DNA methylation and regulatory controls by noncoding RNAs represent mechanisms that may lead to adaptive phenotypes and hence be acted upon by selection. We discuss how this more integrated understanding may be developed in forest trees.

7.1 Genome Resequencing to Uncover Genomic Variations

To date, much of forest genomics research has focussed on SNP variations in or around genes and analyzed relatively small sets of genes (e.g. Eckert et al., 2009, 2010; Prunier et al., 2011). As a result, our understanding of the types of genomic variation is largely incomplete. Furthermore, very little is known of the functional impacts of population-level variations. The early availability of the *P. trichocarpa* genome (Tuskan et al., 2006) has enabled population-level genome resequencing, affording a more in-depth view of genetic variability (Evans et al., 2014; Porth et al., 2013). These studies have primarily reported on SNP discovery and have refined our understanding of genetic diversity (Evans et al., 2014; Porth et al., 2013). For example, Porth et al. (2013) showed that the linkage disequilibrium extended over longer distance than previously described, which has significant implications for adaptation and the development of molecular breeding. Genome resquencing may now take place in eucalyptus (Myburg et al., 2014) and in conifers, and explore other types of genomic variations discussed below.

7.2 Structural Variations: The Case of Gene CNV

Structural polymorphisms such as gene CNVs epitomize the dynamic nature of genomes (Chain et al., 2014). CNVs result when spontaneous gene duplications occur in a population; most gene duplicates are inactivated and lost, but some duplicated gene copies may persist as variable gene copy numbers in the population and even reach fixation depending on fitness impacts (Lynch & Conery, 2003). Although they affect a larger proportion of the genome than SNPs, structural variations including CNVs are the least studied forms of intraspecific genetic variation (Korbel et al., 2008). Genome-wide analyses have associated CNVs with several disease phenotypes in humans (Craddock et al., 2010) and local adaptation among stickleback fish populations (Chain et al., 2014). Many CNVs modify transcript levels (Schlattl, Anders, Waszak, Huber, & Korbel, 2011) and result in protein dosage and other downstream phenotypic effects which may be acted upon by selection.

Studies of CNVs and presence absence variation (PAV) have been initiated in forest trees but their abundance and impacts remain largely unexplored. In *P. sitchensis*, Hall et al., (2011) showed that weevil resistance was associated with CNVs in enzymes involved in (+)-3-carene biosynthesis. An analysis of *P. taeda* L. based on exome capture in 7434 genes identified 408 putative PAVs (Neves et al., 2014). Studies of CNVs have not been reported for hardwood trees; however, gene duplications and retention have been analyzed in detail from an evolution perspective in *Populus* (Evans et al., 2014; Rogers-Melnick et al., 2012) and in *Eucalyptus* (Myburg et al., 2014). Genome resequencing which has been initiated in these species has focussed on SNP discovery and analysis and could now turn to analyzing CNVs on a large-scale by using methods such as CNV-seq (Xie & Tammi, 2009).

It has been suggested that association studies aimed at delineating the genetic architecture of complex traits could gain in resolution and power by including fine-scale CNV information (Schlattl et al., 2011). To this end, complete genome hybridization arrays have been developed in *P. glauca* and used to identify CNVs in several hundreds of genes; much variation in affected genes was observed between full-sib families from the same population (J. Prunier, personal communication).

7.3 Epigenetic Variation

Epigenetic variations encompass mechanisms that result in phenotypic diversity without genetic mutation. The roles of epigenetic variation include the establishment of phenotypic plasticity as well as heritable adaptation in plants (Schmitz et al., 2011). It has been associated with changes in DNA methylation and regulation by noncoding RNAs and generally affects gene expression. DNA methylation (cytosine base modification) is involved in development and ageing in both plants and animals (Bräutigam et al., 2013; Horvath, 2013) and in silencing of transposons and repetitive sequences in plants and fungi (Law & Jacobsen, 2010). In

Arabidopsis, trans-generational epigenetic variation resulting in phenotypic diversity has been directly linked to DNA methylation-altering transcriptions (Schmitz et al., 2011).

Epigenetic variation was proposed to be especially important for long-lived organisms such as forest trees (see reviews: Bräutigam et al., 2013; Yakovlev, Asante, Fossdal, Junttila, & Johnsen, 2011). One of the best documented examples of epigenetic control in trees comes from the discovery of a temperature-dependent epigenetic 'memory' conditioned by the temperature during early embryo development in *P. abies* (Johnsen et al., 2005; Yakovlev et al., 2011). This epigenetic memory was shown to influence the timing of bud phenology in next-generation offspring. Yakovlev et al., (2011) identified specific noncoding microRNAs whose differential expression indicated a putative role in the epigenetic regulation. Conifers accumulate microRNAs that include both shared and distinct sequences compared with angiosperms (Yakovlev et al., 2011) but in contrast to angiosperms, they appear to produce much lower levels of 24 nt small interfering RNAs (Dolgosheina et al., 2008), except in reproductive tissues (Nystedt et al., 2013). In poplar, DNA methylation was associated with ageing and drought responses (Raj et al., 2011). Our understanding of epigenetic control in trees has developed significantly but the underlying mechanisms are only partly identified. Despite this context, Bräutigam et al., (2013) concluded that 'ecological epigenetics' is set to transform our understanding of the way in which organisms such as forest trees function on the landscape.

7.4 Gene Expression as a Focus for Future Research

Several types of genomic variation impact on gene expression either directly or indirectly. These include epigenetic control, CNVs (through gene dosage) as well as regulatory variations in cis-acting sequence elements (e.g. in enhancer elements), at trans-acting loci (e.g. transcriptional regulators, signal transduction proteins, among others) and noncoding regulatory RNAs (e.g. microRNA). These sources of variation and their impacts on phenotypes including gene expression levels have been understudied in forest trees to date. This is thought to be a significant knowledge gap. It has been argued from first principles that mutations that alter the level of gene expression make qualitatively distinct contribution to phenotypic evolution by affecting certain kinds of traits and being acted upon more efficiently by selection (e.g. Jordan, Marino-Ramirez, & Koonin, 2005; Wray, 2007). Of relevance in genetically recombining species (including all forest trees), regulatory changes are more often immediately visible to selection because they are quantitative (additive effects). By contrast, beneficial coding sequence variations tend to be recessive, requiring several generations to increase in frequency within the population.

Recent empirical evidence clearly establishes links between gene expression variations and local adaptation. A study of stress-responsive gene expression comparing *Arabidopsis* accessions showed that genetic variability in responsiveness was a key to adaptation (Lasky et al., 2014). Genes with variable responsiveness were more strongly associated with climatic factors than those with consistent responsiveness (Lasky et al., 2014), implying that interactions occur between plasticity and genetic variability. In stickleback fish, genome resequencing in natural populations revealed the landscape of variation associated with independent local adaptation events (Jones et al., 2012). It was found that 41% variations associated with adaptation to freshwater environments influenced noncoding sequences, i.e. likely regulatory loci, and an additional 42% were potentially regulatory modifications influencing synonymous positions within or near genes, and only 17% of the variations influenced nonsynonymous positions in coding sequences.

Only a few studies have explored genetic variation of gene expression in forest trees experimentally. Expression variability studies have included population analyses showing that up to 50% of genes vary within the population (Palle et al., 2011), hundreds of genes vary between populations adapted to different climates (Holliday et al., 2008) or display allelic variations (Verta, Landry, & Mackay, 2013). Subsets of these genes harboured or were associated with sequence variation (Holliday, Ritland, & Aitken, 2010.) but the extent of results is insufficient to draw inferences regarding their effects on adaptation or fitness. Ultimately, to understand how such expression variation emerges and what is the role of expression variation in adaptation, the field of forest tree genomics needs to continue developing strategies to dissect the genetic and environmental sources of expression variation through either population-based (e.g. Holliday et al., 2008) or progeny-based (e.g. Verta et al., 2013) strategies.

8. CONCLUSION

The potential for deriving benefits from DNA-based tools to enhance tree breeding has been a major driving force for the development of genomics in forest trees including several economically important hardwoods and softwoods over the last two decades (Burdon & Wilcox, 2011; White et al.,

2007). Marker-assisted selection (e.g. see reviews by Burdon & Wilcox, 2011; Neale & Kremer, 2011) and progress in genomic selection (Grattapaglia & Resende, 2011) in both hardwood and softwood trees have shown the potential to shorten genetic selection by several years and thus accelerate breeding (Beaulieu, Doerksen, MacKay, Rainville, & Bousquet, 2014; Resende et al., 2012). Recent developments in NGS technologies and computational analyses promise to lead to other applications in sustainable forest management which may include assisted migration (Aitken et al., 2008), resistance breeding (Mageroy et al., 2015) and conservation of genetic diversity, among others.

In this review of progress, we have argued that a more integrated understanding of genetic diversity and genome function is needed and is possible with NGS. We have proposed that developing an understanding of the functional impacts of different types of diversity in the establishment of phenotypic plasticity and adaptation will enhance our knowledge of fitness determinants in forest trees. NGS technologies can be deployed to reveal variations in gene expression, DNA methylation, regulatory microRNAs, CNVs and other structural variations in addition to coding and regulatory sequence variation, simultaneously. These methodologies will accelerate the analysis of many different species and the development a more unified understanding that spans across the diverse trees species that make up our forests.

REFERENCES

Abril, N., Gion, J. M., Kerner, R., Muller-Starck, G., Cerrillo, R. M., Plomion, C., et al. (2011). Proteomics research on forest trees, the most recalcitrant and orphan plant species. *Phytochemistry, 72*, 1219–1242.

Adomas, A., Heller, G., Olson, A., Osborne, J., Karlsson, M., Nahalkova, J., et al. (2008). Comparative analysis of transcript abundance in *Pinus sylvestris* after challenge with a saprotrophic, pathogenic or mutualistic fungus. *Tree Physiology, 28*, 885–897.

Ahuja, M. R., & Neale, D. B. (2005). Evolution of genome size in conifers. *Silvae Genetica, 54*, 126–137.

Aitken, S. N., Yeaman, S., Holliday, J. A., Wang, T., & Curtis-McLane, S. (2008). Adaptation, migration or extirpation: climate change outcomes for tree populations. *Evolutionary Applications, 1*, 95–111.

Alberto, F. J., Aitken, S. N., Alía, R., González-Martínez, S. C., Hänninen, H., Kremer, A., et al. (2013). Potential for evolutionary responses to climate change – evidence from tree populations. *Global Change Biology, 19*, 1645–1661.

Allen, C. D., & Breshears, D. D. (1998). Drought-induced shift of a forest–woodland ecotone: rapid landscape response to climate variation. *Proceedings of the National Academy of Sciences, 95*, 14839–14842.

Anagnostakis, S. L. (1987). Chestnut blight – the classical problem of an introduced pathogen. *Mycologia, 79*, 23–37.

Archibold, O. W. (1995). *Ecology of World Vegetation*. London: Chapman and Hall.

Arumugasundaram, S., Ghosh, M., Veerasamy, S., & Ramasamy, Y. (2011). Species discrimination, population structure and linkage disequilibrium in *Eucalyptus camaldulensis* and *Eucalyptus tereticornis* using SSR markers. *PLoS One, 6*, e28252.

Barow, M., & Meister, A. (2003). Endopolyploidy in seed plants is differently correlated to systematics, organ, life strategy and genome size. *Plant, Cell & Environment, 26*, 571–584.

Bartholome, J., Mandrou, E., Mabiala, A., Jenkins, J., Nabihoudine, I., Klopp, C., et al. (2014). High-resolution genetic maps of *Eucalyptus* improve *Eucalyptus grandis* genome assembly. *New Phytologist, 206*, 1283–1296.

Bautista, R., Villalobos, D., Díaz-Moreno, S., Cantón, F., Cánovas, F., & Claros, M. G. (2007). Toward a *Pinus pinaster* bacterial artificial chromosome library. *Annals of Forest Science, 64*, 855–864.

Beaulieu, J., Doerksen, T., Boyle, B., Clement, S., Deslauriers, M., Beauseigle, S., et al. (2011). Association genetics of wood physical traits in the conifer white spruce and relationships with gene expression. *Genetics, 188*, 197–214.

Beaulieu, J., Doerksen, T. K., MacKay, J., Rainville, A., & Bousquet, J. (2014). Genomic selection accuracies within and between environments and small breeding groups in white spruce. *BMC Genomics, 15*, 1048.

Bennetzen, J. L. (2002). Mechanisms and rates of genome expansion and contraction in flowering plants. *Genetica, 115*, 29–36.

Birchler, J. A., & Veitia, R. A. (2007). The gene balance hypothesis: from classical genetics to modern genomics. *Plant Cell, 19*, 395–402.

Birol, I., Raymond, A., Jackman, S. D., Pleasance, S., Coope, R., Taylor, G. A., et al. (2013). Assembling the 20 Gb white spruce (*Picea glauca*) genome from whole-genome shotgun sequencing data. *Bioinformatics, 29*, 1492–1497.

Bodénès, C., Chancerel, E., Gailing, O., Vendramin, G. G., Bagnoli, F., Durand, J., et al. (2012). Comparative mapping in the Fagaceae and beyond with EST-SSRs. *BMC Plant Biology, 12*, 153.

Bomal, C., Bedon, F., Caron, S., Mansfield, S. D., Levasseur, C., Cooke, J. E., et al. (2008). Involvement of *Pinus taeda* MYB1 and MYB8 in phenylpropanoid metabolism and secondary cell wall biogenesis: a comparative in planta analysis. *Journal of Experimental Botany, 59*, 3925–3939.

Brasier, C., & Webber, J. (2010). Plant pathology: sudden larch death. *Nature, 466*, 824–825.

Bräutigam, K., Vining, K. J., Lafon-Placette, C., Fossdal, C. G., Mirouze, M., Marcos, J. G., et al. (2013). Epigenetic regulation of adaptive responses of forest tree species to the environment. *Ecology and Evolution, 3*, 399–415.

Brondani, R. P., Williams, E. R., Brondani, C., & Grattapaglia, D. (2006). A microsatellite-based consensus linkage map for species of *Eucalyptus* and a novel set of 230 microsatellite markers for the genus. *BMC Plant Biology, 6*, 20.

Burban, C., & Petit, R. J. (2003). Phylogeography of maritime pine inferred with organelle markers having contrasted inheritance. *Molecular Ecology, 12*, 1487–1495.

Burdon, R. D., & Wilcox, P. L. (2011). Integration of molecular markers in breeding. In C. Plomion, J. Bousquet, & C. Kole (Eds.), *Genetics, genomics and breeding of conifers* (pp. 276–322). New York: Edenbridge Science Publishers and CRC Press.

Burgarella, C., Lorenzo, Z., Jabbour-Zahab, R., Lumaret, R., Guichoux, E., Petit, R. J., et al. (2009). Detection of hybrids in nature: application to oaks (*Quercus suber* and *Q. ilex*). *Heredity, 102*, 442–452.

Butcher, P. A., & Moran, G. F. (2000). Genetic linkage mapping in *Acacia mangium*. 2. Development of an integrated map from two outbred pedigrees using RFLP and microsatellite loci. *Theoretical and Applied Genetics, 101*, 594–605.

Butcher, P. A., Moran, G. F., & Perkins, H. D. (1998). RFLP diversity in the nuclear genome of Acacia mangium. *Heredity, 81*, 205–213.

Callahan, C. M., Rowe, C. A., Ryel, R. J., Shaw, J. D., Madritch, M. D., & Mock, K. E. (2013). Continental-scale assessment of genetic diversity and population structure in quaking aspen (*Populus tremuloides*). *Journal of Biogeography, 40*, 1780–1791.

Camargo, E. L., Nascimento, L. C., Soler, M., Salazar, M. M., Lepikson-Neto, J., Marques, W. L., et al. (2014). Contrasting nitrogen fertilization treatments impact xylem gene expression and secondary cell wall lignification in *Eucalyptus*. *BMC Plant Biology, 14*, 256.

Cappa, E. P., El-Kassaby, Y. A., Garcia, M. N., Acuna, C., Borralho, N. M., Grattapaglia, D., et al. (2013). Impacts of population structure and analytical models in genome-wide association studies of complex traits in forest trees: a case study in *Eucalyptus globulus*. *PLoS One, 8*, e81267.

Casasoli, M., Derory, J., Morera-Dutrey, C., Brendel, O., Porth, I., Guehl, J. M., et al. (2006). Comparison of quantitative trait loci for adaptive traits between oak and chestnut based on an expressed sequence tag consensus map. *Genetics, 172*, 533–546.

Casasoli, M., Pot, D., Plomion, C., Monteverdi, M. C., Barreneche, T., Lauteri, M., et al. (2004). Identification of QTLs affecting adaptive traits in *Castanea sativa* Mill. *Plant, Cell & Environment, 27*, 1088–1101.

Cervera, M. T., Storme, V., Ivens, B., Gusmão, J., Liu, B. H., Hostyn, V., et al. (2001). Dense genetic linkage maps of three *Populus* species *(Populus deltoides, P. nigra and P. trichocarpa)* based on AFLP and microsatellite markers. *Genetics, 158*, 787–809.

Chain, F. J., Feulner, P. G., Panchal, M., Eizaguirre, C., Samonte, I. E., Kalbe, M., et al. (2014). Extensive copy-number variation of young genes across stickleback populations. *PLoS Genetics, 10*, e1004830.

Chen, J., Kallman, T., Ma, X., Gyllenstrand, N., Zaina, G., Morgante, M., et al. (2012). Disentangling the roles of history and local selection in shaping clinal variation of allele frequencies and gene expression in Norway spruce (*Picea abies*). *Genetics, 191*, 865–881.

Chen, J., Uebbing, S., Gyllenstrand, N., Lagercrantz, U., Lascoux, M., & Kallman, T. (2012). Sequencing of the needle transcriptome from Norway spruce (*Picea abies* Karst L.) reveals lower substitution rates, but similar selective constraints in gymnosperms and angiosperms. *BMC Genomics, 13*, 589.

Chou, C.-C., Chen, C.-H., Lee, T.-T., & Peck, K. (2004). Optimization of probe length and the number of probes per gene for optimal microarray analysis of gene expression. *Nucleic acids research, 32*. e99–e99.

Cossu, R. M., Buti, M., Giordani, T., Natali, L., & Cavallini, A. (2012). A computational study of the dynamics of LTR retrotransposons in the *Populus trichocarpa* genome. *Tree Genetics & Genomes, 8*, 61–75.

Craddock, N., Hurles, M. E., Cardin, N., Pearson, R. D., Plagnol, V., Robson, S., et al. (2010). Genome-wide association study of CNVs in 16,000 cases of eight common diseases and 3,000 shared controls. *Nature, 464*, 713–720.

Craven-Bartle, B., Pascual, M. B., Canovas, F. M., & Avila, C. (2013). A MYB transcription factor regulates genes of the phenylalanine pathway in maritime pine. *Plant Journal, 74*, 755–766.

Cullingham, C. I., Cooke, J. E. K., & Coltman, D. W. (2014). Cross-species outlier detection reveals different evolutionary pressures between sister species. *New Phytologist, 204*, 215–229.

Cumbie, W. P., Eckert, A., Wegrzyn, J., Whetten, R., Neale, D., & Goldfarb, B. (2011). Association genetics of carbon isotope discrimination, height and foliar nitrogen in a natural population of *Pinus taeda* L. *Heredity, 107*, 105–114.

Dai, X., Hu, Q., Cai, Q., Feng, K., Ye, N., Tuskan, G. A., et al. (2014). The willow genome and divergent evolution from poplar after the common genome duplication. *Cell Research, 24*, 1274–1277.

Davis, J. C., & Petrov, D. A. (2004). Preferential duplication of conserved proteins in eukaryotic genomes. *PLoS Biology, 2*, 318–326.

De La Torre, A. R., Birol, I., Bousquet, J., Ingvarsson, P. K., Jansson, S., Jones, S. J., et al. (2014). Insights into conifer giga-genomes. *Plant Physiology, 166*, 1724–1732.

De La Torre, A., Ingvarsson, P. K., & Aitken, S. N. (2015). Genetic architecture and genomic patterns of gene flow between hybridizing species of *Picea*. *Heredity*. http://dx.doi.org/10.1038/hdy.2015.19.

Delvas, N., Bauce, É., Labbé, C., Ollevier, T., & Bélanger, R. (2011). Phenolic compounds that confer resistance to spruce budworm. *Entomologia Experimentalis et Applicata, 141*, 35–44.

Dolgosheina, E. V., Morin, R. D., Aksay, G., Sahinalp, S. C., Magrini, V., Mardis, E. R., et al. (2008). Conifers have a unique small RNA silencing signature. *RNA, 14*, 1508–1515.

Douglas, C. J., & DiFazio, S. P. (2010). The *Populus* genome and comparative genomics. In R. Jansson, R. Bhalerao, & A. Groover (Eds.), *Genetics and genomics of Populus* (pp. 67–90). New York: Springer.

Downs, G. S., Bi, Y. M., Colasanti, J., Wu, W., Chen, X., Zhu, T., et al. (2013). A developmental transcriptional network for maize defines coexpression modules. *Plant Physiology, 161*, 1830–1843.

Dubouzet, J. G., Donaldson, L., Black, M. A., McNoe, L., Liu, V., & Lloyd-Jones, G. (2014). Heterologous hybridisation to a *Pinus* microarray: profiling of gene expression in *Pinus radiata* saplings exposed to ethephon. *New Zealand Journal of Forestry Science, 44*, 21.

Duval, I., Lachance, D., Giguère, I., Bomal, C., Morency, M.-J., Pelletier, G., et al. (2014). Large-scale screening of transcription factor–promoter interactions in spruce reveals a transcriptional network involved in vascular development. *Journal of Experimental Botany, 65*, 2319–2333.

Echt, C. S., Saha, S., Krutovsky, K. V., Wimalanathan, K., Erpelding, J. E., Liang, C., et al. (2011). An annotated genetic map of loblolly pine based on microsatellite and cDNA markers. *BMC Genetics, 12*, 17.

Eckert, A. J., Bower, A. D., Gonzalez-Martinez, S. C., Wegrzyn, J. L., Coop, G., & Neale, D. B. (2010). Back to nature: ecological genomics of loblolly pine (*Pinus taeda*, Pinaceae). *Molecular Ecology, 19*, 3789–3805.

Eckert, A., Pande, B., Ersoz, E., Wright, M., Rashbrook, V., Nicolet, C., et al. (2009). High-throughput genotyping and mapping of single nucleotide polymorphisms in loblolly pine (*Pinus taeda* L.). *Tree Genetics & Genomes, 5*, 225–234.

Euring, D., Bai, H., Janz, D., & Polle, A. (2014). Nitrogen-driven stem elongation in poplar is linked with wood modification and gene clusters for stress, photosynthesis and cell wall formation. *BMC Plant Biology, 14*, 391.

Evans, L. M., Slavov, G. T., Rodgers-Melnick, E., Martin, J., Ranjan, P., Muchero, W., et al. (2014). Population genomics of *Populus trichocarpa* identifies signatures of selection and adaptive trait associations. *Nature Genetics, 46*, 1089–1096.

Excoffier, L., Hofer, T., & Foll, M. (2009). Detecting loci under selection in a hierarchically structured population. *Heredity, 103*, 285–298.

Fang, G.-C., Blackmon, B., Staton, M., Nelson, C. D., Kubisiak, T., Olukolu, B., et al. (2013). A physical map of the Chinese chestnut (*Castanea mollissima*) genome and its integration with the genetic map. *Tree Genetics & Genomes, 9*, 525–537.

FAO. (2010). *Global forest resources assessment* (pp. 340).

Farjon, A., & Page, C. N. (1999). Conifers: status survey and conservation action plan. In *ISSC Action Plans for the conservation of biological diversity* (p. 121). IUCN.

Farrar, J. L. (1995). *Trees in Canada. Markham, Ontario: Natural resources*. Canada: Canadian Forest Service and Fitzhenry and Whiteside Limited.

Feder, J. L., Egan, S. P., & Nosil, P. (2012). The genomics of speciation-with-gene-flow. *Trends in Genetics, 28*, 342–350.

Freeman, J. S., O'Reilly-Wapstra, J. M., Vaillancourt, R. E., Wiggins, N., & Potts, B. M. (2008). Quantitative trait loci for key defensive compounds affecting herbivory of *Eucalyptus* in Australia. *New Phytologist, 178*, 846–851.

Friedmann, M., Ralph, S. G., Aeschliman, D., Zhuang, J., Ritland, K., Ellis, B. E., et al. (2007). Microarray gene expression profiling of developmental transitions in Sitka spruce (*Picea sitchensis*) apical shoots. *Journal of Experimental Botany, 58*, 593–614.

Geraldes, A., Difazio, S. P., Slavov, G. T., Ranjan, P., Muchero, W., Hannemann, J., et al. (2013). A 34K SNP genotyping array for *Populus trichocarpa*: design, application to the study of natural populations and transferability to other *Populus* species. *Molecular Ecology Resources, 13*, 306–323.

Geraldes, A., Farzaneh, N., Grassa, C. J., McKown, A. D., Guy, R. D., Mansfield, S. D., et al. (2014). Landscape genomics of *Populus trichocarpa*: the role of hybridization, limited gene flow, and natural selection in shaping patterns of population structure. *Evolution, 68*, 3260–3280.

Gernandt, D., Willyard, A., Syring, J., & Liston, A. (2011). The conifers (Pinophyta). In C. Plomion, J. Bousquet, & C. Kole (Eds.), *Genetics, genomics and breeding of conifers* (pp. 1–39). New York: Edenbridge Science Publishers and CRC Press.

Gesell, A., Blaukopf, M., Madilao, L., Yuen, M. M., Withers, S. G., Mattsson, J., et al. (2015). The gymnosperm cytochrome P450 CYP750B1 catalyzes stereospecific monoterpene hydroxylation of (+)-sabinene in thujone biosynthesis in Western redcedar. *Plant Physiology, 168*, 94–106.

Gion, J.-M., Carouché, A., Deweer, S., Bedon, F., Pichavant, F., Charpentier, J.-P., et al. (2011). Comprehensive genetic dissection of wood properties in a widely-grown tropical tree: *Eucalyptus*. *BMC Genomics, 12*, 301.

Gonzalez-Martinez, S. C., Huber, D., Ersoz, E., Davis, J. M., & Neale, D. B. (2008). Association genetics in *Pinus taeda* L. II. Carbon isotope discrimination. *Heredity, 101*, 19–26.

González-Martínez, S. C., Krutovsky, K. V., & Neale, D. B. (2006). Forest-tree population genomics and adaptive evolution. *New Phytologist, 170*, 227–238. http://dx.doi.org/10.1111/j.1469-8137.2006.01686.x.

González-Martínez, S. C., Wheeler, N. C., Ersoz, E., Nelson, C. D., & Neale, D. B. (2007). Association genetics in *Pinus taeda* L. I. Wood property traits. *Genetics, 175*, 399–409.

Grattapaglia, D., Bertolucci, F. L. G., Penchel, R., & Sederoff, R. R. (1996). Genetic mapping of quantitative trait loci controlling growth and wood quality traits in *Eucalyptus grandis* using a maternal half-sib family and RAPD markers. *Genetics, 144*, 1205–1214.

Grattapaglia, D., & Bradshaw, H. D., Jr. (1994). Nuclear DNA content of commercially important *Eucalyptus* species and hybrids. *Canadian Journal of Forest Research, 24*, 1074–1078.

Grattapaglia, D., & Resende, M. V. (2011). Genomic selection in forest tree breeding. *Tree Genetics & Genomes, 7*, 241–255.

Grattapaglia, D., & Sederoff, R. (1994). Genetic linkage maps of *Eucalyptus grandis* and *Eucalyptus urophylla* using a pseudo-testcross: mapping strategy and RAPD markers. *Genetics, 137*, 1121–1137.

Grattapaglia, D., Vaillancourt, R., Shepherd, M., Thumma, B., Foley, W., Külheim, C., et al. (2012). Progress in Myrtaceae genetics and genomics: *Eucalyptus* as the pivotal genus. *Tree Genetics & Genomes, 8*, 463–508.

Groover, A. T. (2005). What genes make a tree a tree? *Trends in Plant Science, 10*, 210–214.

Guo, H., & Ecker, J. R. (2004). The ethylene signaling pathway: new insights. *Current Opinion in Plant Biology, 7*, 40–49.

Hall, D. E., Robert, J. A., Keeling, C. I., Domanski, D., Quesada, A. L., Jancsik, S., et al. (2011). An integrated genomic, proteomic and biochemical analysis of (+)-3-carene biosynthesis in Sitka spruce (*Picea sitchensis*) genotypes that are resistant or susceptible to white pine weevil. *Plant Journal, 65*, 936–948.

Hamanishi, E. T., Raj, S., Wilkins, O., Thomas, B. R., Mansfield, S. D., Plant, A. L., et al. (2010). Intraspecific variation in the *Populus balsamifera* drought transcriptome. *Plant, Cell & Environment, 33*, 1742–1755.

Hamanishi, E. T., & Campbell, M. M. (2011). Genome-wide responses to drought in forest trees. *Forestry, 84*, 273–283.

Hamann, A., & Wang, T. (2006). Potential effects of climate change on ecosystem and tree species distribution in British Columbia. *Ecology, 87*, 2773–2786.

Harfouche, A., Meilan, R., Kirst, M., Morgante, M., Boerjan, W., Sabatti, M., et al. (2012). Accelerating the domestication of forest trees in a changing world. *Trends in Plant Science, 17*, 64–72.

Hirakawa, H., Nakamura, Y., Kaneko, T., Isobe, S., Sakai, H., Kato, T., et al. (2011). Survey of the genetic information carried in the genome of *Eucalyptus camaldulensis*. *Plant Biotechnology, 28*, 471–480.

Holliday, J. A., Ralph, S. G., White, R., Bohlmann, J., & Aitken, S. N. (2008). Global monitoring of autumn gene expression within and among phenotypically divergent populations of Sitka spruce (*Picea sitchensis*). *New Phytologist, 178*, 103–122.

Holliday, J. A., Ritland, K., & Aitken, S. N. (2010). Widespread, ecologically relevant genetic markers developed from association mapping of climate-related traits in Sitka spruce (*Picea sitchensis*). *New Phytologist, 188*, 501–514.

Horvath, S. (2013). DNA methylation age of human tissues and cell types. *Genome Biology, 14*, R115.

Hu, T. T., Pattyn, P., Bakker, E. G., Cao, J., Cheng, J. F., Clark, R. M., et al. (2011). The *Arabidopsis lyrata* genome sequence and the basis of rapid genome size change. *Nature Genetics, 43*, 476–481.

Ingvarsson, P. K., Garcia, M. V., Luquez, V., Hall, D., & Jansson, S. (2008). Nucleotide polymorphism and phenotypic associations within and around the phytochrome B2 locus in European aspen (*Populus tremula, Salicaceae*). *Genetics, 178*, 2217–2226.

Jaillon, O., Aury, J. M., Noel, B., Policriti, A., Clepet, C., Casagrande, A., et al. (2007). The grapevine genome sequence suggests ancestral hexaploidization in major angiosperm phyla. *Nature, 449*, 463–467.

Jaramillo-Correa, J. P., Rodriguez-Quilon, I., Grivet, D., Lepoittevin, C., Sebastiani, F., Heuertz, M., et al. (2015). Molecular proxies for climate maladaptation in a long-lived tree (*Pinus pinaster* Aiton, *Pinaceae*). *Genetics, 199*, 793–807.

Johnsen, O., Fossdal, C. G., Nagy, N., Molmann, J., Daehlen, O. G., & Skroppa, T. (2005). Climatic adaptation in *Picea abies* progenies is affected by the temperature during zygotic embryogenesis and seed maturation. *Plant Cell and Environment, 28*, 1090–1102.

Jones, F. C., Grabherr, M. G., Chan, Y. F., Russell, P., Mauceli, E., Johnson, J., et al. (2012). The genomic basis of adaptive evolution in threespine sticklebacks. *Nature, 484*, 55–61.

Jordan, I. K., Marino-Ramirez, L., & Koonin, E. V. (2005). Evolutionary significance of gene expression divergence. *Gene, 345*, 119–126.

Kang, B. Y., Mann, I. K., Major, J. E., & Rajora, O. P. (2010). Near-saturated and complete genetic linkage map of black spruce (*Picea mariana*). *BMC Genomics, 11*, 515.

Kim, Y. Y., Choi, H. S., & Kang, B. Y. (2005). An AFLP-based linkage map of Japanese red pine (*Pinus densiflora*) using haploid DNA samples of megagametophytes from a single maternal tree. *Molecules and Cells, 20*, 201–209.

Kirst, M., Johnson, A. F., Baucom, C., Ulrich, E., Hubbard, K., Staggs, R., et al. (2003). Apparent homology of expressed genes from wood-forming tissues of loblolly pine (*Pinus taeda* L.) with *Arabidopsis thaliana*. *Proceedings of the National Academy of Sciences of the United States of America, 100*, 7383–7388.

Kirst, M., Myburg, A. A., De León, J. P. G., Kirst, M. E., Scott, J., & Sederoff, R. (2004). Coordinated genetic regulation of growth and lignin revealed by quantitative trait locus analysis of cDNA microarray data in an interspecific backcross of *Eucalyptus*. *Plant Physiology, 135*, 2368–2378.

Ko, J. H., Kim, H. T., Hwang, I., & Han, K. H. (2012). Tissue-type-specific transcriptome analysis identifies developing xylem-specific promoters in poplar. *Plant Biotechnology Journal, 10*, 587–596.

Kogenaru, S., Qing, Y., Guo, Y., & Wang, N. (2012). RNA-seq and microarray complement each other in transcriptome profiling. *BMC Genomics, 13*, 629.

Komulainen, P., Brown, G. R., Mikkonen, M., Karhu, A., Garcia-Gil, M. R., O'Malley, D., et al. (2003). Comparing EST-based genetic maps between *Pinus sylvestris* and *Pinus taeda*. *Theoretical and Applied Genetics, 107*, 667–678.

Kondo, T., Terada, K., Hayashi, E., Kuramoto, N., Okamura, M., & Kawasaki, H. (2000). RAPD markers linked to a gene for resistance to pine needle gall midge in Japanese black pine (*Pinus thunbergii*). *Theoretical and Applied Genetics, 100*, 391–395.

Korbel, J. O., Kim, P. M., Chen, X., Urban, A. E., Weissman, S., Snyder, M., et al. (2008). The current excitement about copy-number variation: how it relates to gene duplications and protein families. *Current Opinion in Structural Biology, 18*, 366–374.

Kovach, A., Wegrzyn, J. L., Parra, G., Holt, C., Bruening, G. E., Loopstra, C. A., et al. (2010). The *Pinus taeda* genome is characterized by diverse and highly diverged repetitive sequences. *BMC Genomics, 11*, 420.

Kremer, A., Ronce, O., Robledo-Arnuncio, J. J., Guillaume, F., Bohrer, G., Nathan, R., et al. (2012). Long-distance gene flow and adaptation of forest trees to rapid climate change. *Ecology Letters, 15*, 378–392.

Kurz, W. A., Dymond, C. C., Stinson, G., Rampley, G. J., Neilson, E. T., Carroll, A. L., et al. (2008). Mountain pine beetle and forest carbon feedback to climate change. *Nature, 452*, 987–990.

Lander, T. A., Boshier, D. H., & Harris, S. A. (2010). Fragmented but not isolated: contribution of single trees, small patches and long-distance pollen flow to genetic connectivity for *Gomortega keule*, an endangered Chilean tree. *Biological Conservation, 143*, 2583–2590.

Lasky, J. R., Des Marais, D. L., Lowry, D. B., Povolotskaya, I., McKay, J. K., Richards, J. H., et al. (2014). Natural variation in abiotic stress responsive gene expression and local adaptation to climate in *Arabidopsis thaliana*. *Molecular Biology and Evolution, 31*, 2283–2296.

Law, J. A., & Jacobsen, S. E. (2010). Establishing, maintaining and modifying DNA methylation patterns in plants and animals. *Nature Reviews Genetics, 11*, 204–220.

Ledig, F. T., Jacob-Cervantes, V., Hodgskiss, P. D., & Eguiluz-Piedra, T. (1997). Recent evolution and divergence among populations of a rare Mexican endemic, Chihuahua spruce, following Holocene climatic warming. *Evolution, 51*, 1815–1827.

Lepoittevin, C., Harvengt, L., Plomion, C., & Garnier-Géré, P. (2012). Association mapping for growth, straightness and wood chemistry traits in the *Pinus pinaster* Aquitaine breeding population. *Tree Genetics & Genomes, 8*, 113–126.

Lesur, I., Bechade, A., Lalanne, C., Klopp, C., Noirot, C., Leplé, J.-C., et al. (2015). A unigene set for European beech (*Fagus sylvatica* L.) and its use to decipher the molecular mechanisms involved in dormancy regulation. *Molecular Ecology Resources*. http://dx.doi.org/10.1111/1755-0998.12373.

Lesur, I., Le Provost, G., Bento, P., Da Silva, C., Leplé, J. C., Murat, F., et al. (2015). The oak gene expression atlas: insights into Fagaceae genome evolution and the discovery of genes regulated during bud dormancy release. *BMC Genomics, 16*, 112.

Li, C., & Yeh, F. C. (2001). Construction of a framework map in *Pinus contorta* subsp. *latifolia* using random amplified polymorphic DNA markers. *Genome, 44*, 147–153.

Li, S., Chen, Y., Gao, H., & Yin, T. (2010). Potential chromosomal introgression barriers revealed by linkage analysis in a hybrid of *Pinus massoniana* and *P. hwangshanensis*. *BMC Plant Biology, 10*, 37.

Li, X., Wu, H. X., & Southerton, S. G. (2010). Seasonal reorganization of the xylem transcriptome at different tree ages reveals novel insights into wood formation in *Pinus radiata*. *New Phytologist, 187*, 764–776.

Li, X., Wu, H. X., & Southerton, S. G. (2011). Transcriptome profiling of *Pinus radiata* juvenile wood with contrasting stiffness identifies putative candidate genes involved in microfibril orientation and cell wall mechanics. *BMC Genomics, 12*, 480.

Li, X., Yang, X., & Wu, H. X. (2013). Transcriptome profiling of radiata pine branches reveals new insights into reaction wood formation with implications in plant gravitropism. *BMC Genomics, 14*, 768.

Lind, M., Källman, T., Chen, J., Ma, X.-F., Bousquet, J., Morgante, M., et al. (2014). A *Picea abies* linkage map based on SNP markers identifies QTLs for four aspects of resistance to *Heterobasidion parviporum* infection. *PLoS One, 9*, e101049.

Liu, J. J., Sturrock, R. N., & Benton, R. (2013). Transcriptome analysis of *Pinus monticola* primary needles by RNA-seq provides novel insight into host resistance to *Cronartium ribicola*. *BMC Genomics, 14*, 884.

Long, A. D., & Langley, C. H. (1999). The power of association studies to detect the contribution of candidate genetic loci to variation in complex traits. *Genome Research, 9*, 720–731.

Long, Y., Zhang, J., Tian, X., Wu, S., Zhang, Q., Zhang, J., et al. (2014). De novo assembly of the desert tree *Haloxylon ammodendron* (C. A. Mey.) based on RNA-Seq data provides insight into drought response, gene discovery and marker identification. *BMC Genomics, 15*, 1111.

Luikart, G., England, P. R., Tallmon, D., Jordan, S., & Taberlet, P. (2003). The power and promise of population genomics: from genotyping to genome typing. *Nature Reviews Genetics, 4*, 981–994.

Lynch, M., & Conery, J. S. (2003). The origins of genome complexity. *Science, 302*, 1401–1404.

Lynch, M., & Force, A. (2000). The probability of duplicate gene preservation by subfunctionalization. *Genetics, 154*, 459–473.

Ma, L., Sun, N., Liu, X., Jiao, Y., Zhao, H., & Deng, X. W. (2005). Organ-specific expression of *Arabidopsis* genome during development. *Plant Physiology, 138*, 80–91.

Mabberley, D. J. (1987). *The plant-book: A portable dictionary of the higher plants* (1st ed.). Cambridge, UK: Cambridge University Press.

Mackay, J., Dean, J. F., Plomion, C., Peterson, D. G., Canovas, F. M., Pavy, N., et al. (2012). Towards decoding the conifer giga-genome. *Plant Molecular Biology, 80*, 555–569.

Magbanua, Z. V., Ozkan, S., Bartlett, B. D., Chouvarine, P., Saski, C. A., Liston, A., et al. (2011). Adventures in the enormous: a 1.8 million clone BAC library for the 21.7 Gb genome of loblolly pine. *PLoS One, 6*, e16214.

Mageroy, M. H., Parent, G., Germanos, G., Giguère, I., Delvas, N., Maaroufi, H., et al. (2015). Expression of the β-glucosidase gene *Pgβglu-1* underpins natural resistance of white spruce against spruce budworm. *Plant Journal, 81*, 68–80.

Manganaris, G., Rasori, A., Bassi, D., Geuna, F., Ramina, A., Tonutti, P., et al. (2011). Comparative transcript profiling of apricot (*Prunus armeniaca* L.) fruit development and on-tree ripening. *Tree Genetics & Genomes, 7*, 609–616.

Men, L., Yan, S., & Liu, G. (2013). De novo characterization of *Larix gmelinii* (Rupr.) Rupr. transcriptome and analysis of its gene expression induced by jasmonates. *BMC Genomics, 14*, 548.

de Miguel, M., de Maria, N., Guevara, M. A., Diaz, L., Saez-Laguna, E., Sanchez-Gomez, D., et al. (2012). Annotated genetic linkage maps of *Pinus pinaster* Ait. from a Central Spain population using microsatellite and gene based markers. *BMC Genomics, 13*, 527.

Miller, J. T., Seigler, D., & Mishler, B. D. (2014). A phylogenetic solution to the *Acacia* problem. *Taxon, 63*, 653–658.

Mishima, K., Fujiwara, T., Iki, T., Kuroda, K., Yamashita, K., Tamura, M., et al. (2014). Transcriptome sequencing and profiling of expressed genes in cambial zone and differentiating xylem of Japanese cedar (*Cryptomeria japonica*). *BMC Genomics, 15*, 219.

Moraga-Suazo, P., Orellana, L., Quiroga, P., Balocchi, C., Sanfuentes, E., Whetten, R. W., et al. (2014). Development of a genetic linkage map for *Pinus radiata* and detection of pitch canker disease resistance associated QTLs. *Trees, 28*, 1823–1835.

Morgante, M., & Paoli, E. D. (2011). Toward the conifer genome sequence. In C. Plomion, J. Bousquet, & C. Kole (Eds.), *Genetics, genomics and breeding of conifers* (pp. 389–403). New York: Edenbridge Science Publishers and CRC Press.

Morse, A. M., Peterson, D. G., Islam-Faridi, M. N., Smith, K. E., Magbanua, Z., Garcia, S. A., et al. (2009). Evolution of genome size and complexity in Pinus. *PLoS One, 4*, e4332.

Murray, B. G., Leitch, I. J., & Bennett, M. D. (December 2012). *Gymnosperm DNA C-values database*, 5.0. from http://www.kew.org/cvalues.

Myburg, A. A., Grattapaglia, D., Tuskan, G. A., Hellsten, U., Hayes, R. D., Grimwood, J., et al. (2014). The genome of *Eucalyptus grandis*. *Nature, 510*, 356–362. http://www.nature.com/nature/journal/v510/n7505/abs/nature13308.html#supplementary-information.

Nagalakshmi, U., Wang, Z., Waern, K., Shou, C., Raha, D., Gerstein, M., et al. (2008). The transcriptional landscape of the yeast genome defined by RNA sequencing. *Science, 320*, 1344–1349.

Neale, D. B. (2007). Genomics to tree breeding and forest health. *Current Opinion in Genetics & Development, 17*, 539–544.

Neale, D. B., & Ingvarsson, P. K. (2008). Population, quantitative and comparative genomics of adaptation in forest trees. *Current Opinion in Plant Biology, 11*, 149–155.

Neale, D. B., & Kremer, A. (2011). Forest tree genomics: growing resources and applications. *Nature Reviews Genetics, 12*, 111–122.

Neale, D. B., Wegrzyn, J. L., Stevens, K. A., Zimin, A. V., Puiu, D., Crepeau, M. W., et al. (2014). Decoding the massive genome of loblolly pine using haploid DNA and novel assembly strategies. *Genome Biology, 15*, R59.

Nelson, C. D., Nance, W. L., & Doudrick, R. L. (1993). A partial genetic linkage map of slash pine (*Pinus elliottii* Engelm. var. *elliottii*) based on random amplified polymorphic DNAs. *Theoretical and Applied Genetics, 87*, 145–151.

Neves, L. G., Davis, J. M., Barbazuk, W. B., & Kirst, M. (2014). A high-density gene map of loblolly pine (*Pinus taeda* L.) based on exome sequence capture genotyping. *G3, 4*, 29–37.

Novaes, E., Drost, D. R., Farmerie, W. G., Pappas, G. J., Jr., Grattapaglia, D., Sederoff, R. R., et al. (2008). High-throughput gene and SNP discovery in *Eucalyptus grandis*, an uncharacterized genome. *BMC Genomics, 9*, 312.

Nystedt, B., Street, N. R., Wetterbom, A., Zuccolo, A., Lin, Y. C., Scofield, D. G., et al. (2013). The Norway spruce genome sequence and conifer genome evolution. *Nature, 497*, 579–584.

Ohno, S. (1970). *Evolution by gene duplication*. New York, NY: Springer.

Padovan, A., Lanfear, R., Keszei, A., Foley, W. J., & Kulheim, C. (2013). Differences in gene expression within a striking phenotypic mosaic *Eucalyptus* tree that varies in susceptibility to herbivory. *BMC Plant Biology, 13*, 29.

Paiva, J. A., Garnier-Gere, P. H., Rodrigues, J. C., Alves, A., Santos, S., Graca, J., et al. (2008). Plasticity of maritime pine (*Pinus pinaster*) wood-forming tissues during a growing season. *New Phytologist, 179*, 1080–1094.

Pakull, B., Groppe, K., Meyer, M., Markussen, T., & Fladung, M. (2009). Genetic linkage mapping in aspen (*Populus tremula* L. and *Populus tremuloides* Michx.). *Tree Genetics & Genomes, 5*, 505–515.

Palle, S. R., Seeve, C. M., Eckert, A. J., Cumbie, W. P., Goldfarb, B., & Loopstra, C. A. (2011). Natural variation in expression of genes involved in xylem development in loblolly pine (*Pinus taeda* L.). *Tree Genetics & Genomes, 7*, 193–206.

Paolucci, I., Gaudet, M., Jorge, V., Beritognolo, I., Terzoli, S., Kuzminsky, E., et al. (2010). Genetic linkage maps of *Populus alba* L. and comparative mapping analysis of sex determination across *Populus* species. *Tree Genetics & Genomes, 6*, 863–875.

Parchman, T. L., Geist, K. S., Grahnen, J. A., Benkman, C. W., & Buerkle, C. A. (2010). Transcriptome sequencing in an ecologically important tree species: assembly, annotation, and marker discovery. *BMC Genomics, 11*, 180.

Parchman, T. L., Gompert, Z., Mudge, J., Schilkey, F. D., Benkman, C. W., & Buerkle, C. A. (2012). Genome-wide association genetics of an adaptive trait in lodgepole pine. *Molecular Ecology, 21*, 2991–3005.

Paterson, A. H. (1998). *Molecular dissection of complex traits*. New York, NY: CRC Press.

Patzlaff, A., Newman, L., Dubos, C., Whetten, R., Smith, C., McInnis, S., et al. (2003). Characterisation of PtMYB1, an R2R3-MYB from pine xylem. *Plant Molecular Biology, 53*, 597–608.

Pavy, N., Pelgas, B., Laroche, J., Rigault, P., Isabel, N., & Bousquet, J. (2012). A spruce gene map infers ancient plant genome reshuffling and subsequent slow evolution in the gymnosperm lineage leading to extant conifers. *BMC Biology, 10*, 84.

Pelgas, B., Beauseigle, S., Acheré, V., Jeandroz, S., Bousquet, J., & Isabel, N. (2006). Comparative genome mapping among *Picea glauca, P. mariana* × *P. rubens* and *P. abies*, and correspondence with other Pinaceae. *Theoretical and Applied Genetics, 113*, 1371–1393.

Pelgas, B., Bousquet, J., Meirmans, P. G., Ritland, K., & Isabel, N. (2011). QTL mapping in white spruce: gene maps and genomic regions underlying adaptive traits across pedigrees, years and environments. *BMC Genomics, 12*, 145.

Plomion, C., Aury, J.-M., Amselem, J., Alaeitabar, T., Barbe, V., Belser, C., et al. (2015). Decoding the oak genome: public release of sequence data, assembly, annotation and publication strategies. *Molecular Ecology Resources*. http://dx.doi.org/10.1111/1755-0998.12425.

Porth, I., Klapste, J., Skyba, O., Hannemann, J., McKown, A. D., Guy, R. D., et al. (2013). Genome-wide association mapping for wood characteristics in *Populus* identifies an array of candidate single nucleotide polymorphisms. *New Phytologist, 200*, 710–726.

Pot, D., Rodrigues, J.-C., Rozenberg, P., Chantre, G., Tibbits, J., Cahalan, C., et al. (2006). QTLs and candidate genes for wood properties in maritime pine (*Pinus pinaster* Ait.). *Tree Genetics & Genomes, 2*(1), 10–24.

Prunier, J., Gerardi, S., Laroche, J., Beaulieu, J., & Bousquet, J. (2012). Parallel and lineage-specific molecular adaptation to climate in boreal black spruce. *Molecular Ecology, 21*, 4270–4286.

Prunier, J., Laroche, J., Beaulieu, J., & Bousquet, J. (2011). Scanning the genome for gene SNPs related to climate adaptation and estimating selection at the molecular level in boreal black spruce. *Molecular Ecology, 20*, 1702–1716.

Prunier, J., Pelgas, B., Gagnon, F., Desponts, M., Isabel, N., Beaulieu, J., et al. (2013). The genomic architecture and association genetics of adaptive characters using a candidate SNP approach in boreal black spruce. *BMC Genomics, 14*, 368.

Pullat, J., Fleischer, R., Becker, N., Beier, M., Metspalu, A., & Hoheisel, J. D. (2007). Optimization of candidate-gene SNP-genotyping by flexible oligonucleotide microarrays; analyzing variations in immune regulator genes of hay-fever samples. *BMC genomics, 8*, 282.

Qiu, Q., Ma, T., Hu, Q., Liu, B., Wu, Y., Zhou, H., et al. (2011). Genome-scale transcriptome analysis of the desert poplar, *Populus euphratica*. *Tree Physiology, 31*, 452–461.

Qiu, Z., Wan, L., Chen, T., Wan, Y., He, X., Lu, S., et al. (2013). The regulation of cambial activity in Chinese fir (*Cunninghamia lanceolata*) involves extensive transcriptome remodeling. *New Phytologist, 199*, 708–719.

Quesada, T., Gopal, V., Cumbie, W. P., Eckert, A. J., Wegrzyn, J. L., Neale, D. B., et al. (2010). Association mapping of quantitative disease resistance in a natural population of Loblolly pine (*Pinus taeda* L.). *Genetics, 186*, 677–686.

Raffa, K. F., Powell, E. N., & Townsend, P. A. (2013). Temperature-driven range expansion of an irruptive insect heightened by weakly coevolved plant defenses. *Proceedings of the National Academy of Sciences, 110*, 2193–2198.

Raherison, E. S., Giguere, I., Caron, S., Lamara, M., & MacKay, J. J. (2015). Modular organization of the white spruce (*Picea glauca*) transcriptome reveals functional organization and evolutionary signatures. *New Phytologist, 207*, 172–178.

Raherison, E., Rigault, P., Caron, S., Poulin, P. L., Boyle, B., Verta, J. P., et al. (2012). Transcriptome profiling in conifers and the PiceaGenExpress database show patterns of diversification within gene families and interspecific conservation in vascular gene expression. *BMC Genomics, 13*, 434.

Raj, S., Brautigam, K., Hamanishi, E. T., Wilkins, O., Thomas, B. R., Schroeder, W., et al. (2011). Clone history shapes *Populus* drought responses. *Proceedings of the National Academy of Sciences, 108*, 12521–12526.

Ralph, S. G., Yueh, H., Friedmann, M., Aeschliman, D., Zeznik, J. A., Nelson, C. C., et al. (2006). Conifer defence against insects: microarray gene expression profiling of Sitka spruce (*Picea sitchensis*) induced by mechanical wounding or feeding by spruce budworms (*Choristoneura occidentalis*) or white pine weevils (*Pissodes strobi*) reveals large-scale changes of the host transcriptome. *Plant Cell and Environment, 29*, 1545–1570.

Ranade, S., Abrahamsson, S., Niemi, J., & García-Gil, M. (2013). *Pinus taeda* cDNA microarray as a tool for candidate gene identification for local red/far-red light adaptive response in *Pinus sylvestris*. *American Journal of Plant Sciences, 4*, 479–493.

Ren, L. L., Liu, Y. J., Liu, H. J., Qian, T. T., Qi, L. W., Wang, X. R., et al. (2014). Subcellular relocalization and positive selection play key roles in the retention of duplicate genes of *Populus* class III peroxidase family. *Plant Cell, 26*, 2404–2419.

Resende, M. D., Resende, M. F., Jr., Sansaloni, C. P., Petroli, C. D., Missiaggia, A. A., Aguiar, A. M., et al. (2012). Genomic selection for growth and wood quality in *Eucalyptus*: capturing the missing heritability and accelerating breeding for complex traits in forest trees. *New Phytologist, 194*, 116–128.

Rigault, P., Boyle, B., Lepage, P., Cooke, J. E., Bousquet, J., & MacKay, J. J. (2011). A white spruce gene catalog for conifer genome analyses. *Plant Physiology, 157*, 14–28.

Rodgers-Melnick, E., Mane, S. P., Dharmawardhana, P., Slavov, G. T., Crasta, O. R., Strauss, S. H., et al. (2012). Contrasting patterns of evolution following whole genome versus tandem duplication events in *Populus*. *Genome Research, 22*, 95–105.

Sato, S., Yoshida, M., Hiraide, H., Ihara, K., & Yamamoto, H. (2014). Transcriptome analysis of reaction wood in gymnosperms by next-generation sequencing. *American Journal of Plant Sciences, 5*, 2785–2798.

Scalfi, M., Troggio, M., Piovani, P., Leonardi, S., Magnaschi, G., Vendramin, G. G., et al. (2004). A RAPD, AFLP and SSR linkage map, and QTL analysis in European beech (*Fagus sylvatica* L.). *Theoretical and Applied Genetics, 108*, 433–441.

Schlattl, A., Anders, S., Waszak, S. M., Huber, W., & Korbel, J. O. (2011). Relating CNVs to transcriptome data at fine resolution: assessment of the effect of variant size, type, and overlap with functional regions. *Genome Research, 21*, 2004–2013.

Schmitz, R. J., Schultz, M. D., Lewsey, M. G., O'Malley, R. C., Urich, M. A., Libiger, O., et al. (2011). Transgenerational epigenetic instability is a source of novel methylation variants. *Science, 334*, 369–373.

Schnurr, J., Cheng, Z., & Boe, A. (1996). Effects of plant growth regulators on sturdiness of Jack pine seedlings. *Journal of Environmental Horticulture, 14*, 228–230.

Scotti-Saintagne, C., Mariette, S., Porth, I., Goicoechea, P. G., Barreneche, T., Bodénès, C., et al. (2004). Genome scanning for interspecific differentiation between two closely related oak species [*Quercus robur* L. and *Q. petraea* (Matt.) Liebl.]. *Genetics, 168*, 1615–1626.

Sena, J. S., Giguere, I., Boyle, B., Rigault, P., Birol, I., Zuccolo, A., et al. (2014). Evolution of gene structure in the conifer *Picea glauca*: a comparative analysis of the impact of intron size. *BMC Plant Biology, 14*, 95.

Shafer, A. B., Cullingham, C. I., Cote, S. D., & Coltman, D. W. (2010). Of glaciers and refugia: a decade of study sheds new light on the phylogeography of northwestern North America. *Molecular Ecology, 19*, 4589–4621.

Simpson, J. T., Wong, K., Jackman, S. D., Schein, J. E., Jones, S. J., & Birol, I. (2009). ABySS: a parallel assembler for short read sequence data. *Genome Research, 19*, 1117–1123.

Siol, M., Wright, S. I., & Barrett, S. C. H. (2010). The population genomics of plant adaptation. *New Phytologist, 188*, 313–332.

Sisco, P. H., Kubisiak, T. L., Casasoli, M., Barreneche, T., Kremer, A., Clark, C., et al. (2005). An improved genetic map for *Castanea mollissima/Castanea dentata* and its relationship to the genetic map of *Castanea sativa*. *Acta Horticulturae, 693*, 491–496.

Sjodin, A., Street, N. R., Sandberg, G., Gustafsson, P., & Jansson, S. (2009). The *Populus* genome integrative explorer (PopGenIE): a new resource for exploring the *Populus* genome. *New Phytologist, 182*, 1013–1025.

Soltis, P. S., & Soltis, D. E. (2013). A conifer genome spruces up plant phylogenomics. *Genome Biology, 14*, 122.

Sork, V. L., Aitken, S. N., Dyer, R. J., Eckert, A. J., Legendre, P., & Neale, D. B. (2013). Putting the landscape into the genomics of trees: approaches for understanding local adaptation and population responses to changing climate. *Tree Genetics & Genomes, 9*, 901–911.

Stackpole, D., Vaillancourt, R., de Aguigar, M., & Potts, B. (2010). Age trends in genetic parameters for growth and wood density in *Eucalyptus globulus*. *Tree Genetics & Genomes, 6*, 179–193.

ter Steege, H., Pitman, N. C., Sabatier, D., Baraloto, C., Salomao, R. P., Guevara, J. E., et al. (2013). Hyperdominance in the Amazonian tree flora. *Science, 342*, 1243092.

Sterky, F., Regan, S., Karlsson, J., Hertzberg, M., Rohde, A., Holmberg, A., et al. (1998). Gene discovery in the wood-forming tissues of poplar: analysis of 5, 692 expressed sequence tags. *Proceedings of the National Academy of Sciences, 95*, 13330–13335.

Stevens, P.F. (2012, Version 12). Angiosperm phylogeny website. Retrieved July, 2012, from http://www.mobot.org/MOBOT/research/APweb/

Street, N. R., Skogström, O., Sjödin, A., Tucker, J., Rodríguez-Acosta, M., Nilsson, P., et al. (2006). The genetics and genomics of the drought response in *Populus*. *Plant Journal, 48*, 321–341.

Tani, N., Takahashi, T., Iwata, H., Mukai, Y., Ujino-Ihara, T., Matsumoto, A., et al. (2003). A consensus linkage map for sugi (*Cryptomeria japonica*) from two pedigrees, based on microsatellites and expressed sequence tags. *Genetics, 165*, 1551–1568.

Thamarus, K. A., Groom, K., Murrell, J., Byrne, M., & Moran, G. F. (2002). A genetic linkage map for *Eucalyptus globulus* with candidate loci for wood, fibre, and floral traits. *Theoretical and Applied Genetics, 104*, 379–387.

Thavamanikumar, S., Southerton, S. G., Bossinger, G., & Thumma, B. R. (2013). Dissection of complex traits in forest trees – opportunities for marker-assisted selection. *Tree Genetics & Genomes, 9*, 627–639.

Thumma, B. R., Matheson, B. A., Zhang, D., Meeske, C., Meder, R., Downes, G. M., et al. (2009). Identification of a cis-acting regulatory polymorphism in a eucalypt COBRA-like gene affecting cellulose content. *Genetics, 183*, 1153–1164.

Thumma, B. R., Sharma, N., & Southerton, S. G. (2012). Transcriptome sequencing of *Eucalyptus camaldulensis* seedlings subjected to water stress reveals functional single nucleotide polymorphisms and genes under selection. *BMC Genomics, 13*, 364.

Tschaplinski, T. J., Tuskan, G. A., Sewell, M. M., Gebre, G. M., Todd, D. E., & Pendley, C. D. (2006). Phenotypic variation and quantitative trait locus identification for osmotic potential in an interspecific hybrid inbred F2 poplar pedigree grown in contrasting environments. *Tree Physiology, 26*, 595–604.

Tuskan, G. A., Difazio, S., Jansson, S., Bohlmann, J., Grigoriev, I., Hellsten, U., et al. (2006). The genome of black cottonwood, *Populus trichocarpa* (Torr. & Gray). *Science, 313*, 1596–1604.

Ueno, S., Klopp, C., Leplé, J. C., Derory, J., Noirot, C., Léger, V., et al. (2013). Transcriptional profiling of bud dormancy induction and release in oak by next-generation sequencing. *BMC Genomics, 14*, 236.

Verne, S., Jaquish, B., White, R., Ritland, C., & Ritland, K. (2011). Global transcriptome analysis of constitutive resistance to the white pine weevil in spruce. *Genome Biology and Evolution, 3*, 851–867.

Verta, J. P., Landry, C. R., & Mackay, J. J. (2013). Are long-lived trees poised for evolutionary change? Single locus effects in the evolution of gene expression networks in spruce. *Molecular Ecology, 22*, 2369–2379.

Villalobos, D. P., Diaz-Moreno, S. M., Said el, S. S., Canas, R. A., Osuna, D., Van Kerckhoven, S. H., et al. (2012). Reprogramming of gene expression during compression wood formation in pine: coordinated modulation of S-adenosylmethionine, lignin and lignan related genes. *BMC Plant Biology, 12*, 100.

Villar, S., Plomion, C., & Gion, J.-M. (2011). Integrative approach involving RNA-seq, foliar traits and growth measurements revealed genotype-specific plasticity on *Eucalyptus* subjected to seasonal water shortage. *BMC Proceedings, 5*(Suppl 7), O28.

Vining, K. J., Romanel, E., Jones, R. C., Klocko, A., Alves-Ferreira, M., Hefer, C. A., et al. (2014). The floral transcriptome of *Eucalyptus grandis*. *New Phytologist, 206*, 1406–1422.

Wang, J., Abbott, R. J., Ingvarsson, P. K., & Liu, J. (2014). Increased genetic divergence between two closely related fir species in areas of range overlap. *Ecology and Evolution, 4*, 1019–1029.

Wang, Z., Chen, J., Liu, W., Luo, Z., Wang, P., Zhang, Y., et al. (2013). Transcriptome characteristics and six alternative expressed genes positively correlated with the phase transition of annual cambial activities in Chinese Fir (*Cunninghamia lanceolata* (Lamb.) Hook). *PLoS One, 8*, e71562.

Wasternack, C. (2007). Jasmonates: an update on biosynthesis, signal transduction and action in plant stress response, growth and development. *Annals of Botany, 100*, 681–697.

Wegrzyn, J. L., Eckert, A. J., Choi, M., Lee, J. M., Stanton, B. J., Sykes, R., et al. (2010). Association genetics of traits controlling lignin and cellulose biosynthesis in black cottonwood (*Populus trichocarpa, Salicaceae*) secondary xylem. *New Phytologist, 188*, 515–532.

Wegrzyn, J. L., Lee, J. M., Tearse, B. R., & Neale, D. B. (2008). TreeGenes: a forest tree genome database. *International Journal of Plant Genomics, 2008*.

Wegrzyn, J. L., Liechty, J. D., Stevens, K. A., Wu, L.-S., Loopstra, C. A., Vasquez-Gross, H. A., et al. (2014). Unique features of the Loblolly pine (*Pinus taeda* L.) megagenome revealed through sequence annotation. *Genetics, 196*, 891–909.

Wen, J. (1999). Evolution of eastern Asian and eastern North American disjunct distributions in flowering plants. *Annual Review of Ecology and Systematics, 30*, 421–455.

White, T. L., Adams, W. T., & Neale, D. B. (2007). *Forest genetics*. Wallingford, UK: CABI.

Wong, M. M. L., Cannon, C. H., & Wickneswari, R. (2011). Identification of lignin genes and regulatory sequences involved in secondary cell wall formation in *Acacia auriculiformis* and *Acacia mangium* via de novo transcriptome sequencing. *BMC Genomics, 12*, 342.

Woodward, F. I., & Williams, B. G. (1987). Climate and plant distribution at global and local scales. *Vegetation, 69*, 189–197.

Wray, G. A. (2007). The evolutionary significance of cis-regulatory mutations. *Nature Reviews Genetics, 8*, 206–216.
Xie, C., & Tammi, M. T. (2009). CNV-seq, a new method to detect copy number variation using high-throughput sequencing. *BMC Bioinformatics, 10*, 80.
Yakovlev, I. A., Asante, D. K. A., Fossdal, C. G., Junttila, O., & Johnsen, O. (2011). Differential gene expression related to an epigenetic memory affecting climatic adaptation in Norway spruce. *Plant Science, 180*, 132–139.
Yakovlev, I. A., Lee, Y., Rotter, B., Olsen, J. E., Skroppa, T., Johnsen, O., et al. (2014). Temperature-dependent differential transcriptomes during formation of an epigenetic memory in Norway spruce embryogenesis. *Tree Genetics & Genomes, 10*, 355–366.
Yang, S. H., & Loopstra, C. A. (2005). Seasonal variation in gene expression for loblolly pines (*Pinus taeda*) from different geographical regions. *Tree Physiology, 25*, 1063–1073.
Yeaman, S., Hodgins, K. A., Suren, H., Nurkowski, K. A., Rieseberg, L. H., Holliday, J. A., et al. (2014). Conservation and divergence of gene expression plasticity following c. 140 million years of evolution in lodgepole pine (*Pinus contorta*) and interior spruce (*Picea glauca* x *Picea engelmannii*). *New Phytologist, 203*, 578–591.
Yin, T. M., DiFazio, S. P., Gunter, L. E., Riemenschneider, D., & Tuskan, G. A. (2004). Large-scale heterospecific segregation distortion in *Populus* revealed by a dense genetic map. *Theoretical and Applied Genetics, 109*, 451–463.
Zhang, J., Feng, J., Lu, J., Yang, Y., Zhang, X., Wan, D., et al. (2014). Transcriptome differences between two sister desert poplar species under salt stress. *BMC Genomics, 15*, 337.
Zhao, S., Fung-Leung, W. P., Bittner, A., Ngo, K., & Liu, X. (2014). Comparison of RNA-Seq and microarray in transcriptome profiling of activated T cells. *PLoS One, 9*, e78644.
Zhou, F., & Xu, Y. (2009). RepPop: a database for repetitive elements in *Populus trichocarpa*. *BMC Genomics, 10*, 14.
Zimin, A. V., Marcais, G., Puiu, D., Roberts, M., Salzberg, S. L., & Yorke, J. A. (2013). The MaSuRCA genome assembler. *Bioinformatics, 29*, 2669–2677.
Zobel, B., & Talbert, J. (1984). *Applied forest tree improvement*. New York, NY: John Wiley & Sons.

CHAPTER THREE

Acceleration of Forest and Fruit Tree Domestication by Genomic Selection

Fikret Isik*[,1], Satish Kumar[§], Pedro J. Martínez-García[¶], Hiroyoshi Iwata[||], Toshiya Yamamoto[#]

*Department of Forestry and Environmental Resources, North Carolina State University, Raleigh, NC, USA
[§]The New Zealand Institute for Plant & Food Research Limited, Havelock North, New Zealand
[¶]Department of Plant Sciences University of California, Davis, CA, USA
[||]Department of Agricultural and Environmental Biology, Graduate School of Agricultural and Life Sciences, The University of Tokyo, Bunkyo, Tokyo, Japan
[#]NARO Institute of Fruit Tree Science, Tsukuba, Ibaraki, Japan
[1]Corresponding author: E-mail: Fikret_isik@ncsu.edu

Contents

Advances in Botanical Research, Volume 74
ISSN 0065-2296
http://dx.doi.org/10.1016/bs.abr.2015.05.002

Abstract

Trees are important sources of food, fibre, fodder, fuel, and timber. Fruit and forest trees provide habitat for wildlife, sequester carbon, and protect soil and water resources. Despite a very long history of domestication of some fruit tree species, such as apple, domestication and breeding of many tree species is still in their early stages. The domestication of trees can be accelerated by modern tree breeding practices. Advances in genomic technologies are providing valuable tools to tree breeders. Breeders can be more efficient in improving the traits of interest to increase productivity, and develop genetic resources that can adapt to changing environment and have resistance to disease and pests. In this chapter, we first briefly examined the current status of major forest (conifers, eucalypts) and fruit trees (apple, peach, pear) breeding across the world. We then described the breeding strategies employed to improve the populations. A summary of genomic resources available for major tree species is also covered. The challenges and opportunities of genomic selection applications are finally summarized.

1. INTRODUCTION

Domestication of plant species is a complex modification of evolutionary processes and considered one of humankind's most important technical innovations. Domestication is believed to have started in Neolithic period (13,000−10,000 years ago), which gave rise to current human culture (Purugganan & Fuller, 2009; Simons & Leakey, 2004). In a narrow-sense, domestication is defined as the change in plant morphology and genetics to adapt to human-controlled environments (Harlan, 1975; Jamnadass et al., 2011). In a more broad sense, domestication is defined as a process to change the plant exploitation practices, with effects on the morphology and genetics (Wiersum, 1997). From an evolutionary perspective, domestication is defined as a rapid evolutionary change that modifies the morphological and physiological characteristics of taxa to distinguish them from wild relatives (Purugganan & Fuller, 2009).

Trees are important sources of food, fibre, fodder, fuel, and timber in the world. Plantations of fruit and forest trees provide habitat for wildlife, sequester carbon, and protect soil and water resources. The domestication of some trees species, such as apple, goes back thousands of years (Cornille, Giraud, Smulders, Roldán-Ruiz, & Gladieux, 2014). However, a large number of tree species (more than 80,000) are still essentially wild, or domestication is in its infancy (Ofori et al., 2014). Although few examples of domestication of indigenous African food tree species have been reported (Jamnadass et al., 2011). Given the very recent history of tree domestication in tropical and subtropical regions, there is a great potential to domesticate new tree species to improve human wellbeing in Asia, Africa and South America (Ofori et al., 2014). Wild food trees in Africa and in South America are under threat due to deforestation (Dawson et al., 2011). Therefore, conservation of indigenous tree species and cultivation need to be accelerated through modern tree breeding strategies.

In this chapter, we first review the current breeding practices of major fruit and forest trees in various regions of the world. For fruit trees, we focus our discussion to apple, pear and peach, while for forest trees, we cover major conifers (*Pinus* spp. and *Picea* spp.) and eucalypts (*Eucalyptus* spp.). We address bottlenecks of traditional breeding practices and try to address the following questions:

1. What are the major challenges facing tree breeding, and how can they be addressed to make tree breeding more efficient?
2. Are there successful examples of marker-based breeding in major tree breeding programs of the above mentioned tree species?
3. What are the major difficulties for bringing genomic-based breeding to forest and fruit trees?

Advances in genomic technologies are expected to bring new opportunities for accelerating tree breeding. New methods are not free of challenges, and they vary from one species to another. For each species covered in this Chapter, the prospect of marker-assisted selection (MAS) and genome selection (GS) are reviewed in light of the opportunities provided by high-throughput genotyping and next-generation sequencing technologies. MAS is using DNA markers to study the inheritance of quantitative trait loci with large effects, and use the marker-linked loci to make selection decisions in breeding. We hope that this review will help next-generation tree breeders to better understand the scope of tree domestication using genomics.

2. CURRENT STATUS OF BREEDING AND SELECTION IN MAJOR FOREST TREE SPECIES

Domestication of forest trees using modern breeding methods started in the late 1950s with the goals of increasing wood for construction, firewood and bioenergy production, fibre for paper and engineered wood products, and food from a given land base while reducing the exploitation of natural forests. The domestication process moved slowly for most species mainly due to lack of resources and complex logistics of breeding. Advances in DNA sequencing technologies and development of high-throughput genotyping technologies are expected to bring dramatic changes in forest tree breeding. In this section, we take a snapshot of current breeding practices for the major conifer species. A short review of genomic resources and likely applications of genomic selection methods follow. The discussions are limited to major forest trees that have advanced breeding programs. These breeding programs are likely to be used for genomic selection in the near future. Readers should refer to Mullin et al. (2011) for an extensive review of breeding status of conifers and eucalypts. They should also refer to Neale, Langley, Salzberg, and Wegrzyn (2013) for an extensive review on genomic resources and to Isik (2014) for a review of genomic selection in forest trees.

There is a large number of forest tree species subjected to breeding. A high majority of species being bred belongs to the *Pinus*, *Picea* and *Eucalyptus* genera. These species have different reproductive biology, growth patterns and genetic structures. Sitka spruce (*Picea sitchensis* Bong. Carr) for example can take up to two decades to produce strobili and reproduce (Philipson, 1983). The speed of reproduction (i.e. recombined selected genotypes) is obviously one of the major limiting factors to accelerate tree domestication.

Unlike fruit trees, domestication of forest trees started in the 1950s, and many forest tree breeding programs are only in their first or second cycles (Mullin et al., 2011). The most advanced tree breeding programs, such as loblolly pine (*Pinus taeda* L.) breeding program in the southern USA are in their fourth cycle. The relatively slow advancement is partly due to the fact that forest tree breeders started with thousands of trees from wild populations or plantations, and it took major effort, time, and resources to accomplish the breeding and testing steps. Despite slow progress, significant achievements were obtained for growth and disease resistance. In loblolly pine for instance, the best families provide 60% more growth over unimproved planting stock in the southern USA (McKeand et al., 2006).

For conifer trees, the objective of the breeding is to increase the frequency of favourable alleles with additive effects on the traits of interest by capturing general combining ability (GCA) (White, Adams, & Neale, 2007). Breeders have employed various strategies to achieve their goals, such as the complementary breeding strategy used by the North Carolina State University Cooperative Tree Improvement Program in the southern USA (McKeand & Bridgwater, 1998). This strategy uses wind pollination or mixed pollen of a set of individuals (polymix) to carry out mating of candidate trees (Figure 1). Half-sib progeny of the candidate trees are tested in field trials to estimate parental GCA. Once the trees are ranked for their GCA values, they are mated using controlled pollination for forward selection. Full-sib progeny from crosses are tested in field trials, and the best full-sib progeny from crosses are selected to form the next-generation breeding population (McKeand & Bridgwater, 1998). In order to control deleterious effects of inbreeding, multiple breeding sublines strategy has been suggested (Burdon & Namkoong, 1983; Burdon, Shelbourne, & Wilcox, 1977). Sublines are a subset of trees (from 4 to 30) of the main breeding population. The strategy requires that all matings be carried out within sublines (McKeand & Bridgwater, 1998). Trees from different sublines are then put into seed orchards to avoid relatedness in plantations. The subline strategy has been incorporated into several conifer breeding programs (Li & Wyckoff, 1994; McKeand & Bridgwater, 1998; White, Hodge, & Powell, 1993) to increase genetic diversity and genetic gain. Each subline is managed as a discrete unit, and mating among individuals across sublines

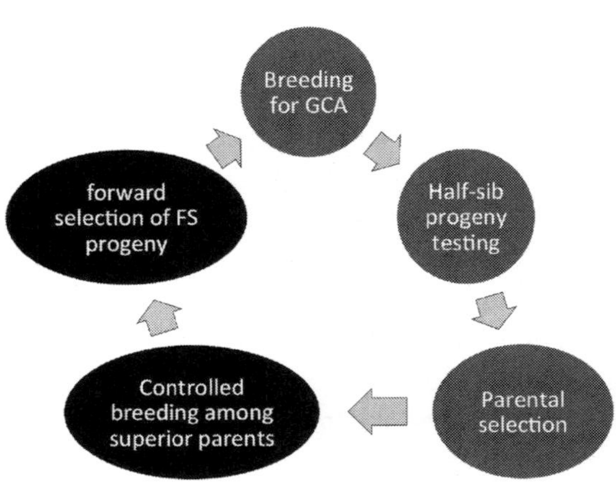

Figure 1 A schematic representation of recurrent selection based on complementary mating design used for some conifers. (See colour plate)

is not allowed, except to produce seed for plantations. Although the subline strategy has helped to control inbreeding in deployment populations and helped to manage breeding and testing efforts, it has major drawbacks. Sublines had usually small number of trees with 4–20 trees (McKeand & Bridgwater, 1998; White et al., 2007). After one cycle of breeding, it was realized that relatedness (inbreeding) would increase rapidly if the progeny of same trees in the subline were used for another round of mating. A high majority of forest trees, particularly conifers, cannot tolerate increased level of inbreeding and mating would not be successful. Since mating of the trees with high index breeding values was not considered across sublines, and selection intensity within sublines was low, genetic gain was sacrificed. Complementary mating designs proved to be slow to finish a breeding cycle because of two stage mating and testing schemes.

In *Eucalyptus*, species comparisons trials were carried out in early 1960s and 1970s to match species to suitable environments across the world (Assis & Resende, 2011). In more advanced cycles, rolling-front types of strategies were implemented. These strategies are characterized by nondiscrete cycles. In many countries, interspecies hybrid breeding schemes were adopted to capture hybrid vigour and to combine desired characteristics of multiple species (Borralho & Dutkowski, 1998). Open-pollinated breeding strategies are still common in some eucalyptus species, because it is difficult to produce seed from controlled breeding in some species such as *Eucalyptus nitens* (Hamilton, Joyce, Williams, Dutkowski, & Potts, 2008). In Brazil, reciprocal recurrent selection between divergent populations and intrapopulation recurrent selection are common (Assis & Resende, 2011). Taking advantage of rooting abilities, the selected hybrid progeny are cultivated by vegetative propagation techniques. This allows cloning hybrid progeny and testing them in replicated field trials. A subset of superior clones is further evaluated in field trials for verification and the best individuals selected for further breeding and deployment.

Growth, disease resistance, stem form (straightness, forking) and wood quality have been the major traits improved by tree breeders. For pine and spruce species, fungal disease and pest resistance are among the most important traits to improve for plantations. For example in loblolly pine fusiform rust disease caused by the endemic fungus *Cronartium quercuum* (Berk.) Miyabe ex Shirai f. sp. *fusiforme* in the southern USA is the major threat to plantations, and extensive selection has been done to improve disease resistance (Bridgwater, Kubisiak, Byram, & McKeand, 2005). Improving stem form is important for most pines, while stem straightness is not an issue for spruces. Because many eucalyptus pure species and their hybrids outgrow conifers, there has been great interest to introduce them to temperate climates (Turnbull, 1999). Most species cannot tolerate frost, so cold tolerance (as well as pest resistance) has been a primary trait for improvement. Pulp and paper characteristics are also important selection criteria in *Eucalyptus* (Borralho, Cotterill, & Kanowski, 1993; Terdwongworakul, Punsuwan, Thanapase, & Tsuchikawa, 2005).

2.1 Genomic Resources for Breeding and Population Management of Forest Trees

Large and complex genomes of conifers, particularly pines (with a typical physical size ranging from 20 to 40 Gb), and a lack of well-characterized genome sequence for close relatives have been challenging for developing a reference genome sequence (Wegrzyn et al., 2014). Large-scale development of cost-effective single nucleotide polymorphism (SNP) markers is time-consuming. With the transition to next-generation sequencing technologies, the cost of DNA sequencing has fallen dramatically since 2008 (Wetterstrand, 2015). This advance has made it possible to sequence whole genomes of several forest trees in recent years (Birol et al., 2013; Muchero, Labbé, Ranjan, DiFazio, & Tuskan, 2014; Myburg et al., 2011; Neale et al., 2014; Nystedt et al., 2013). The first draft assembly of loblolly pine was released in 2014 (Neale et al., 2014). The second draft assembly is expected to be released in late 2015. Draft assembly of *Eucalyptus grandis*, *Picea abies* and *Picea glauca* have been published (Birol et al., 2013; Myburg et al., 2011; Nystedt et al., 2013). The whole genome assembly of loblolly pine will help to better understand gene regulation and how the genotype gives rise to the phenotype. Although genome assemblies of forest trees are far from being complete, they are expected to provide invaluable resources to develop SNP markers for various applications in forest tree breeding and breeding population management.

Thousands of markers have become available for genetic mapping to help us understand the genetic basis of complex traits in forest trees. High-throughput SNP genotyping has been carried out in several conifers and high-density genetic linkage maps have been developed to understand genome structure and evolution of conifers (Chancerel et al., 2011, 2013; Eckert et al., 2010; Hudson et al., 2012; Neves, Davis, Barbazuk, & Kirst, 2013). Such maps, constructed by using larger number of DNA markers, are essential to identify genes involved in complex traits by linkage analysis.

SNP arrays are also crucial to enable researchers to uncover the molecular basis of quantitative traits by linkage disequilibrium (LD) mapping. LD mapping is more robust than linkage analysis for identifying the causal polymorphisms given the rapid decay of LD (Neale & Kremer, 2011). We have already observed significant reduction in genotyping cost using custom SNP arrays.

2.2 Genomic Selection in Forest Trees

2.2.1 Summary of Empirical Studies

We should expect accelerated discovery and development of genomic tools for tree breeding in the near future. As observed for cattle breeding, using large number of DNA markers that are reproducible, cost-efficient and reliable should bring a paradigm shift in forest tree breeding. In contrast to efforts on discovering quantitative trait loci (QTLs) with large effects to use for selection, we should expect a major shift using large number of markers to cover the whole genome and trace possibly all the genes with large and small effects on the traits of interest.

Marker assisted selection will likely be applied for traits with oligogenetic effects, such as disease resistance genes (Muranty, Jahier, Tanguy, Worland, & Law, 2002). For such traits, a low-density SNP chip could be used to screen the population for disease resistance genes and make selections without challenging trees with the pathogen. Further refinements of genome assembly of forest trees coupled with increasing marker density and improved genetic maps should enable breeders to identify resistance genes in the near future. For example, in loblolly pine, the top five SNP markers explained about 30% of the variance in gall score caused by the fusiform rust fungus (Quesada et al., 2014).

Isik (2014) developed a strategy to use genomic selection in loblolly pine for a proof of concept breeding plan in the southern USA. In this strategy, the phenotype of trees in the population is known. A subset of the phenotype in the population is removed for validation. The effects of the makers on the trait are estimated in a larger subset of the population (testing population), which is independent from the validation population. The cumulative effects of the large number of markers in the validation population are the genomic estimated breeding values. Once the population goes through a breeding cycle, we would expect most marker—trait associations to weaken because of recombination in the genome, unless very large number of markers are used (Isik, 2014; Zapata-Valenzuela et al., 2012). The real benefit of genomic selection would be the genotyping of young seedlings without known phenotypic characteristics and the use of markers to predict traits of interest. For most conifers, the progeny testing takes more than 10 years to complete (White et al., 2007). If genomic selection is proven to be successful, we would expect substantial reduction in progeny testing efforts in long-lived species. Genomic selection strategies have thus the potential to cut the progeny testing time in half and increase genetic gain onefold (Isik, 2014).

Research on using SNP markers to predict the genetic merit of forest trees has increased since 2011. Studies to date are considered a proof of concept because the same populations were split into testing and validation sets to test the predictive ability of markers (Isik, 2014). In this case, the linkage phase of markers and alleles controlling traits do not change until the population goes through breeding(s).

The very first proof of concept genomic selection studies were carried out in loblolly pine in the USA. Zapata-Valenzuela et al. (2012) estimated genomic breeding values of 149 cloned progeny from 13 crosses in loblolly pine by fitting 3406 markers simultaneously. The accuracy of genomic estimated breeding values ranged from 0.61 to 0.83 for lignin and cellulose content while the predictive ability of markers for growth was lower (a range of 0.30–0.68). In another study by Zapata-Valenzuela, Whetten, Neale, Mckeand, and Isik (2013), 165 cloned progeny of loblolly pine individuals were genotyped at 3461 SNP loci and genomic estimated breeding values were obtained using genomic relationship matrices. On average, the accuracy of predictions based on genomic relationship matrices ranged between 0.37 and 0.74. The authors pointed out that using genomic relationship matrices in linear mixed models provided certain advantages, e.g. less computing resources, over fitting large number of markers simultaneously. Resende, Muñoz, Resende et al. (2012) genotyped 952 cloned individuals at 4853 SNP loci in loblolly pine. The predictive abilities of various statistical models were evaluated for 17 traits with distinct heritabilities. The predictive ability of markers for growth traits ranged from 0.38 to 0.49. Authors reported lower accuracy of predictions for wood quality, developmental traits and disease resistance. The statistical methods differed marginally in accuracy of genomic estimated breeding values. Regressing the predictive ability of markers for 17 traits on their heritability values suggested that as heritability increases the accuracy of predictions increases linearly ($R^2 = 0.79$). Resende, Muñoz, Acosta et al. (2012) studied the efficiency of genomic selection in loblolly pine using 926 cloned progeny from 61 full-sib families. They reported selection efficiency of 53–112% with the genomic selection, assuming breeding cycle is reduced in half.

The efficiency of genomic selection was assessed in two unrelated hybrid eucalyptus breeding populations with contrasting effective population sizes (Ne = 11 and Ne = 51) with about 3000 DArT markers Resende, Resende et al. (2012). The accuracies of predictions ranged between 0.55 and 0.88, similar to accuracies obtained by the traditional progeny testing approach. Markers captured a substantial portion (74—97%) of the heritability of the traits. However, prediction models performed poorly between two populations.

In white spruce (*P. glauca*), 1694 trees from 214 open-pollinated families were genotyped at 6385 SNP loci (Beaulieu, Doerksen, Clément, MacKay, & Bousquet, 2014). The estimated accuracy of predictions for various wood quality and growth traits ranged between 0.33 and 0.44. These accuracies are encouraging for reducing the progeny testing time and efforts despite using a wild population sampled from 43 natural stands.

In a recent study on maritime pine (*Pinus pinaster* L.), 661 individuals sampled from two generations were genotyped for 2500 SNP markers (Isik et al., 2015). Authors used linear mixed models with realized genomic relationships matrix (genomic Best Linear Unbiased Prediction (BLUP)), Bayesian Ridge and Bayesian LASSO regression to estimate predicted ability of markers for growth and stem sweep. Multiple testing and validation scenarios suggested that markers could estimate genetic merit of trees with accuracies as high as 0.85. The average predictive ability across statistical models and validation scenarios was about 0.49 for stem sweep, and 0.47 and 0.43 for total height and tree diameter, respectively. The authors concluded that the results are encouraging despite low marker coverage of genome (1.39 markers/cM) and low LD (r = 0.006). It was suggested that a larger sample size that includes third generation individuals should be analyzed to test the efficiency of marker prediction and such analyses are underway.

Despite small sample of trees genotyped in most studies, with limited breeding history and small number of markers used to cover very large genomes (>20 Gb for conifers), the accuracy of genomic predicted breeding values are encouraging. We should expect sharp increases in the number of studies on genomic selection in forest trees in the near future as researchers develop cost-efficient, reliable and repeatable genotyping platforms for many more species, such as multiple species SNP chip developed for eucalyptus (Silva-Junior et al., 2015).

2.3 Challenges and Recommendations

There are different ways to improve forest trees for desirable properties. The availability of financial resources, land, and public/private sector support can dictate breeding practices. The goal of the breeder is to increase the frequency of favourable mutations in a population using the most cost and time efficient methods. Most tree improvement programs in developing countries are still at early stages. Breeders should keep in mind that if the program is starting from wild populations, the highest possible genetic gain in a breeding program can be achieved by starting with the right species and the right provenances within species. Thus, tree breeders and forest geneticists have responsibility to make sure that the species and provenance trials are carried out first. Limited financial resources should not be wasted on high-tech, popular and expensive technologies unless there are clear returns on investment.

The major challenge in applications of genomics in tree breeding is still a lack of reliable, repeatable and cost-efficient genotyping platforms. This is particularly the case for many conifer species since this gap has now been filled in *Eucalyptus* (Silva-Junior et al., 2015). The second challenge is the lack of infrastructure (e.g. databases) to store, retrieve, and analyze large number of markers, although significant progress is being made (Neale et al., 2013; Wegryzn et al., 2014). Routine applications of genomic tools in tree breeding will require trained personnel with analytical skills. The third challenge is the lack of well-designed and well-tested breeding populations for many species.

Genomic selection would work at the population level. The ultimate goal is to increase favourable alleles controlling the traits of interest in the population through selection. If a breeding program is based on cultivar development using a few crosses, then genomic selection may not produce desired outcome.

Although it is difficult to make general recommendations for very diverse species and breeding programs around the world, some general guidelines can be followed in genomic selection applications;

- Reliable marker—trait associations and genomic predictions require phenotypes with high accuracies. Tree breeders need to make sure that the populations are properly tested in replicated progeny tests such that phenotypes are accurate and comparable. For example, if two groups of trees are to be genotyped but they were not tested together in progeny tests, phenotypes will not be on the same scale. The phenotypes will be confounded with the environmental noise. Markers may not be successful to predict the genomic merit of individuals because of the large environmental noise in the phenotypes of the two groups. Progeny tests connected with checklots are not recommended. Connections between different test series (e.g. across years) should be established with a subset of families of the same population.

- A genomic prediction model based on unrelated individuals or based on the same population split into testing validation sets would have a low power once the population goes through recombination. Breeders should develop populations with at least a few generations of mating and selection to test the predictive ability of markers across different generations. In a population with some history of breeding, associations between markers and traits can be better established (with LD).
- Finally, tree breeders usually have limited resources, so a cost-benefit analysis should be done before investing in expensive genotyping projects. For some species with short breeding cycles (e.g. *Eucalyptus*) genomic selection may not produce the desired impacts if the strategy is to cross a few individuals to select superior clones for short-term genetic gain.

3. CURRENT STATUS OF BREEDING AND SELECTION IN APPLE

The origin of controlled breeding of apples (*Malus* × *domestica* Borkh.) is attributed to Thomas Andrew Knight (1759–1838) who developed the first cultivars of known parentage (Janick, Cummins, Brown, & Hemmat, 1996). Apple is now one of the most widely cultivated tree fruits in the temperate regions of the world, and is the third most internationally traded fruit, behind only bananas and grapes. The advent of affordable and high-throughput genotyping and sequencing technologies provides grounds for advances in the field of apple breeding. Apple breeders now commonly use genome sequence information to uncover clues to the genetic basis of economically important traits, and are using this information to improve fruit yield, resistance to various pests and diseases and quality.

3.1 Breeding Strategy

The traditional apple breeding strategy has involved crossings among a few top commercial cultivars and elite selections, and planting full-sib families in order to forward-select (based on phenotypic performance) individuals for further clonal testing. Selection of parents with complementary characteristics is essential in order to produce seedlings that have the desired attribute inherited from each parent. Crossing of two commercial cultivars produces progeny exhibiting a wide variation for any one quality trait, with only a very small proportion showing trait improvements over the parents. In apple, selected plants are used as potential cultivars and/or as breeding parents, so that frequencies of desirable alleles are increased in successive populations.

3.2 Selection Traits

A large number of selection criteria need to be evaluated before an apple seedling is considered for commercialization. Some plant traits, such as resistance to apple scab, can be assessed on juvenile plants, with phenotypes similar to those observed on adult trees in the orchard, provided inoculum and environmental conditions are optimum. On the other hand, the evaluation of powdery mildew resistance based on one-year-old seedling assessments does not correlate with assessments made following infection of more mature trees in the orchard (Janse, Verhaegh, & den Nijs, 1994). Clearly fruit characteristics can only be assessed once trees are past the juvenile phase and appraisal of traits that influence yield, such as biennial bearing, requires several years of continuous evaluation (Guitton et al., 2012). For most fruit quality traits, successive measurements over 2—3 years are made in order to select seedlings for commercial testing.

The most important objective across most apple breeding programs has been to increase the marketability of the fruit by improving their eating quality and/or appearance (Laurens, 1999). In response to consumer-based research (e.g. Daillant-Spinnler, MacFie, Beyts, & Hedderley, 1996; Jaeger, Andani, Wakeling, & MacFie, 1998), considerable effort is expended on breeding for desirable flesh textural attributes (especially high crispness and juiciness), sweetness and acidity. Breeding against non-desired astringency and bitterness, as well as bland low-flavour types, is also important. While fruit size preferences in individual countries vary considerably (e.g. consumers in the UK prefer smaller apples than the Japanese), generally apples need to weigh at least 150 g. Fruit skin needs to be unblemished, with an absence of russet and other markings, together with an attractive and consistent colour and shape. With the requirement for new cultivars to store for longer times, freedom from storage disorders such as bitter pit, superficial scald and rots has become increasingly important as a breeding objective. Tree productivity and resistance to pests and diseases are also among important traits for most apple breeding programs (Laurens, 1999).

3.3 Breeding Cycle

Flowering may begin as early as four years of age in some seedlings, with the majority fruiting after 6 years. This period can be reduced by promoting seedling growth in glasshouses and employing flower bud-inducing cultural practices (Volz, Rikkerink, Austin, Lawrence, & Bus, 2009). Dwarfing rootstocks induce earlier flowering of seedlings than those grown on their own roots. However fruiting is somewhat delayed, simply because of the requirement to grow from the propagated single bud or 2–3 bud graft of the seedling, a structure on the rootstock capable of bearing fruit. Generally scion-wood is taken from top (>1.7 m) of two-year-old nursery-grown seedlings and grafted onto 'M.9' cuttings from stoolbeds. These are planted in the nursery for a further year, before planting in the orchard. In New Zealand conditions, it generally takes 7 years before an individual can be identified for further use as a parent or as a potential stage-2 cultivar (Figure 2).

3.4 Genomic Resources for Breeding

3.4.1 Applications of MAS in Apple Breeding Programs

The earliest use of MAS in apple employed genetic markers flanking major gene resistance loci and involved the development and selection of parents with pyramided resistances to infection by a single pathogen, with the goal of eventually developing new cultivars with more durable resistance (Luby & Shaw, 2001). Selection of seedlings for fruit traits such as red skin (Zhu, Evans, & Peace, 2011) or fruit softening (Zhu & Barritt, 2008), has also been reported. Some breeding program in the US employ MAS principally for parental selection for loci that include crispness, acidity, fruit colour, firmness, and apple scab (personal communication with Dr Luby). Impediments

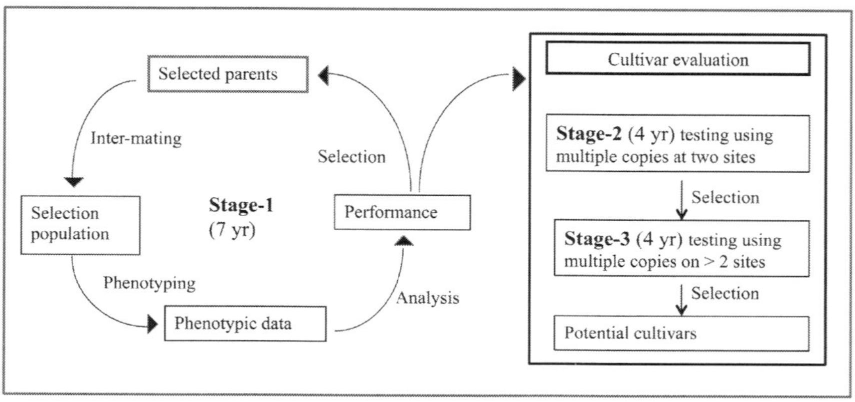

Figure 2 A schematic presentation of the timeline for apple cultivar breeding in New Zealand conditions.

to the wider use of MAS include the relatively small number of traits for which genetic markers (flanking loci or candidate gene/allele markers) are available. The solution to the limitations of traditional MAS perhaps lies not in seeking single markers associated with single large effects, but in capitalizing on developing the capacity for identifying and scoring tens-of-thousands of markers at low cost (Jannink, Lorenz, & Iwata, 2010).

3.4.2 Genome Sequencing and Genotyping Platforms

The sequencing of the apple genome by an international consortium was performed on 'Golden Delicious' (GD) (Velasco et al., 2010), as a common founder of most breeding programs. The GD genome assembly that spanned a total of 598 Mb, anchored on 17 pseudo-chromosomes, yielded more than 2 million SNPs. Following this landmark success, the US-based international research program 'RosBREED' led the development of the 8K apple Infinium® SNP chip (Chagné et al., 2012). The European program 'FruitBreedomics', which focuses on bridging the gap between breeding and genomics, has recently developed a 20K SNP array (Bianco et al., 2014).

The cost per data point is constantly decreasing, not only for highly paralleled targeted genotyping (SNP assays) but also for whole genome sequencing. Genotyping-by-sequencing (GBS), which involves simultaneous detection and scoring of SNPs, is becoming a common platform for genotyping large populations (Deschamps, Llaca, & May, 2013; Elshire et al., 2011). GBS approach uses data directly from the populations being genotyped, thus removing ascertainment bias towards a particular population. DNA sequences from samples distantly related to the reference genome will sometimes not align properly to the reference sequence and will result in incorrect or missing genotype calls (Myles, 2013). Also, GBS does not produce an equal amount of sequence for each sample or site in the genome and this would result in large amounts of missing genotype data, which requires innovative data imputation techniques.

The era of utilizing GBS for breeding has indeed begun for apple. The first evaluation of GBS in apple germplasm was reported by Myles (2013), and recently Deng (2014) demonstrated that of the three enzymes *BamHI*, *ApeKI* and *MspI* tested, *ApeKI* is the most suitable, with the most SNP sites being detected and the lowest missing data rate. Use of GBS has been reported in various agricultural crops (reviewed by Deschamps et al., 2013), and this technique is now also being evaluated to implement GS in Plant and Food Research (PFR) apple breeding program.

3.5 Genomic Selection in Apple

3.5.1 Summary of Empirical Studies

Kumar et al. (2012) conducted an empirical study for evaluating accuracy of GS in the PFR apple cultivar breeding program. The training dataset used for this purpose was obtained from an existing population generated in 2004 using a factorial mating design involving four female parents with two male parents. The number of offspring per full-sib family varied from 40 to 353, with a total sample size of about 1200. These seedlings were assessed for a range of fruit quality traits (e.g. fruit firmness, soluble solids, russet, astringency, titratable acidity) using instrumental, sensory or visual assessment methods (Kumar et al., 2012). The training population was genotyped using the apple 8K SNP array v1 (Chagné et al., 2012). Predicted accuracies, using cross-validation method, were very similar for the two contrasting GS models (RR-BLUP and Bayesian LASSO), and varied from 0.70 to 0.90 for various fruit quality traits (Kumar et al., 2012). The selection response per unit time using GS compared with the traditional BLUP-based selection was very high (>100%), especially for low-heritability traits. Currently the FruitBreedomics program in Europe and the RosBREED program in the USA are evaluating the potential of GS in their apple breeding program.

3.5.2 GS Implementation Strategy

The current application of GS in the New Zealand apple breeding program focuses on selecting candidates for stage-2 testing based on SNP profiles of stage-1 seedlings – hence obviating the need to phenotype stage-1 seedlings, and as a result knocking out 5 years off the cultivar development timeline (Figure 3). The pollen parents of the selection population were chosen from the training population, which provided strong genetic link between the training and selection candidates. About 10% of the available selection candidates were identified based on their genomic breeding value (GEBV) and were then planted out in the orchards for stage-2 testing.

3.6 Challenges and Recommendations

- Genetic relatedness between the training individuals and selection candidates is a critical factor affecting accuracy of genomic predictions. In other words, accuracies in across-family scenario (where training population does not contain the families being predicted) would be lower than within-family scenario. Such risks could, however, be minimized

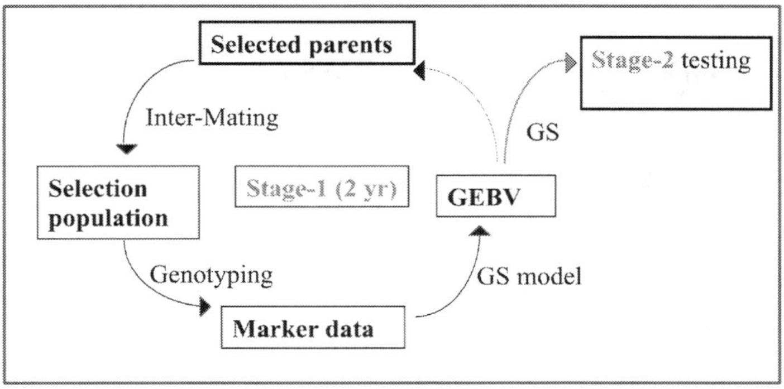

Figure 3 GS implementation strategy in the New Zealand apple breeding program.

by using high marker density, which ensures that the marker phase is consistent across families. Additionally, combining training individuals from families that are related to populations being predicted can achieve better accuracy.

- For a number of apple fruit quality traits, the genotype-by-environment interaction is an important factor, which means that estimates of SNP effects could be different across environmental conditions and prediction accuracies could be lower across sites compared to within-sites. Experiments using clonal replications of training population seedlings at different sites could be conducted to study the magnitude of SNP × environment interactions.
- The cost of genotyping a large number of SNPs is still an impediment for most apple breeding programs, but new techniques such as GBS could be used to reduce these costs significantly.
- Now that gene technologies are advancing rapidly, a huge challenge for apple breeders is the high-throughput phenotyping of large populations in order to further our understanding of which genes influence which physical traits. Lack of such technology would constitute a 'phenotyping bottleneck' that would limit our ability to capitalize on substantial investments made in genome sequencing.

4. CURRENT STATUS OF BREEDING AND SELECTION IN PEACH

Peach (*Prunus persica* (L.) Bastch) is one the most produced stone fruits in the world. In 2012, peach and nectarine production was around 21.08 Mt

(FAOSTAT, http://faostat3.fao.org/). The largest producers in the world are China, Italy, USA and Greece. The specific characteristics of peach make this species as a model stone fruit in the Rosaceae family. A small diploid genome, self-compatibility and a short juvenile period (Abbott, Arús, & Scorza, 2008), can summarize some of these characteristics. Peach, domesticated over 4000 years ago (Huang, Cheng, Zhang, & Wang, 2008) exhibits high phenotypic variability but reduced genetic diversity (Aranzana, Illa, Howad, & Arús, 2012). This pattern of low diversity seems to be driven by three genetic bottleneck events which occurred during the whole domestication process from their native origin, China, to its extensive western propagation (Verde et al., 2013).

4.1 Current Status of Breeding Efforts

In the history of peach breeding, a very limited gene pool was used one century ago to develop the main cultivars currently commercialized in Europe and America (Scorza, Mehlenbacher, & Lightner, 1985). Peach breeding is a common practise in the top peach producing countries (Abbott, Arús, & Scorza, 2007), and is usually separated for fresh-market and canning cultivars. More than 2000 cultivars have been released worldwide, which can be divided into freestone and clingstone (Lurie & Crisosto, 2005), depending on whether there is a separation of the pit from the flesh. Developing a new peach cultivar usually takes 15–20 years and requires at least eight basic steps, described previously by Abbot et al. (2007). Important cultivar development goals are to obtain cultivars with higher fruit quality, better postharvest fruit characteristics, greater resistance to disease and pests and with high levels of health-protective properties and phytochemical contents. In addition, developing cultivars for a targeted market, increases the range of harvest times, modified tree architecture to facilitate fruit culture and adaptation to low-chill zones which are also important objectives pursued by breeders (Abbott et al., 2007; Byrne, 2012). Two main breeding methods have been used in the development of new peach cultivars: the use of self-pollinations (or recurrent self-pollinations) and the use of intraspecific hybridization. The latter is the most common method for peach breeding, supplying the vast majority of the new cultivars worldwide.

4.2 Challenges and Prospects of Genomic Breeding and Selection

The availability of a large number of genomics resources for peach confirms this species is a model for comparative genomics studies of Rosaceae species.

A preliminary high-quality reference genome was sequenced from the double haploid genotype of the peach cv. 'Lovell', Lovel v1.0 (Verde et al., 2013), which facilitated the genome assembly. The peach genome size is approximately 227.3 million base pairs (227.3 MB), and its eight main scaffolds were aligned against the update version of the *Prunus* reference map (Howad et al., 2005). Recently an improved version, version 2.0, was obtained and is available for the peach research community (personal communication with Dr I. Verde). The new version presents a better genome contiguity and genome annotation. Additionally, to the presence of a reference genome, several approaches have been developed to obtain a large set of SNPs markers (Ahmad et al., 2011; Verde et al., 2012). The sequences of three additional peach genomes were aligned to the 'Lovell' genome to obtain a set of 6654 high-quality SNPs (Ahmad et al., 2011). The existence of these sequences allowed a pioneering comparative genomic approach in peach to study and quantify genomic variants between these genotypes and the peach reference genome (Fresnedo-Ramírez, Martínez-García, Parfitt, Crisosto, & Gradziel, 2013). An international Peach SNP Consortium was created to resequence 56 breeding-relevant accessions and to develop a 9K SNP array for peach, which is commercially available and used worldwide for genetic studies in peach (Da Silva Linge et al., 2015) and related species (Verde et al., 2012). A preliminary study using GBS (Elshire et al., 2011) has been developed to sequence 280 peach accessions (cultivars, landraces, breeder's lines and wild relatives from all over the world) from the National Clonal Germplasm Repository (NCGR) in Davis, California. A more saturated genetic map was created and QTLs for chilling requirement (CR) and bloom date were obtained, using an F2 mapping population with 57 genotypes from a selfed F1 progeny of a cross between 'Hakuho' (high CR) and 'UFGold' (low CR). As a result, 20 QTL were detected for these traits on linkage groups (LG) 1, LG4 and LG7 (personal communication with Dr D.G. Bielenberg). Together with the strong effort to develop genomic data for peach, efforts to understand complex traits in this species have been carried out over last two decades, with a large amount of phenotypic data obtained from the different breeding program around the world. Breeders have crossed around 42 different parents (selections or cultivars) and they have obtained around 24 mapping populations, mainly F1, and a large number of QTLs (498), associated with ~75 different traits, were discovered. These QTLs represent 65% of the total (760) of the discovered QTLs in *Prunus* species (Salazar et al., 2014). As a result of some of those QTLs mapping approaches, a large number of interesting genes were associated with important traits, such as chilling injuring symptoms or disease resistance (Martínez-García, Fresnedo-Ramírez, Parfitt, Gradziel, & Crisosto, 2013).

4.3 Implementation of GS in Peach

With the ineffective use of phenotypic selection for low heritable traits and the long generation time in peach for progeny testing, other approaches have been consider as alternative selection methods and as better methods to estimate variance components. The prediction of breeding values is one of these alternatives and preliminary studies supported the use of this method to increase the response to selection, affecting positively the breeding efficiency in peach (de Souza, Byrne, & Taylor, 2000). Generalized linear mixed model was used to predict breeding values for different parents in the Texas A&M University peach breeding program, and also to estimate genetic and phenotypic covariances among traits (de Souza, Byrne, & Taylor, 1998a, 1998b; de Souza et al., 2000). Within the RosBREED initiative, a pioneer approach called Pedigree-Based Analysis (PBA) has been developed. PBA uses multiple full-sib families that are related through common ancestors and exploit the identity-by-descent principle in known pedigrees. This method can predict the genetic potential of any breeding parent examined and to perform a more efficient QTL discovery and characterization (Bink et al., 2014), which will increase the accuracy of the breeder's decisions. In addition, approaches based on pedigree correction have been implemented in peach to facilitate an accurate calculation of realized additive relationship matrices to precisely estimate narrow-sense heritability and breeding values for fruit weight through Bayesian and frequentist approaches (personal communication with Dr J. Fresnedo-Ramírez). With the manifest increment of number of markers in peach, covering the entire genome, the implementation of genomic selection as a regular application in peach breeding will be a prominent and an exciting routine in the future. The massive genomic data generated for peach, the historical pedigree and large phenotype dataset commonly available for this tree crop, will be combined to effectively establish GS in this species. The use of the historical and corrected pedigree offers a more efficient population design in GS, with a potential improvement of the accuracy of the models (Jacobson, Lian, Zhong, & Bernardo, 2014). Other considerations to boost the effective use of GS in this species are the use of large breeding populations and the study of traits with medium–high heritability. High levels of LD have been observed in peach (Aranzana, Abbassi, Howad, & Arús, 2010) but with a clear variation between accessions of different origin (Li et al., 2013), in this sense a previous study of LD will be needed to decide the amount of markers used for GS. However, as plant breeders we should consider to increase the accuracy of phenotyping assays (less subjectivity more precision) and a better understanding of genotype-by-environment interaction, the latter can be a hard approach to reach mainly due to the high cost of extensive land required for tree crops.

5. CURRENT STATUS OF BREEDING AND SELECTION IN PEAR

Pears belong to the tribe Pyreae, subfamily Spiraeoideae, of the Rosaceae, along with apple and loquat, all of which share a basic chromosome number of $x = 17$, which indicates a polyploid origin. Four major species (Japanese pear, *Pyrus pyrifolia* Nakai; European pear, *Pyrus communis* L.; Chinese pear, *Pyrus bretschneideri* Rehd.; and *Pyrus ussuriensis* Maxim.) are grown commercially. Pears rank second to apples in the worldwide production of deciduous tree fruits; production approached 24 million tonnes in 2012 (FAOSTAT, http://faostat3.fao.org/). Total production has doubled over the past 20 years, of which 70% is Asian pears, mainly in China (FAOSTAT, http://faostat3.fao.org/). Italy, the USA, Argentina and Spain are the major producers of European pears. Most of the crop is sold fresh, but some is processed for canning, juice, purée or other products.

5.1 Current Status of Breeding and Selection in Pear

5.1.1 Breeding Objectives

Breeding for the improvement of both European and Asian pears has a long history. The objectives are complex, but most pear breeders share some common goals such as fruit quality, fruit shelf life, yield stability, and disease and pest resistance (Bell, Quamme, Layne, & Skirvin, 1996). Pear breeding methods generally involve the enlargement of genetic variation, the selection of elite material, and the evaluation of promising selections for market release. Considerable efforts have also been made in disease resistance in European pear, particularly to fire blight (caused by *Erwinia amylovora*) and pear scab (caused by *Venturia pirina*).

5.1.2 Breeding of Japanese Pear

In Japan, Japanese pear is an important fruit species and has the third greatest share of fruit production (MAFF, http://www.maff.go.jp/e/tokei/kikaku/monthly_e/). The cultivars 'Kosui' and 'Hosui', which account for more than 60% of production, were released in 1959 and 1972, respectively. Improved cultivars that can compensate for the shortcomings of these two cultivars are needed. The major method used in Japanese pear breeding is cross-breeding. But because of the long lifespan and large size of pear trees, large amounts of time, space and labour are required: it takes about 20 years from the crossing of parental lines to the release and registration of new cultivars of Japanese pear. During this time, large areas of land are tied up; at the National Institute of Fruit Tree Science (Ibaraki, Japan), the first stage of a selection program requires 0.4 ha to test 1000 trees in a field trial, and the last stage requires 0.33 ha to test 100 trees. Although intensive labour is also necessary for plant maintenance and fruit evaluation, a high proportion of inferior seedlings are destined for culling by the end of the breeding process.

5.2 Genomic Resources for Breeding

5.2.1 Genome Sequences and Markers

Draft genome sequences are now available for *P. bretschneideri* 'Dangshansuli' (512 Mb, 42,712 genes; Wu et al., 2013) and *P. communis* 'Bartlett' (577 Mb, 43,419 putative genes; Chagné et al., 2014). *P. pyrifolia* 'Hosui' has also been used for Expressed Sequence Tag (EST) sequencing (185 Mb) and genome sequencing (529 Mb; Terakami et al., 2014). High-throughput SNP markers have also been developed using the latest molecular genetics technologies: Montanari et al. (2013) designed an Infinium II 9K apple and pear array to evaluate a set of 1096 European pear SNPs and 7692 apple SNPs, and mapped 857 and 1031 SNPs, respectively, on pear genetic maps. Terakami et al. (2014) used an Illumina GoldenGate Assay to evaluate 1536 SNPs detected in EST and genome sequences of 'Hosui', and mapped 609 SNPs on a linkage map. Wu et al. (2014) used restriction-associated DNA sequencing, which is a fractional genomic sequencing technology that enables genotyping of SNPs by next-generation sequencing, and mapped 3143 SNPs on a linkage map. Simple sequence repeat (SSR) markers have also been developed (e.g. Nishitani, Terakami, Sawamura, Takada, & Yamamoto, 2009; Yamamoto et al., 2002). Most recently, Chen et al. (2014) designed 1341 SSRs based on draft genome sequences of Chinese pear; they evaluated a total of 1756 SSR markers, including the newly designed ones, and mapped 734 of these on a linkage map. Some of these SNP and SSR markers are transferable within the subfamily Spiraeoideae, proving useful in investigating syntenic relationships between pear and apple chromosomes (e.g. Chen et al., 2014;

Montanari et al., 2013). The genome resources that have been developed in pear will contribute to the practical application of MAS and GS to pear breeding programs.

5.2.2 Linkage/QTL Mapping for Marker-Assisted Selection

In pear, linkage/QTL mapping has been conducted mainly for disease and pest resistance. Loci controlling resistance to fire blight disease (Dondini et al., 2004), pear scab disease (e.g. Bouvier et al., 2012; Pierantoni et al., 2007; Terakami et al., 2006; Won et al., 2014), black spot disease (Terakami et al., 2007) and aphid pests (Evans, Govan, & Fernández-Fernández, 2008) have been located on genetic linkage maps. Previously, the only report of linkage/QTL mapping of fruit traits concerned redness in European pear (Dondini et al., 2008). Recently, Zhang et al. (2013) reported a QTL analysis of six fruit traits (soluble solids content, fruit weight, fruit length, fruit diameter, fruit shape index and fruit maturity date) using a population derived from a cross between a European × Chinese pear hybrid and Chinese pear, and detected 19 QTLs. Saeed et al. (2014) performed QTL analysis for fruit friction discolouration and related traits (fruit firmness, polyphenol oxidase activity, concentrations of phenolic compounds and ascorbic acid) using two populations derived from Asian × European pears, and identified a total of 221 QTLs for 22 phenotyped variables. Yamamoto et al. (2014) performed QTL mapping of seven fruit traits (harvest time, fruit skin colour, flesh firmness, fruit weight, acid content, total soluble solids content, and preharvest fruit drop) using a population derived from a Japanese pear cross, and detected 12 QTLs. Some of these 12 QTLs were detected in genome-wide association studies (GWAS) described below.

5.2.3 Genome-Wide Association Studies

GWAS are especially suitable for long-lived perennials because they do not require a segregating population derived from a biparental cross (Oraguzie, Rikkerink, Gardiner, & De Silva, 2007). The potential of GWAS was examined in pear with a population of 76 Japanese pear cultivars: significant associations detected in the GWAS agreed well with the results of biparental linkage/QTL mapping (Iwata, Hayashi, Terakami, Takada, Sawamura et al., 2013). Two significant associations of marker genotypes with harvest time detected in GWAS were also detected in biparental QTL mapping (Yamamoto et al., 2014). A significant association of marker genotypes with resistance to black spot disease was located close to the disease susceptibility genes *Ani* and *Ana*, which had been detected in biparental linkage mapping (Terakami et al., 2007). The results of GWAS and biparental mapping also agreed in fruit skin colour (Yamamoto et al., 2014). These results suggest that GWAS will be useful for the detection of QTLs for important agronomic traits, and will provide an effective alternative to the traditional time-consuming QTL mapping.

5.2.4 Genomic Selection in Pear

Because pear is a long-lived perennial, selection methods that require no phenotypic evaluation have great potential for accelerating genetic improvement. For this selection, GS is more suitable than MAS because it can take into account minor genes as well as major genes. The accuracy of GS predictions was evaluated through leave-one-out cross-validation of phenotype and marker genotype data of 76 Japanese pear cultivars (Iwata, Hayashi, Terakami, Takada, Sawamura et al., 2013). The accuracy was the highest for harvest time; medium for fruit size, flesh firmness, number of spurs, fruit shape in longitudinal section, resistance to black spot and acid content; and low for soluble solids content and tree vigour. GWAS detected two significant markers for harvest time and one for the number of spurs, but predictions based on these markers were much less accurate than GS predictions; this result suggests that the traits are determined by several minor and medium QTLs as well as the major QTLs detected by GWAS. Saeed et al. (2014) also suggested a possible benefit of the application of GS to the genetic improvement of friction discolouration controlled by multiple small-effect QTLs.

5.2.5 Selection of Good Parental Combinations

In cross-breeding, it is important to choose a parental combination that has a high probability of generating offspring with desired characteristics. GS prediction models can be used for predicting trait segregation in a progeny population. Iwata, Hayashi, Terakami, Takada, Saito et al. (2013) proposed a method for predicting the segregation of target traits and for selecting promising parental combinations on the basis of the prediction. The method combines segregation simulation and prediction modelling for genomic selection. Marker segregation in a progeny population is simulated on the basis of parental genotypes. Marker effects estimated for the GS prediction model are used to predict the segregation patterns of target traits. The method enables the prediction of promising parental combinations, the population size required for obtaining feasible genotypes and the genetic gain achievable with selection. Analysis showed that the predicted segregation of harvest time and fruit weight in a Japanese pear breeding population agreed well with actual segregation in two years. The result suggests that the method is useful to provide objective and quantitative criteria for choosing a parental combination and the breeding population size.

5.3 Future Challenges for Breeding

The potential of GWAS and GS in pear breeding is discussed in Iwata, Hayashi Terakami, Takada, Sawamura et al. (2013). A major challenge is the collection of data for performing GWAS and building GS prediction models. Because marker genotype data can now be obtained relatively easy and cheap, the extent of data collection is limited mainly by phenotypic data. Despite inefficiencies in phenotyping, phenotypic data should be collected for a number of genotypes (individuals or cultivars) in a number of environments, because phenotype is influenced by environmental as well as genetic factors. One possible strategy for collecting phenotypic data for a number of genotypes and environments is to use breeding populations, which are routinely developed and evaluated in breeding programs. We have begun collecting genome-wide SNP data for populations used in Japanese pear breeding programs. If data can be gathered routinely, the resultant collection will boost the detection power of GWAS and the accuracy of GS prediction. Because many trees are culled and thus not repeatedly measured in multiple environments, genotype-by-environment interaction would affect the power of GWAS and the accuracy of GS prediction. As suggested by Heffner, Sorrells, and Jannink (2009), however, the genotype of any tree in a breeding program is composed of alleles that have been evaluated in a large number of target environments, and thus it may be possible to keep the accuracy of GS high even in the presence of high genotype-by-environment interaction. As the throughput and cost of genotyping systems further improve, the routine collection of genome-wide SNPs in pear breeding programs will be facilitated.

REFERENCES

Abbott, A. G., Arús, P., & Scorza, R. (2007). Peach. In C. Kole (Ed.), *Fruits and nuts: Vol. 4. Genome mapping and molecular breeding in plants* (pp. 137–156). Heidelberg, Germany: Springer-Verlag Berlin Heidelberg.

Abbott, A. G., Arús, P., & Scorza, R. (2008). Genetic engineering and genomics. In D. Layne, & D. Bassi (Eds.), *The peach: Botany, production and uses* (pp. 85–105). Wallingford, UK: CABI.

Ahmad, R., Parfitt, D., Fass, J., Ogundiwin, E., Dhingra, A., Gradziel, T., et al. (2011). Whole genome sequencing of peach (*Prunus persica* L.) for SNP identification and selection. *BMC Genomics, 12*, 569.

Aranzana, M., Abbassi, E.-K., Howad, W., & Arús, P. (2010). Genetic variation, population structure and linkage disequilibrium in peach commercial varieties. *BMC Genetics, 11*, 69.

Aranzana, M. J., Illa, E., Howad, W., & Arús, P. (2012). A first insight into peach [*Prunus persica* (L.) batsch] SNP variability. *Tree Genetics and Genomes, 8*(6), 1359–1369.

Assis, T. F., & Resende, M. D. V. (2011). Genetic improvement of forest trees. *Crop Breeding and Applied Biotechnology, S1*, 44–49.

Beaulieu, J., Doerksen, T., Clément, S., MacKay, J., & Bousquet, J. (2014). Accuracy of genomic selection models in a large population of open-pollinated families in white spruce. *Heredity, 113*(4), 343–352.

Bell, R. L., Quamme, H. A., Layne, R. E. C., & Skirvin, R. M. (1996). Pears. In J. Janick, & J. N. Moore (Eds.), *Fruit breeding, volume I: Tree and tropical fruits* (pp. 441–514). UK: John Wiley and Sons Inc.

Bianco, L., Cestaro, A., Sargent, D. J., Banchi, E., Derdak, S., Guardo, M. D., et al. (2014). Development and validation of a 20K single nucleotide polymorphism (SNP) whole genome genotyping array for apple (*Malus* × *domestica* Borkh.). *PLoS One, 9*, e110377. http://doi.org/10.1093/bioinformatics/btt178.

Bink, M. C. A. M., Jansen, J., Madduri, M., Voorrips, R. E., Durel, C.-E., Kouassi, A. B., et al. (2014). Bayesian QTL analyses using pedigreed families of an outcrossing species, with application to fruit firmness in apple. *Theoretical and Applied Genetics, 127*, 1073–1090.

Birol, I., Raymond, A., Jackman, S. D., Pleasance, S., Coope, R., Taylor, G. A., et al. (2013). Assembling the 20 Gb white spruce (*Picea glauca*) genome from whole-genome shotgun sequencing data. *Bioinformatics, 29*(12), 1492–1497.

Borralho, N. M., & Dutkowski, G. W. (1998). Comparison of rolling front and discrete generation breeding strategies for trees. *Canadian Journal of Forest Research, 28*(7), 987–993. http://dx.doi.org/10.1139/x98-075.

Borralho, N. M. G., Cotterill, P. P., & Kanowski, P. J. (1993). Breeding objectives for pulp production of *Eucalyptus globulus* under different industrial cost structures. *Canadian Journal of Forest Research, 23*(4), 648–656.

Bouvier, L., Bourcy, M., Boulay, M., Tellier, M., Guerif, P., Denance, C., et al. (2012). A new pear scab resistance gene *Rvp1* from the European pear cultivar 'Navara' maps in a genomic region syntenic to an apple scab resistance gene cluster on linkage group 2. *Tree Genetics and Genomes, 8*, 53–60.

Bridgwater, F., Kubisiak, T., Byram, T., & McKeand, S. (2005). Risk assessment with current deployment strategies for fusiform rust-resistant loblolly and slash pines. *Southern Journal of Applied Forestry, 29*(2), 80–87.

Burdon, R., Shelbourne, C. J., & Wilcox, M. D. (1977). Advanced selection strategies. In *Presented at the proceedings of third world consultation on forest tree breeding* (Vol. 2, pp. 1133–1147). Canberra, Australia: IUFRO.

Burdon, R. D., & Namkoong, G. (1983). Multiple populations and sublines. *Silvae Genetica, 32*(5–6), 221–222.

Byrne, D. (2012). Trends in fruit breeding. In M. L. Badenes, & D. H. Byrne (Eds.), *Fruit breeding* (pp. 623–662). New York, USA: Springer Science + Business Media.

Chagné, D., Crowhurst, R. N., Pindo, M., Thrimawithana, A., Deng, C., Ireland, H., et al. (2014). The draft genome sequence of European pear (*Pyrus communis* L. 'Bartlett'). *PLoS One, 9*, 392644.

Chagné, D., Crowhurst, R. N., Troggio, M., Davey, M. W., Gilmore, B., Lawley, C., et al. (2012). Genome-wide SNP detection, validation, and development of an 8K SNP array for apple. *PLoS One, 7*, e31745.

Chancerel, E., Lamy, J.-B., Lesur, I., Noirot, C., Klopp, C., Ehrenmann, F., et al. (2013). High-density linkage mapping in a pine tree reveals a genomic region associated with inbreeding depression and provides clues to the extent and distribution of meiotic recombination. *BMC Biology, 11*(1), 50.

Chancerel, E., Lepoittevin, C., Provost, G. L., Lin, Y.-C., Jaramillo-Correa, J. P., Eckert, A. J., et al. (2011). Development and implementation of a highly-multiplexed SNP array for genetic mapping in maritime pine and comparative mapping with loblolly pine. *BMC Genomics, 12*(1), 368. http://dx.doi.org/10.1186/1471-2164-12-368.

Chen, H., Song, Y., Li, L.-T., Khan, M. A., Li, X.-G., Korban, S. S., et al. (2014). Construction of a high-density simple sequence repeat consensus genetic map for pear (*Pyrus* spp.). *Plant Molecular Biology Reporter, 33*(2), 316–325. http://dx.doi.org/10.1007/s11105-014-0745-x.

Cornille, A., Giraud, T., Smulders, M. J. M., Roldán-Ruiz, I., & Gladieux, P. (2014). The domestication and evolutionary ecology of apples. *Trends in Genetics, 30*(2), 57–65. http://dx.doi.org/10.1016/j.tig.2013.10.002.

Da Silva Linge, C., Bassi, D., Bianco, L., Pacheco, I., Raul Pirona, I., & Rossini, L. (2015). Genetic dissection of fruit weight and size in an F2 peach (*Prunus persica* (L.) Batsch) progeny. *Molecular Breeding, 35*, 71.

Daillant-Spinnler, B., MacFie, H. J. H., Beyts, P. K., & Hedderley, D. (1996). Relationships between perceived sensory properties and major preference directions of 12 varieties of apples from the Southern Hemisphere. *Food Quality Preference, 7*, 113–126.

Dawson, I. K., Vinceti, B., Weber, J. C., Neufeldt, H., Russell, J., Lengkeek, A. G., et al. (2011). Climate change and tree genetic resource management: maintaining and enhancing the productivity and value of smallholder tropical agroforestry landscapes. A review. *Agroforestry Systems, 81*(1), 67–78. http://dx.doi.org/10.1007/s10457-010-9302-2.

Deng, C. H. (2014). Genotyping by sequencing in fruit tree species. In *Plant and animal genome XXII conference*. Plant and Animal Genome.

Deschamps, S., Llaca, V., & May, G. D. (2013). Genotyping-by-sequencing in plants. *Biology, 1*, 460–483.

Dondini, L., Pierantoni, L., Ancarani, V., D'Angelo, M., Cho, K.-H., Shin, I.-S., et al. (2008). The inheritance of the red colour character in European pear (*Pyrus communis*) and its map position in the mutated cultivar 'Max Red Bartlett'. *Plant Breeding, 127*, 524–526.

Dondini, L., Pierantoni, L., Gaiotti, F., Chiodini, R., Tartarini, S., Bazzi, C., et al. (2004). Identifying QTLs for fire-blight resistance via a European pear (*Pyrus communis* L.) genetic linkage map. *Molecular Breeding, 14*, 407–418.

Eckert, A. J., van Heerwaarden, J., Wegrzyn, J. L., Nelson, C. D., Ross-Ibarra, J., González-Martínez, S. C., et al. (2010). Patterns of population structure and environmental associations to aridity across the range of loblolly pine (*Pinus taeda* L., Pinaceae). *Genetics, 185*(3), 969–982. http://dx.doi.org/10.1534/genetics.110.115543.

Elshire, R. J., Glaubitz, J. C., Sun, Q., Poland, J. A., Kawamoto, K., Buckler, E. S., et al. (2011). A robust, simple genotyping-by-sequencing (GBS) approach for high diversity species. *PLoS One, 6*, e19379.

Evans, K. M., Govan, C. L., & Fernández-Fernández, F. (2008). A new gene for resistance to *Dysaphis pyri* in pear and identification of flanking microsatellite markers. *Genome, 51*, 1026–1031.

Fresnedo-Ramírez, J., Martínez-García, P. J., Parfitt, D. E., Crisosto, C. H., & Gradziel, T. M. (2013). Heterogeneity in the entire genome for three genotypes of peach (*Prunus persica* (L.) Batsch) as distinguished from sequence analysis of genomic variants. *BMC Genomics, 14*, 750.

Guitton, B., Kelner, J.-J., Velasco, R., Gardiner, S. E., Chagné, D., Costes, E., et al. (2012). Genetic control of biennial bearing in apple. *Journal of Experimental Botany, 63*, 131–149.

Hamilton, M., Joyce, J., Williams, D., Dutkowski, G., & Potts, B. (2008). Achievements in forest tree improvement in Australia and New Zealand. Genetic improvement of *Eucalyptus nitens* in Australia. *Australian Forestry, 71*(2), 82–93.

Harlan, J. R. (1975). *Crops and man*, 306 pp.

Heffner, E. L., Sorrells, M. E., & Jannink, J.-L. (2009). Genomic selection for crop improvement. *Crop Science, 49*, 1–12.

Howad, W., Yamamoto, T., Dirlewanger, E., Testolin, R., Cosson, P., Cipriani, G., et al. (2005). Mapping with a few plants: using selective mapping for microsatellite saturation of the *Prunus* reference map. *Genetics, 171*, 1305–1309.

Huang, H., Cheng, Z., Zhang, Z., & Wang, Y. (2008). History of cultivation and trends in China. In D. Layne, & D. Bassi (Eds.), *The peach: Botany, production and uses* (pp. 37–60). Wallingford, UK: CABI.

Hudson, C. J., Kullan, A. R., Freeman, J. S., Faria, D. A., Grattapaglia, D., Kilian, A., et al. (2012). High synteny and colinearity among Eucalyptus genomes revealed by high-density comparative genetic mapping. *Tree Genetics and Genomes, 8*(2), 339–352.

Isik, F. (2014). Genomic selection in forest tree breeding: the concept and an outlook to the future. *New Forests, 45*(3), 379–401. http://dx.doi.org/10.1007/s11056-014-9422-z.

Isik, F., Bartholome, J., Farjat, A., Chancerel, E., Raffin, A., Sanchez, L., et al. (2015). Genomic selection in maritime pine. Plant Science (under review).

Iwata, H., Hayashi, T., Terakami, S., Takada, N., Saito, T., & Yamamoto, T. (2013). Genomic prediction of trait segregation in a progeny population: a case study of Japanese pear (*Pyrus pyrifolia*). *BMC Genetics, 14*, 81.

Iwata, H., Hayashi, T., Terakami, S., Takada, N., Sawamura, Y., & Yamamoto, T. (2013). Potential assessment of genome-wide association study and genomic selection in Japanese pear *Pyrus pyrifolia*. *Breeding Science, 63*, 125–140.

Jacobson, A., Lian, L., Zhong, S., & Bernardo, R. (2014). General combining ability model for genomewide selection in a biparental cross. *Crop Science, 54*, 895–905.

Jaeger, S. R., Andani, Z., Wakeling, I. N., & MacFie, H. J. H. (1998). Consumer preferences for fresh and aged apples: a crosscultural comparison. *Food Quality Preference, 9*, 355–366.

Jamnadass, R. H., Dawson, I. K., Franzel, S., Leakey, R. R. B., Mithöfer, D., Akinnifesi, F. K., et al. (2011). Improving livelihoods and nutrition in sub-Saharan Africa through the promotion of indigenous and exotic fruit production in smallholders' agroforestry systems: a review. *International Forestry Review, 13*(3), 338–354. http://dx.doi.org/10.1505/146554811798293836.

Janick, J., Cummins, J. N., Brown, S. K., & Hemmat, M. (1996). Apples. In J. Janick, & J. N. Moore (Eds.), *Fruit breeding, volume I, tree and tropical fruits* (pp. 1–77). New York: John Wiley & Sons Inc.

Jannink, J. L., Lorenz, A. J., & Iwata, H. (2010). Genomic selection in plant breeding: from theory to practice. *Briefing in Functional Genomics, 9*, 166–177.

Janse, J., Verhaegh, J. J., & den Nijs, A. P. M. (1994). Early selection of partial resistance to powdery mildew, *Podosphaera leucotricha* (Ell. et EV.) Salm. in apple progenies. *Euphytica, 77*, 7–9.

Kumar, S., Chagné, D., Bink, M. C. A. M., Volz, R. K., Whitworth, C., & Carlisle, C. (2012). Genomic selection for fruit quality traits in apple (*Malus* × *domestica* Borkh.). *PLoS One*, 7(5), e36674.

Laurens, F. (1999). Review of the current apple breeding programs in the world: objectives for scion cultivar improvement. *Acta Horticulturae, 484*, 163–170.

Li, X., Meng, X., Jia, H., Yu, M., Ma, R., Wang, L., et al. (2013). Peach genetic resources: diversity, population structure and linkage disequilibrium. *BMC Genetics, 14*, 84.

Li, B., & Wyckoff, G. W. (1994). Breeding strategies for *Larix decidua, L. leptolepis* and their hybrids in the United States. *Forest Genetics, 1*(2), 65–72.

Luby, J. J., & Shaw, D. V. (2001). Does marker-assisted selection make dollars and sense in a fruit breeding program? *HortScience, 36*, 872–879.

Lurie, S., & Crisosto, C. H. (2005). Chilling injury in peach and nectarine. *Postharvest Biology and Technology, 37*(3), 195–208.

Martínez-García, P., Fresnedo-Ramírez, J., Parfitt, D., Gradziel, T., & Crisosto, C. (2013). Effect prediction of identified SNPs linked to fruit quality and chilling injury in peach (*Prunus persica* (L.) Batsch). *Plant Molecular Biology, 81*(1–2), 161–174.

McKeand, S. E., & Bridgwater, F. E. (1998). A strategy for the third breeding cycle of loblolly pine in the southeastern US. *Silvae Genetica, 47*(4), 223–234.

McKeand, S. E., Jokela, E. J., Huber, D. A., Byram, T. D., Allen, H. L., Li, B., et al. (2006). Performance of improved genotypes of loblolly pine across different soils, climates, and silvicultural inputs. *Forest Ecology and Management, 227*(1–2), 178–184. http://dx.doi.org/10.1016/j.foreco.2006.02.016.

Montanari, S., Saeed, M., Knäbel, M., Kim, Y.-K., Troggio, M., Malnoy, M., et al. (2013). Identification of *Pyrus* single nucleotide polymorphisms (SNPs) and evaluation for genetic mapping in European pear and interspecific *Pyrus* hybrids. *PLoS One, 8*, e77022.

Muchero, W., Labbé, J., Ranjan, P., DiFazio, S., & Tuskan, G. A. (2014). Genome resequencing in *Populus*: revealing large-scale genome variation and implications on specialized-trait genomics. In *Challenges and opportunities for the world's forests in the 21st century* (pp. 587–595). Netherlands: Springer.

Mullin, T., Andersson, B., Bastien, J.-C., Beaulieu, J., Burdon, R. D., Dvorak, W. S., et al. (2011). Economic importance, breeding objectives and achievements. In *Genetics, genomics and breeding of conifers* (Vols 1–0, pp. 40–127). Science Publishers. Retrieved from http://www.crcnetbase.com/doi/abs/10.1201/b11075-3.

Muranty, H., Jahier, J., Tanguy, A. M., Worland, A. J., & Law, C. (2002). Inheritance of resistance of wheat to eyespot at the adult stage. *Plant breeding, 121*(6), 536–538.

Myburg, A., Grattapaglia, D., Tuskan, G., Jenkins, J., Schmutz, J., Mizrachi, E., et al. (2011). The *Eucalyptus grandis* Genome Project: genome and transcriptome resources for comparative analysis of woody plant biology. *BMC Proceedings, 5*(Suppl. 7), I20. http://dx.doi.org/10.1186/1753-6561-5-S7-I20.

Myles, S. (2013). Improving fruit and wine: what does genomics have to offer? *Trends in Genetics, 29*, 190–196.

Neale, D. B., & Kremer, A. (2011). Forest tree genomics: growing resources and applications. *Nature Reviews Genetics, 12*(2), 111–122.

Neale, D. B., Langley, C. H., Salzberg, S. L., & Wegrzyn, J. L. (2013). Open access to tree genomes: the path to a better forest. *Genome Biology, 14*(120), 1–8.

Neale, D. B., Wegrzyn, J. L., Stevens, K. A., Zimin, A. V., Puiu, D., Crepeau, M. W., et al. (2014). Decoding the massive genome of loblolly pine using haploid DNA and novel assembly strategies. *Genome Biology, 15*(3), R59. http://dx.doi.org/10.1186/gb-2014-15-3-r59.

Neves, L. G., Davis, J. M., Barbazuk, W. B., & Kirst, M. (2013). A high-density gene map of loblolly pine (*Pinus taeda* L.) based on exome sequence capture genotyping. *G3: Genes, Genomes, Genetics*, g3–113.

Nishitani, C., Terakami, S., Sawamura, Y., Takada, N., & Yamamoto, T. (2009). Development of novel EST-SSR markers derived from Japanese pear. *Breeding Science, 59*, 319–400.

Nystedt, B., Street, N. R., Wetterbom, A., Zuccolo, A., Lin, Y.-C., Scofield, D. G., et al. (2013). The Norway spruce genome sequence and conifer genome evolution. *Nature, 497*(7451), 579–584. http://dx.doi.org/10.1038/nature12211.

Ofori, D. A., Gyau, A., Dawson, I. K., Asaah, E., Tchoundjeu, Z., & Jamnadass, R. (2014). Developing more productive African agroforestry systems and improving food and nutritional security through tree domestication. *Current Opinion in Environmental Sustainability, 6*, 123–127. http://dx.doi.org/10.1016/j.cosust.2013.11.016.

Oraguzie, N. C., Rikkerink, E. H. E., Gardiner, S. E., & De Silva, H. N. (2007). *Association mapping in plants*. New York: Springer.

Philipson, J. J. (1983). The role of gibberellin $A_{4/7}$, heat and drought in the induction of flowering in Sitka spruce. *Journal of Experimental Botany, 34*(3), 291–302. http://dx.doi.org/10.1093/jxb/34.3.291.

Pierantoni, L., Dondini, L., Cho, K.-H., Shin, I.-S., Gennari, F., Chiodini, R., et al. (2007). Pear scab resistance QTL via a European pear (*Pyrus communis*) linkage map. *Tree Genetics and Genomes, 3*, 311–317.

Purugganan, M. D., & Fuller, D. Q. (2009). The nature of selection during plant domestication. *Nature, 457*(7231), 843–848. http://dx.doi.org/10.1038/nature07895.

Quesada, T., Resende, M. F. R., Jr., Muñoz, P., Wegrzyn, J. L., Neale, D. B., Kirst, M., et al. (2014). Mapping fusiform rust resistance genes within a complex mating design of loblolly pine. *Forests, 5*(2), 347–362. http://dx.doi.org/10.3390/f5020347.

Resende, M. F. R., Jr., Muñoz, P., Resende, M. D. V., Garrick, D. J., Fernando, R. L., Davis, J. M., et al. (2012). Accuracy of genomic selection methods in a standard data set of loblolly Pine (*Pinus taeda* L.). *Genetics, 190*(4), 1503–1510.

Resende, M. F. R., Jr., Muñoz, P., Acosta, J. J., Peter, G. F., Davis, J. M., Grattapaglia, D., et al. (2012). Accelerating the domestication of trees using genomic selection: accuracy of prediction models across ages and environments. *New Phytologist, 193*(3), 617–624. http://onlinelibrary.wiley.com/doi/10.1111/j.1469-8137.2011.03895.x/full.

Resende, M. D. V., Resende, M. F. R., Jr., Sansaloni, C. P., Petroli, C. D., Missiaggia, A. A., Aguiar, A. M., et al. (2012). Genomic selection for growth and wood quality in *Eucalyptus*: capturing the missing heritability and accelerating breeding for complex traits in forest trees. *New Phytologist, 194*(1), 116–128. http://onlinelibrary.wiley.com/doi/10.1111/j.1469-8137.2011.04038.x/full.

Saeed, M., Brewer, L., Johnston, J., McGhie, T. K., Gardiner, S. E., Hayes, J. A., et al. (2014). Genetic, metabolite and developmental determinism of fruit friction discolouration in pear. *BMC Plant Biology, 14*, 241.

Salazar, J. A., Ruiz, D., Campoy, J. A., Sánchez-Pérez, R., Crisosto, C. H., Martínez-García, P. J., et al. (2014). Quantitative trait loci (QTL) and Mendelian trait loci (MTL) analysis in *Prunus*: a breeding perspective and beyond. *Plant Molecular Biology Reporter, 32*(1), 1–18.

Scorza, R., Mehlenbacher, S. A., & Lightner, G. W. (1985). Inbreeding and coancestry of freestone peach cultivars of the eastern United States and implications for peach germplasm improvement. *Journal of the American Society for Horticultural Science, 110*, 547–552.

Silva-Junior, O. B., Faria, D. A., & Grattapaglia, D. (2015). A flexible multi-species genome-wide 60K SNP chip developed from pooled resequencing of 240 *Eucalyptus* tree genomes across 12 species. *New Phytologist, 206*(4), 1527–1540. http://dx.doi.org/10.1111/nph.13322.

Simons, A. J., & Leakey, R. R. B. (2004). Tree domestication in tropical agroforestry. *Agroforestry Systems, 61-62*(1–3), 167–181. http://dx.doi.org/10.1023/B:AGFO.0000028997.74147.f9.

de Souza, V. A. B., Byrne, D. H., & Taylor, J. F. (1998a). Heritability, genetic and phenotypic correlations, and predicted selection response of quantitative traits in peach: I. An analysis of several reproductive traits. *Journal of the American Society for Horticultural Science, 123*(4), 598–603.

de Souza, V. A. B., Byrne, D. H., & Taylor, J. F. (1998b). Heritability, genetic and phenotypic correlations, and predicted selection response of quantitative traits in peach: II. An analysis of several fruit traits. *Journal of the American Society for Horticultural Science, 123*(4), 604–611.

de Souza, V. A. B., Byrne, D. H., & Taylor, J. F. (2000). Predicted breeding values for nine plant and fruit characteristics of 28 peach genotypes. *Journal of the American Society for Horticultural Science, 125*(4), 460–465.

Terakami, S., Adachi, Y., Iketani, H., Sato, Y., Sawamura, Y., Takada, N., et al. (2007). Genetic mapping of genes for susceptibility to black spot disease in Japanese pear. *Genome, 50*, 735–741.

Terakami, S., Nishitani, C., Kunihisa, M., Shirasawa, K., Sato, S., Tabata, S., et al. (2014). Transcriptome-based single nucleotide polymorphism markers for genome mapping in Japanese pear (*Pyrus pyrifolia* Nakai). *Tree Genetics and Genomes, 10*, 853–863.

Terakami, S., Shoda, M., Adachi, Y., Gonai, T., Kasumi, M., Sawamura, Y., et al. (2006). Genetic mapping of the pear scab resistance gene *Vnk* of Japanese pear cultivar Kinchaku. *Theoretical and Applied Genetics, 113*, 743–752.

Terdwongworakul, A., Punsuwan, V., Thanapase, W., & Tsuchikawa, S. (2005). Rapid assessment of wood chemical properties and pulp yield of *Eucalyptus camaldulensis* in Thailand tree plantations by near infrared spectroscopy for improving wood selection for high quality pulp. *Journal of Wood Science, 51*(2), 167–171.

Turnbull, J. W. (1999). Eucalypt plantations. In *Planted forests: Contributions to the quest for sustainable societies* (pp. 37–52). Netherlands: Springer.

Velasco, R., Zharkikh, A., Affourtit, J., Dhingra, A., Cestaro, A., Kalyanaraman, A., et al. (2010). The genome of the domesticated apple (*Malus* x *domestica* Borkh.). *Nature Genetics, 42*, 833–839.

Verde, I., Abbott, A. G., Scalabrin, S., Jung, S., Shu, S., Marroni, F., et al. (2013). The high-quality draft genome of peach (*Prunus persica*) identifies unique patterns of genetic diversity, domestication and genome evolution. *Nature Genetics, 45*, 487–494.

Verde, I., Bassil, N., Scalabrin, S., Gilmore, B., Lawley, C. T., Gasic, K., et al. (2012). Development and evaluation of a 9K SNP array for peach by internationally coordinated SNP detection and validation in breeding germplasm. *PLoS One*, 7(4), e35668, 13p.

Volz, R. K., Rikkerink, E., Austin, P., Lawrence, T., & Bus, V. G. M. (2009). "Fast Breeding" in apple: a strategy to accelerate introgression of new traits into elite germplasm. *Acta Horticulturae, 814*, 163–168.

Wegrzyn, J. L., Liechty, J. D., Stevens, K. A., Wu, L.-S., Loopstra, C. A., Vasquez-Gross, H. A., et al. (2014). Unique features of the loblolly Pine (*Pinus taeda* L.) megagenome revealed through sequence annotation. *Genetics, 196*(3), 891–909. http://doi.org/10.1534/genetics.113.159996.

Wetterstrand, K. A. (2015). *DNA sequencing costs: data from the NHGRI genome sequencing program (GSP)*. Available at: www.genome.gov/sequencingcosts. Accessed March 2015.

White, T. L., Adams, T. W., & Neale, D. B. (2007). *Forest genetics*. Cambridge, MA, USA: CABI.

White, T. L., Hodge, G. R., & Powell, G. L. (1993). An advanced-generation tree improvement plan for slash pine in the southeastern United States. *Silvae Genetica, 42*, 359.

Wiersum, K. F. (1997). From natural forest to tree crops, co-domestication of forests and tree species, an overview. *NJAS Wageningen Journal of Life Sciences, 45*(4), 425–438.

Won, K., Bastiaanse, H., Kim, Y. K., Song, J. H., Kang, S. S., Lee, H. C., et al. (2014). Genetic mapping of polygenic scab (*Venturia pirina*) resistance in an interspecific pear family. *Molecular Breeding, 34*, 2179–2189.

Wu, J., Li, L.-T., Li, M., Khan, M. A., Li, X.-G., Chen, H., et al. (2014). High-density genetic linkage map construction and identification of fruit-related QTLs in pear using SNP and SSR markers. *Journal of Experimental Botany*. http://dx.doi.org/10.1093/jxb/eru311.

Wu, J., Wang, Z., Shi, Z., Zhang, S., Ming, R., Zhu, S., et al. (2013). The genome of the pear (*Pyrus bretschneideri* Rehd.). *Genome Research, 23*, 396–408.

Yamamoto, T., Kimura, T., Sawamura, Y., Manabe, T., Kotobuki, K., Hayashi, T., et al. (2002). Simple sequence repeats for genetic analysis in pear. *Euphytica, 124*, 129–137.

Yamamoto, T., Terakami, S., Takada, N., Nishio, S., Onoue, N., Nishitani, C., et al. (2014). Identification of QTLs controlling harvest time and fruit skin color in Japanese pear (*Pyrus pyrifolia* Nakai). *Breeding Science, 64*, 351–361.

Zapata-Valenzuela, J., Isik, F., Maltecca, C., Wegrzyn, J., Neale, D., McKeand, S., et al. (2012). SNP markers trace familial linkages in a cloned population of *Pinus taeda*—prospects for genomic selection. *Tree Genetics and Genomes, 8*(6), 1307–1318. http://dx.doi.org/10.1007/s11295-012-0516-5.

Zapata-Valenzuela, J., Whetten, R. W., Neale, D. B., McKeand, S. E., & Isik, F. (2013). Genomic estimated breeding values using genomic relationship matrices in a cloned population of loblolly Pine. *G3: Genes, Genomes, Genetics, 3*(5), 909–916. http://dx.doi.org/10.1534/g3.113.005975. April.

Zhang, R.-P., Wu, J., Li, X.-G., Khan, M. A., Chen, H., Korban, S. S., et al. (2013). An AFLP, SRAP, and SSR genetic linkage map and identification of QTLs for fruit traits in pear (*Pyrus* L.). *Plant Molecular Biology Reporter, 31*, 678–687.

Zhu, Y., & Barritt, B. H. (2008). *Md-ACS1* and *Md-ACO1* genotyping of apple (*Malus* × *domestica* Borkh.) breeding parents and suitability for marker-assisted selection. *Tree Genetics and Genomes, 4*, 555–562.

Zhu, Y., Evans, K., & Peace, C. (2011). Utility testing of an apple skin colour MdMYB1 marker in two progenies. *Molecular Breeding, 27*, 525–532.

CHAPTER FOUR

Population and Conservation Genomics in Forest and Fruit Trees

Filippos A. Aravanopoulos[*,§,1], **Ioannis Ganopoulos**[*,§], **Athanasios Tsaftaris**[§,¶]

[*]Laboratory of Forest Genetics and Tree Breeding, Faculty of Forestry and Natural Environment, Aristotle University of Thessaloniki, Thessaloniki, Greece
[§]Centre for Research and Technology Hellas (CERTH), Institute of Applied Biosciences, Thessaloniki, Greece
[¶]Department of Genetics and Plant Breeding, Aristotle University of Thessaloniki, Thessaloniki, Greece
[1]Corresponding author: E-mail: aravanop@for.auth.gr

Contents

Abstract

Population and conservation genomics in forest and fruit trees are advancing at an accelerated rate due to the recent developments in high-throughput next-generation sequencing capabilities. Our ability to understand the wealth and structure of genetic variation in nature, plantations and orchards, and the processes of selection, adaptation and speciation will be limited in the future more by conceptual and analytical weaknesses of biostatistical tools, than by the amount of sequence data available. Population genomics are discussed considering the still elusive elucidation of quantitative and adaptive variation at the molecular level. This chapter focuses on the use of genome-wide sampling to discern evolutionary processes and the spatially explicit patterns of genomic variation. The transition from conservation genetics to conservation genomics will increase manifold our ability to select and prioritize conservation targets. Conservation genomics will lead to the identification of particular genomic regions that

Advances in Botanical Research, Volume 74
ISSN 0065-2296
http://dx.doi.org/10.1016/bs.abr.2015.04.001

may have undergone natural selection and are of adaptive significance, in addition to improving the precision of genetic and demographic inferences by significantly raising the number of putatively neutral markers assayed. Comparative analysis with cultivated trees will facilitate our understanding of domestication and dispersal effects on tree genomes and the effect of tree breeding on gene selected and genome effects.

1. INTRODUCTION

Trees constitute more than 80% of continental biomass, while forests harbour more than 50% of terrestrial biodiversity. Forest ecosystems, agroforestry communities and tree orchards, if managed properly, offer a sustainable source of primary production that includes wood, paper and its products, firewood and bioenergy, tree-crop foods, pharmaceuticals and a multitude of ecological services. A large number of trees are keystone species for many ecosystems and form a cornerstone for the support of different biological communities. Tree orchards sustain the rural economies in many areas of the developing world and in the mountainous areas of the developed countries. Trees have become an important and invaluable component of the urban environment. Trees represent an integral part of the human cultural heritage (Aravanopoulos, 2010).

Trees are long-living, perennial sessile organisms that sustain biotic and abiotic stress and environmental change potentially for eons during their lifetime. Gymnosperms are the most ancient living plants having appeared 360—400 M year ago. Angiosperms, that include most fruit trees, were separated from gymnosperms about 130—190 M year (Magallo & Sanderson, 2005). Contemporary gymnosperms number fewer than 1000 species (Farjon, 2008), compared to about 250,000 extant angiosperm species (Kenrick, 1999). Gymnosperm genome sizes are vast, e.g. in *Cupressus*, *Pinus* and *Picea* the size ranges from 10,500 to more than 35,000 M bp, being on the average more than 200× the size of the *Arabidopsis thaliana* genome, and close to 249×, 109× and 79× the genomes of rice, maize and human, respectively (Mackay et al., 2012; Manescu et al., 2011). Perennial woody angiosperms present more moderately sized genomes, for example *Quercus* (about 650—900 M bp, Zoldos, Papes, Brown, Panaud, & Siljak-Yakovlev, 1998) and *Populus* (about 550—650 M bp, Tuskan et al., 2006), while fruit trees sized genomes varied from 227 M bp in *Prunus persica* (IPGI, 2011) to 742 M bp in *Malus domestica* (Velasco et al., 2010). Despite the large differences in genome size, numbers of coding genes are comparable, likely in the vicinity of about 30,000–60,000 genes (IPGI, 2011; Velasco et al., 2010; White, Adams, & Neale, 2007).

Trees spend their pluriannual lifetime in one location, in stark contrast to the highly mobile animal species, or herbaceous plants with high annual dispersal. Trees are unique among seed plants as many species can be found in both a domesticated and an undomesticated state (Ganopoulos, Aravanopoulos, & Tsaftaris, 2013). A wealth of different situations emerges; populations with no, little or strong human disturbance can be found. Extant tree populations are the result of different degrees of natural evolutionary forces and questions pertaining to speciation, adaptation, and demography may have different levels of confounding anthropogenic influence (Ganopoulos, Aravanopoulos, Argiriou, Kalivas, & Tsaftaris, 2011). Fruit trees have lower rates of evolution as they may propagate clonally, while their long juvenile phases further decrease the number of sexual cycles (Zohary & Spiegel-Roy, 1975). The same is true for some forest tree species such as poplar. Unlike most crop and livestock species, trees form large continuous natural population in highly heterogeneous environments harbour significant genetic diversity that ensures adaptation to changing environmental conditions and provides the genetic fuel for both evolution and breeding. Trees are characterized by high gene flow that spreads adaptive variants across landscapes. Nevertheless, selection for local adaptation is apparently strong, as indicated by strong population differentiation for key adaptive traits (Savolainen et al., 2007).

Genomic approaches have become important since the introduction of next-generation sequencing (NGS; Mardis, 2013; Visendi, Batley, & Edwards, 2014) because NGS allows for the low-cost/high-throughput sequencing of multiple genotypes (Ekblom & Galindo, 2010; Manel & Holderegger, 2013; Schoville et al., 2012). With the advent of molecular marker-based technologies in plant breeding, researchers in fruit and forest tree genetics have explored their application to genetically map and genotype key tree species genomes. Much of the evolution of these technologies progressed in parallel between forest and fruit tree genetics with little exploration of the interface between the two. As there are common breeding targets for forest and fruit trees, merging genomics research of each will lead to a better understanding of tree species fundamental biology and the development of molecular breeding (Abbott, 2014).

The study of genes at the population level (population genomics) may lead to the identification of genes function responsible for species adaptation and evolution. Population genomics of fruit trees in particular are facilitated by the large amount of sequence information now available for several species, coupled with the implementation of high-throughput expression analyses (Ramawat, Mérillon, & Ahuja, 2014). Genome and transcriptome sequencing (both de novo sequencing and resequencing) using NGS technologies is advancing in trees, forming a quantum leap in the volume of genomic data available for species with insofar scarce genomic resources available. An extension of population genomics is conservation genomics. Population genomics facilitate disentangling the distribution and evolution of ecologically relevant and evolutionary important adaptive diversity. Population genomics can support genetic improvement programs and develop diagnostic tools (Neale & Kremer, 2011) for the conservation, restoration and management of natural populations, plantations and orchards. Conservation genomics, the application of genomic analysis to the preservation of the viability of populations and the biodiversity of forest and fruit trees, involves the use of population genomics in the management, sustainable use and especially protection and conservation of undomesticated and long-lived natural tree populations of numerous species and mildly or partially domesticated fruit tree species. This review presents and critically analyzes recent advances in population and conservation genomics of forest and fruit trees.

2. POPULATION GENOMICS

Genetic variation in natural populations is the fuel of adaptation and evolution. Moreover, it forms the foundation for artificial selection and breeding. Population differentiation originates mainly from adaptive responses to local environmental changes that occurred since the last glacial period and was not blurred by demographic patterns or historical trends (Neale & Kremer, 2011). Trees have often shown geographical patterns of adaptive genetic variation. Field experiments and molecular markers have been used to unravel the genetic basis of adaptation in trees. None of them has proved to be fully informative. Time-consuming and expensive field tests are solely based on phenotypes and the statistical power of genetic parameter estimation is dependent on the number of different test environments, field test duration, the diversity of experimental populations and the number of the genetic entries used. Explaining quantitative-adaptive variation at the gene and molecular level has remained elusive. Field experiments fail to provide information on which particular genes and how many of them are involved in adaptation, or how much phenotypic variation can be explained by genetic variation in these genes (Abril et al., 2011). On the other hand, most of the molecular markers measure either neutral or highly conservative genetic variation. The adaptive value of molecular markers may only be indirect (Aravanopoulos, 2011).

NGS technologies show a great promise in unravelling the genetic base of complex quantitative traits in natural tree populations. The availability of genome-wide sequences, even for non-model species, creates the opportunity for a population genomic approach that combines traditional population genetics with a genome-wide level of analysis and can simultaneously examine the effects of demographic history, migration, and selection (Sork et al., 2013). By population genome-wide analysis factors that affect individual genes, alleles, or nucleotides can be discerned from factors affecting the entire genome (Abril et al., 2011). The power of population genomics lies in the ability to combine information on phenotype, genotype and local environment in a high number of spatially referenced samples collected across a multitude of landscape scales (Sork et al., 2013).

Trees are in fact well placed to be good models for population genomic applications. They (1) are long-lived sessile organisms, (2) occupy diverse environments with potentially very extensive (but in many cases fragmented) distribution, (3) are predominantly outcrossing with high gene dispersal, (4) present late reproductive maturity, (5) have faced range contractions and expansions associated with glacial cycles, (6) exhibit clinal variation across environmental gradients and (7) do not present extensive domestication and artificial selection as they have not been subjected to intensive and prolonged breeding.

Population genomics combine genome-wide sampling with traditional population genetic approaches (González-Martínez, Krutovsky, & Neale, 2006), supported the high genetic variation that is present in woody plants. Trees present high genetic variation compared to other plants and animals (Hamrick, Godt, & Sherman-Broyles, 1992; González-Martínez et al., 2006; Savolainen & Pyhajarvi, 2007). Trees possess high levels of individual heterozygosity and high genetic diversity within populations (Hamrick et al., 1992), nevertheless rather low differentiation among populations at biparentally inherited nuclear gene markers (Austerlitz, Mariette, Machon, Gouyon, & Godelle, 2000).

Two unique characteristics of most tree genomes are low linkage disequilibrium (LD) at two nucleotide sites and rapid LD decay. Evidently LD is affected in genes located near the pericentromeric area where chromosomal crossovers are restricted. At the population level, after many generations of recombination and random mating, only tightly linked loci will show significant association to quantitative traits (González-Martínez, Krutovsky, et al., 2006). Natural tree populations exhibit continuous variation for phenological traits along geographical patterns (clinal variation) in response to latitudinal, altitudinal, climate, or soil gradients (White et al., 2007). It is therefore feasible to conduct simultaneous association testing and clinal signature detection at the genomic level (Neale & Kremer, 2011). Sampling the edges of a steep cline, for example can increase the chance of elucidating the molecular basis of the divergent adaptive trait, as extreme genotypes might be sampled (González-Martínez, Krutovsky, et al., 2006). LD decay lies in the vicinity of 200−1500 bp (in contrast, for instance to >20,000 bp in *Arabidopsis*). Low LD and rapid LD decay facilities the identification of sequences associated with specific phenotypes, since the causal variants responsible for phenotypic traits can be located specifically within a sequence (Parchman et al., 2012).

Marker loci (mostly single nucleotide polymorphisms (SNPs) originating from candidate genes or from whole-genome scans) are genotyped in individuals from different populations or environments and tested for patterns of variation that signal selection. The detection of loci with unusually high or low levels of variation and differentiation (outlier loci) for multiple-population genetic differentiation estimates, is a powerful method to find loci under selection and to separate genome-wide effects that are caused by demographic processes from adaptive locus-specific effects (Luikart, England, Tallmon, Jordan, & Taberlet, 2003). The identification of outliers with respect to population structure by employing, e.g. an F_{ST} outlier analysis for putatively neutral molecular markers (usually nuclear Simple Sequence Repeats (SSRs)) and candidate gene markers (e.g. SNPs or Expressed Sequence Tags (EST)-based markers), is a common approach to study local adaptation and monitor conservation. Markers that show higher (or lower) differentiation than putatively neutral ones can be considered as being under-diversifying (or stabilizing) selection (González-Martínez, Krutovsky, et al., 2006). F_{ST} outlier analyses to identify loci showing strong differentiation across populations do not require spatially explicit data because they do not associate allele frequencies with environmental gradients, but they can identify loci under selection when compared either to the entire genome or to a set of noncandidate markers (Sork et al., 2013). There is no requirement for large-scale, costly and time-consuming common garden tests, or for prior knowledge of phenotypic traits involved in local adaptation.

Although natural selection leaves its marks in the genome, separating these marks from those of other relaxations of the panmictic model and demographic events can be an intricate task. Currently, the best way to detect the effects of selection on a locus is to demonstrate that its pattern of polymorphism differs significantly from the genome-wide polymorphism (Savolainen & Pyhajarvi, 2007; Wright & Gaut, 2005). Such analyses are hindered by the absence of databases with extensive multilocus population datasets, such as those, e.g. available for humans (Biswas & Akey, 2006). However, it appears that evidence for either balancing or directional selection can be inferred. Savolainen and Pyhajarvi (2007) have compiled the scattered available information on the selection on individual loci in trees and found that 15% have been reported to be under some kind of selection, while a more recent and more comprehensive review of approximately 290 genes identified 55 (20%) showing departures from neutrality (Neale & Ingvarsson, 2008). Based on these preliminary data it appears that a rather small number of one to two in 10 genes in tree genomes may be under some form of natural selection. Nevertheless, selection can be highly efficient and easily detectable in a large tree population mainly due to the low LD in forest trees and evolution in response to natural selection is very rapid (Savolainen & Pyhajarvi, 2007).

Population genomics are highly powerful in detecting marker–phenotype associations and in discerning the genetic basis of quantitative traits in trees compared to earlier approaches, but they are not free of caveats. One drawback is that environmental association analyses do not necessarily relate a variant under selection to a particular phenotype. Furthermore, most phenotypic traits involved in local adaptation are highly polygenic. High gene flow and locally variable selection result in large population differences in mean adaptive traits, in spite of small changes in underlying Quantitative Trait Loci (QTL) frequencies (Le Corre & Kremer, 2003), suggesting epistasis, epigenomics or transposable element activity (Grattapaglia, Plomion, Kirst, & Sederoff, 2009). Kremer and Le Corre (2011) showed through simulations that if phenotypic traits involve more than a few genes, the underlying loci may show only modest levels of population differentiation (F_{ST}), similar to that of selectively neutral loci, even when there is strong population differentiation for the phenotypic trait (Q_{st}). This can be expected in cross-fertilized trees where only alleles with modest effects are selected for, in response to avoiding major disturbances at the phenotype level. Moreover, SNP discovery and candidate gene selection need functional information about the genes' role in order to validate candidates, a process that may introduce ascertainment bias (Grattapaglia et al., 2009). An additional problem is the detection of signatures of selection in tree genomes (Savolainen & Pyhajarvi, 2007). For example selective sweeps are accompanied by a variability reduction in areas surrounding the selected site (Nordborg et al., 2005) as it has been found, for instance in maize (Wang, Stec, Hey, Lukens, & Doebley, 1999). Nevertheless, reduction may well be because of a demographic departure from panmixis. Selective sweeps can result in an excess of rare alleles as well; however, the latter may also be the result of a population expansion phase. A high frequency of SNP loci that present intermediate allele frequencies might be due to negative frequency-dependent selection (Ganopoulos, Aravanopoulos, Argiriou, & Tsaftaris, 2012), but it could also be due to a recent bottleneck. Natural selection can cause an increase in the local LD level (Nordborg et al., 2005), but LD also varies because of variation in recombination rates or due to population structure. High genetic population differentiation can be due to differential local selection between populations, but it can also result from population isolation. If past demographic changes have led to systematic departures from neutrality, the probability that a gene shows such departures is increased, which will inflate the number of false positives (Neale & Ingvarsson, 2008). However, demographic processes are expected to affect the entire genome of a species, whereas selection is likely to affect only small segments of the genome, i.e. a few alleles and the vicinity were LD has not decayed and thus demography can in theory be distinguished from selection.

2.1 Population Genomics in Forest Trees

Tree genomes neutral evolution (nucleotide substitution rate at synonymous sites) is slow and close to the rates estimated for many annual plant species if generation time is taken into account (Savolainen & Pyhajarvi, 2007). In *Populus*, the evolutionary (mutation) rate has been estimated to be about 0.25×10^{-8}/site/year (Tuskan et al., 2006) and in *Pinus* about 0.7×10^{-9}/site/year (Willyard, Syring, Gernandt, Liston, & Cronn, 2007).

Most nucleotide diversity resides within tree populations (>95% for European populations; Savolainen & Pyhajarvi, 2007), and even in the highly divergence South-Eastern European populations the variation found between populations is not much higher (Malliarou & Aravanopoulos, 2012). By contrast, a considerably higher divergence was found, for instance in Central European populations of *Arabidopsis lyrata* where 17% of the variation resides between populations (Clauss & Mitchell-Olds, 2006).

The study of population genomics becomes very important for both evolution, phylogeography, as well as breeding, when we consider that neutral loci across the genome will be similarly affected by the demographic and evolutionary history of populations (Sork et al., 2013), while loci under selection will often behave differently and reveal "outlier" patterns of variation (Grattapaglia et al., 2009). Therefore, population genomics can be employed in a jointly informative sense to study both gene movement and local adaptation in natural environments (Sork et al., 2013).

The standard panmictic neutral model predicts that given the above mutation rates, large N_es and high inherent genetic variation, contemporary levels of genetic diversity in an equilibrium tree population should be high (Dvornyk, Sirviö, Mikkonen, & Savolainen, 2002). Nevertheless, diversity estimates of trees are highly variable and not higher than those for other plants (Savolainen & Pyhajarvi, 2007). This suggests that many tree populations might not be at equilibrium, for instance due to fluctuations of historical population size, or selection (Evans et al., 2014; Müller, Freund, Wildhagen, & Schmid, 2015).

For most tree (non-model) species the use of the association between environmental gradients and genotypes may reveal 'outlier' patterns of variation and thus identify potential loci underlying selection. Associations using high-resolution genome-wide scans are now becoming feasible for non-model species. In contrast to mapping where the dissection of complex traits is based on the simultaneous assessment of genotypes and phenotypes, population genomics requires scoring genotypes, but not necessarily phenotypes. Studies in trees successfully identified polymorphisms involved in adaptive population differentiation and detected genes with contrasting patterns of variation. Population genomic representative studies are presented in Table 1. Two recent studies show the importance of this approach. McKown et al. (2014) uncovered the genetic basis of phenotypic trait variation in 448 accessions of *Populus trichocarpa* and suggested that phenology had the largest number of associated genes (240 genes), followed by biomass (53 genes) and ecophysiology traits (25 genes). Moreover, Tsumura et al. (2007, 2014) studied the genetic structure of *Cryptomeria japonica* natural populations using almost 4000 SNPs. While genetic differentiation was low ($F_{ST} = 0.05$), it accounted for the presence of important genes associated with adaptation to specific environments. Moreover, they detected 208 outlier-loci, of which 43 were associated with environmental variables. On the contrary, in *Populus* no evidence was found for population genetic differentiation at SNPs in phenology candidate genes that had

Table 1 Representative Population Genomics Studies in Forest and Fruit Trees

Species	Methodology/Approach	Genomic Data	Major Results	References
Alnus glutinosa	Genotyping-by-sequencing; 24 European populations	A total of 1990 SNPs	Very low genetic differentiation among populations; high intrapopulation genetic variation	Kort et al. (2014)
Pinus halepensis	Sequencing; 7 populations from Greece, Spain, Morocco, Algeria, Italia, Israel and South Africa	Ten drought-response candidate genes	Marked loss of genetic diversity from the relictual Greek population towards the western range of the species, as well as molecular signatures of intense bottlenecks	Grivet, Sebastiani, González-Martínez, and Vendramin (2009)
Pinus pinaster Ait	Sequencing; 24 populations from France, Spain, Morocco and Tunisia	Eleven drought-response candidate genes	Two genes presented higher F_{ST} values than expected (PR-AGP4 and erd3), suggesting that they could have been affected by the action of diversifying selection among populations; 3 genes Presented lower F_{ST} values than expected (dhn-1, dhn2, and lp3-1)	Eveno et al. (2008)

Picea abies	Sequencing; 18 populations from Europe	A total of 137 SNPs from 18 candidate genes putatively affecting bud set.	Population genetic structure was low ($F_{ST} = 0.05$), but latitudinal patterns were apparent among Scandinavian populations Twenty-nine SNPs were outliers	Chen et al. (2012)
Picea mariana [Mill.] B.S.P	Sequencing; boreal populations	A total of 583 SNPs from 313 genes potentially playing adaptive roles	Average differentiation estimates for gene SNPs were null, with F_{ST} values of 0.005 and 0.006, respectively, among temperature and precipitation population groups A total of 26 SNPs from 25 genes distributed among 11 of the 12 linkage groups of black spruce were detected as outliers with F_{ST} as high as 0.078.	Prunier, Laroche, Beaulieu, and Bousquet (2011)

(Continued)

Table 1 Representative Population Genomics Studies in Forest and Fruit Trees—cont'd

Species	Methodology/Approach	Genomic Data	Major Results	References
Prunus persica	Large-scale resequencing of 10 wild and 74 cultivated peach varieties	4.6 million SNPs	A single domestication event, the speciation of *P. persica* from wild peach. Ornamental and edible peach both belong to *P. persica*, along with another geographically separated subgroup, *P. ferganen*	Cao et al. (2014)
Citrus spp.	Sanger sequencing; 45 genotypes of *Citrus* and relatives of *Citrus*	A total of 1097 SNPs and 50 indels were identified from 18 genes involved in metabolite biosynthesis pathways and nine putative genes for salt tolerance	Nuclear phylogenetic analysis revealed that *Citrus reticulate* and *Fortunella* form a cluster that is differentiated from the clade that includes three other basic taxa of cultivated citrus (*C. maxima*, *C. medica* and *C. micrantha*)	Garcia-Lor et al. (2013)

Vitis spp.	9000-SNP genotyping array (Vitis9kSNP); 950 accessions of *Vitis vinifera* and 59 accessions wild relative *Vitis sylvestris*	5387 SNPs	The decay of LD is very rapid in *vinifera* and appears unchanged between the wild ancestor and the domesticated grape	Myles et al. (2011)
Vitis spp.	9000-SNP genotyping array (Vitis9kSNP); 1173 accessions representing 19 taxa (16 unique species, two hybrid taxa, one species with two intraspecific groups)	4073 SNPs after filtering	They reconstructed the phylogeny of the genus *Vitis* using hybridization data. North American subgenus *Vitis* species are monophyletic	Miller et al. (2013)

previously shown significant clinal variation (Hall et al., 2007), but Ingvarsson, García, Hall, Luquez, and Jansson (2006) reported SNP clinal variation at the phytochrome locus PhyB2 in *Populus tremula*.

2.2 Population Genomics in Fruit Trees

Studies in the population genetics of fruit trees are less extensive than those on forest trees presumably because fruit trees do not form large natural populations. The focus of genomics research is more associated with the understanding of the process of domestication, the pressure of the environment, as well as the manifestation of breeding objectives. Nevertheless, from the studies available it becomes apparent that they also possess notable amounts of genetic diversity. For instance in *Malus*, the average polymorphism rate ranges between 4.8 and 9.6 SNPs/kb. In *Prunus*, the nucleotide diversity (π) ranges between 1.6 and 4.8×10^{-3} (IPGI, 2011) and in *Citrus* between 3.0 and 5.0×10^{-3} (Garcia-Lor et al., 2013). Furthermore, Myles et al. (2011) characterized genome-wide patterns of genetic variation in over 1000 samples of the domesticated grape, *Vitis vinifera* subsp. *vinifera*, and its wild relative, *V. vinifera* subsp. *sylvestris* and found high levels of genetic diversity and rapid LD decay in the former. Recently, Miller et al. (2013) used the Vitis9kSNP array for *Vitis* phylogenomics purposes and suggested that the North American subgenus *Vitis* species are monophyletic.

Fruit tree population genomics have also recently expanded in studies relevant to the history and evolution of wild and cultivated species and the anthropogenic influences leading to domestication. For instance Cao et al. (2014) performed large-scale resequencing of wild and cultivated peach (*P. persica*) varieties and identified 4.6 million SNPs, a large number of which could explain the phenotypic variation in cultivated peach. Their population genomic analysis showed a single domestication event, the speciation of *P. persica* from wild peach. Furthermore, in *Citrus* species Wu et al. (2014) sequenced and analyzed the genomes of a diverse collection of cultivated pummels, mandarins and oranges to test the pummel-mandarin species hypothesis and to uncover the origins of several important citrus cultivars. They suggested that edible varieties of this species complex are usually hybrids of varying complexity (Wu et al., 2014).

3. ASSOCIATION MAPPING

Association mapping refers to the testing of the functional importance of SNP markers by associating nucleotide variation in genes with phenotypic variation in adaptive traits in natural or segregating populations (Gailing, Vornam, Leinemann, & Finkeldey, 2009). It provides much higher level map resolution and can potentially reveal individual genes underlying complex traits (Neale & Ingvarsson, 2008). Association mapping uses LD in populations (natural populations, but also populations of varieties, cultivars or clones) to find statistical associations between molecular markers and phenotypes. Trees are amenable organisms for detecting traces of selection, compared to many other species, as LD declines fast, while large tree populations appear to respond rapidly to natural selection, in contrast to the slow rate of neutral evolution (Savolainen & Pyhajarvi, 2007).

Association mapping in trees have confirmed the rapid LD decay within candidate genes for various adaptive traits (Brown, Gill, Kuntz, Langley, & Neale, 2004; Ingvarsson, 2005; Krutovsky & Neale, 2005; Myles et al., 2011; Rafalski & Morgante, 2004), as R^2 often declines to <0.20 within 1000—2000 bp (Neale & Savolainen, 2004). The extent of LD within genes is very short (from approximately 200—400 bp in *Picea abies* and *Pseudostuga menziensii* to approximately 800—1500 bp in *Pinus taeda* and <500 bp in *Populus tremuloides*). There is a general consensus that LD within genes declines within less than 1000 bp (Savolainen & Pyhajarvi, 2007). In contrast, the rate of LD decay in self-pollinated plants, or in more evolutionarily recent outbreeding organisms such as humans, is in the vicinity of many thousand base pairs if not million base pairs (Neale, 2007). In the highly inbred *A. thaliana*, it is over 20,000 bp in worldwide samples (Nordborg et al., 2005). Cao et al. (2014) found that the LD decay of breeding varieties in *P. persica* (peach) was 19,000 bp. These values showed that peach had a medium LD level compared with other self-compatible plants (Cao et al., 2014). Furthermore, in *Malus* × *domestica* the averaged LD between adjacent SNPs pairs in a training population consisting of 1200 seedlings was 0.32 (Kumar et al., 2012). The major advantage of rapid LD decay is that once a marker—trait association has been discovered and validated, it is likely that such a marker is at a close physical distance to the functional variant (probably within the gene itself), or even is the functional variant itself (González-Martínez, Ersoz, Brown, Wheeler, & Neale, 2006). This is important for the application of marker-assisted selection because the marker—trait association is less likely to break down with successive rounds of sexual reproduction and recombination. Hence, marker—phenotype associations can be applied both for within and among family marker (gene)-assisted selection (Plomion et al., 2007).

The absence of population stratification has been regarded as one of the advantages of the association mapping approach; however, this has to be shown beforehand. In fact, population structure (such as that resulting from historical migration patterns) is the most common systematic bias producing false positives in association studies (Hirschhorn & Daly, 2005). To increase the power of detection suitable populations are restricted to those that minimize the effect of demography and/or family relatedness. Therefore, population structure should be tested in advance and methods that correct for population structure should be used (González-Martínez, Krutovsky, et al., 2006). Moreover, most markers used are neutral and therefore they may not be within the actual genes that play a causal role in the determination of quantitative traits. Of course, the lower the LD the higher the chances such neutral markers such as SNPs may be within a causal gene itself, but this may not be that apparent if the gene effect is small, or if it is not subjected to fairly strong selection. Furthermore, an absence of differences in neutral marker genotypes does not preclude significant adaptive differences between individuals or populations. For instance common garden experiments in *Pinus sylvestris* exhibited distinct genetic differences in the time of bud flush among populations sampled along a climatic gradient; however, molecular markers failed to distinguish these differences (Neale, 2007). Additional problems in association mapping are (1) low sample size, (2) need to perform multiple testing, (3) preference for the use of common alleles, need for tight linkage (only markers very close to an SNP causing phenotypic differences are expected to show a significant association between phenotypic trait expressions and marker variation), (4) difficulty in mutation detection, (5) low level of variation explained (individual SNP—trait associations only account for a small proportion of the variation, generally less than 2—3% of the total phenotypic variance and the total variation explained by all markers is generally less than 50%) (Aravanopoulos, 2014).

3.1 Association Mapping of Forest Trees

SNPs in candidate genes associated with traits have already been detected, e.g. in drought tolerance (González-Martinez, Huber, Ersoz, Davis, & Neale, 2008), cold hardiness (Eckert et al., 2009) and wood properties (González-Martinez, Wheeler, Ersoz, Nelson, & Neale, 2007; Thumma, Nolan, Evans, & Moran, 2005; Yu et al., 2006). Genes associated to phenotypes appear to be more often correlated to environmental variables and more often to depart from neutrality than randomly chosen genes (Eckert,

2011). Nevertheless, the magnitude of the effect of each individual association has been very small, rarely exceeding 5% of the phenotypic variance (Grattapaglia & Resende, 2011). In congruence to earlier QTL mapping results, individual SNP—trait associations only account for a small proportion of the total phenotypic variance (usually <2—3%) and in some cases weak associations are mostly lower than 1% (González-Martinez, Huber, Ersoz, Davis, & Neale, 2008). Much larger and significantly more powerful genome-wide association studies in crop plants (Buckler et al., 2009), domestic animals (Goddard & Hayes, 2009), and humans (Visscher, 2008) have found similar results.

3.2 Association Mapping of Fruit Trees

Association mapping has become the method of choice for fruit trees of high economic value, such as apple, peach, pear, cherries and grapevine, where creating and preserving high generation filial structures (that would be amenable to direct QTL mapping) remains a constant major challenge. In apple, after identifying candidate orthologous FRUITFULL-like genes linked to fruit flesh firmness, Cevik et al. (2010) have used association mapping and linked one of the markers and flesh firmness in a population of 168 apple accessions. Furthermore, Khan, Zhao, and Korban (2013) discovered the genetic basis of fire blight resistance in apple using both QTL and genome-wide association mapping. They used an F_1 pedigree and an association mapping population of *Malus* accessions and found significant associations of three QTLs with resistance to fire blight.

In cherry (*Prunus avium* L.), NGS sequencing was used to discover SNPs, providing markers for genetic and genomic studies. High-density genetic maps were built using a cherry 6K SNP array (Klagges et al., 2013) which was used to identify genes that may be involved in fruit quality. De Franceschi et al. (2013) found two SNPs located within the confidence intervals of major QTLs previously discovered in sweet cherry (*P. avium*).

The first study of association mapping in fruit trees and particularly in grape was conducted by Fournier-Level et al. (2009). They selected a set of five putative quantitative trait nucleotides, corresponding to one retrotransposon insertion, one 2-bp insertion/deletion (indel), and three SNPs which accounted for 84% of the variation in anthocyanin content of a highly diversified collection of grape cultivars. Fruit development is an important economic trait of grapevine (*V. vinifera*) and Houel et al. (2010) studied the genetic and molecular controls of fleshy fruit development. They identified markers from a gene (*flb*) proposed to be linked to a fleshless berry mutation by resequencing genomic regions of two grape cultivars and then sequenced these gene fragments in a highly diverse set of cultivated and wild *V. vinifera* genotypes to identify possible signatures of domestication in cultivated *V. vinifera*. SNPs significantly associated to grape weight variation have been identified in the *flb* region. A major QTL for muscat flavour determination (Emanuelli et al., 2010), when resequenced in an association population of grape cultivars, resulted in the identification of three SNPs in moderate LD, significantly associated with muscat-flavoured cultivars (Myles et al., 2011).

Representative studies on the association mapping of forest and fruit trees are presented in Table 2.

4. COMPARATIVE GENOMICS

Comparative genomics are used in plant and animal systems to fully understand the evolutionary basis of gene function, developmental processes and adaptation (Neale & Ingvarsson, 2008). Trees can take full advantage of comparative genomics as in contrast to agricultural plants, not a single tree species is extensively studied throughout the world. In trees, comparative mapping has shown that macrosynteny is conserved above the species, or even above the family level (Aravanopoulos, 2001). Abrouk et al. (2010) and Salse (2012) have employed archeogenomics and identified common and lineage-specific patterns of conservation indicating that plants have evolved from common ancestors with a basic number of five/seven chromosomes, mainly through whole-genome duplications, as well as translocations followed by lineage specific segmental duplications, chromosome fusions and translocations. These findings are important because they show the promise of comparative genomics to leap from one well-studied genome to another relatively unknown one.

Co-localization of QTLs and especially how gene function may be transferable across species form a major facet of comparative genomics. Wider comparative studies, of the already sequenced long-lived perennial *Populus* and extremely short-lived annual *Arabidopsis*, showed that the photoperiodic pathway regulates photoperiodic control of flowering in *Arabidopsis*, while in *Populus* it has also been implicated to control seasonal growth cessation, a process absent in *Arabidopsis* (Jansson & Douglas, 2007). Hence, there are indications that at least some aspects of gene expression are conserved between different major groups of plants.

Table 2 Representative Association Mapping Studies in Forest and Fruit Trees

Species	Sample Size Populations/ Individuals	Genomic Data	Major Results	References
Pinus taeda L.	A total of 32 seed Megagametophytes Seed donors included 22 unrelated, first-generation selections of *P. taeda* (plus trees) from undisturbed natural stands covering the south-eastern range.	A total of 18 candidate genes for drought-stress response.	Candidate genes had moderate average nucleotide diversity at silent sites ($\pi_{sil} = 0.00853$), varying 100-fold among single genes. The level of within-gene LD was low, with an average pairwise r^2 of 0.30, decaying rapidly from ~0.50 to ~0.20 at 800 bp.	González-Martínez, Ersoz, et al. (2006)
Pseudotsuga menziesii var. *menziesii*	Sequencing; 700 unrelated families.	A total of 384 SNPs from 117 candidate genes and 21 cold-hardiness-related traits.	A Set of 12 candidate genes that is associated significantly with cold hardiness. Four of the 15 polymorphisms within these genes are non-synonymous substitutions, with the remainder being located largely in synonymous positions.	Eckert et al. (2009)
Eucalyptus spp.	Sequencing of cinnamoyl CoA reductase (*CCR*) gene; 290 trees from an *E. nitens* natural population.	A total of 25 SNPs in *CCR gene.*	Two haplotypes significantly associated with microfibril angle were found.	Thumma et al. (2005)

(Continued)

Table 2 Representative Association Mapping Studies in Forest and Fruit Trees—cont'd

Species	Sample Size Populations/ Individuals	Genomic Data	Major Results	References
Malus domestica	Sequencing of *FRUITFULL* (*FUL*) and *SHATTERPROOF* (*SHP*) genes Prima × Fiesta' progeny.	Two candidate orthologous *FUL*-like genes, *MdMADS2.1* and *MdMADS2.2*.	Significant association between the *MdMADS2.1* and fruit flesh firmness at both allelic ($\chi^2 = 34$, $df = 9$, $P < 0.001$) and genotypic ($\chi^2 = 57$, $df = 32$, $P < 0.01$) levels.	Cevik et al. (2010)
Vitis vinifera	Sequencing of 1 Mb grapevine genome region at the top of chromosome 18 carrying the *fleshless berry* mutation (*flb*); 38 grapevine varieties.	A total of 62 regions (17 SSR, 3 SNP, 1 CAPS and 41 re-sequenced gene fragments).	SNPs significantly associated to berry weight variation were identified in the *flb* region.	Houel et al. (2010)
Vitis vinifera	Sequencing of *MybA1* locus; Syrah × Grenache (S × G) was a F_1 progeny of 191.	Five QTNs corresponding to one retrotransposon insertion, one 2-bp insertion/deletion (indel), and three single nucleotide polymorphisms (SNPs).	QTNs significantly associated with anthocyanin content of a highly diversified collection of grape cultivars.	Fournier-Level et al. (2009)

Prunus avium	Sequencing of candidate fruit weight *CNR*s; 557 individuals of N (New York 54) × E (Emperor Francis) progeny and 133 individuals of R (Regina) × L (Lapins) progeny.	A total of 23 *FW2.2/CNR* family members were identified in the peach genome. Two of these *CNRs* were located on linkage groups 2 and 6, named *PavCNR1 2* and *PavCNR20*, respectively.	Association with fruit size strongly supports a role of *PavCNR12* in the sweet cherry linkage group 2 fruit size QTL.	De Franceschi et al. (2013)

In fruit trees, relevant work has focused in Rosaceae and in particular in *Prunus*. Le Dantec et al. (2010) revealed the co-localization of candidate genes and QTLs for fruit quality traits in both peach (*P. persica*) and strawberry (*Fragaria*). Primers designed from candidate genes of the former were amplified in the latter and a congruent pattern synteny was also found. A gene-based genome-wide set of markers has successfully been employed in comparative genomics studies of *P. avium*, *P. persica*, *Malus* × *domestica* and *Fragaria* spp. (Cabrera et al., 2012).

Overall, comparative genomics hold a great promise as a major component of genomic studies in forest and fruit trees. Comparative genomics should be complemented by ultra-deep sequencing of all closely related species to enable powerful genomic scans for functional sequences (Lawrie & Petrov, 2014).

5. CONSERVATION GENOMICS

In an era of strong and fast environmental change imposed on long-lived tree species, genetic conservation and monitoring is of the utmost importance for the perpetual survival of natural tree populations (Aravanopoulos, 2011). The holy grail of genetic conservation is estimating and preserving adaptive genetic potential and therefore securing the long-term adaptation through evolution of natural tree populations. The inherent limitations of common garden experiments, the use of a small number of putatively neutral molecular markers covering a very limited subset of the genome and the need for immediate action by natural resource managers form a Gordian knot.

This can now be solved by estimating adaptive genomic potential based on DNA sequencing and molecular genetic marker information from population samples. Population genetic parameters such as measures of nucleotide diversity and divergence can be estimated in a straightforward manner and departures from neutrality can be tested (Neale & Kremer, 2011). This approach can be combined with the identification of allele-specific effects for thousands of genes through association mapping. In forest and fruit trees in particular, where typical genetic conservation studies used about 10 variable loci, the analytical power of conservation genomics involving tens of thousands of loci is unprecedented. It will allow for high-precision estimates of genetic and demographic parameters (N_e, gene flow, latent genetic potential, etc.), as well as for the characterization and higher resolution of adaptive genetic variation (identification of selection signals) *in natura*

(Aravanopoulos, 2011; Graudal et al., 2014). Hence, it is possible to identify particular genomic regions that may have undergone natural selection, in addition to improving the precision of demographic inferences by significantly rising the number of putatively neutral markers assayed (Narum, Buerkle, Davey, Miller, & Hohenlohe, 2013). Conservation of both neutral and adaptive genetic diversity can finally be tuned. Genomics permit genome-wide studies and will facilitate the study of the genetic basis of local adaptation and its genetic architecture, even in non-model species, expanding this field beyond traditional model (usually non-tree) species (Savolainen, Lascoux, & Merilä, 2013).

Regarding in situ conservation, genomic data will usually be better than genetic marker data for delineating gene conservation units, as the former are generally inadequate for characterizing and quantifying adaptive patterns, but not the latter. Hence, NGS data would be more powerful especially in cases where significant adaptive differentiation is expected among evolutionarily significant units that are considered as candidates for gene conservation (Funk, McKay, Hohenlohe, & Allendorf, 2012), will improving the accuracy of the estimation of genetic and demographic parameters. The higher added value of genomics will be primarily in determining which parts of the genome are responsible for local adaptation and therefore important to preserve. Genomic tools will be especially important for small populations of (non-model) forest tree species and wild fruit trees as they will also facilitate studying comprehensively the genomic mechanisms that translate the effects of small population size into effects on adaptation (Ouborg, Pertoldi, Loeschcke, Bijlsma, & Hedrick, 2010). The leap from studying the effects on sequence variation to investigating the effects on gene expression will pave the way for an integrated conservation genomics approach towards assessing the relative impact of genetic and environmental threats (Ouborg et al., 2010), and especially for fruit trees the effects of domestication, assisted migration through cultivar dispersal, and anthropogenic influences in general.

However, conservation genomic studies in trees are still scarce, despite an extensive relevant discussion (Allendorf, Hohenlohe, & Luikart, 2010; Avise, 2010; Ekblom & Galindo, 2011; Funk et al., 2012, Hoban et al., 2013; Luikart et al., 2003; McMahon, Teeling, & Höglund, 2014; Ouborg et al., 2010), and in spite of reports indicating the vulnerability of in situ conservation due to climatic change (Schueler et al., 2014). Few recent studies exist. Genome-wide SNP data were used to infer demographic history and conservation units in *C. japonica* (Uchiyama, Miyamoto, Takahashi,

Watanabe, & Tsumura, 2014) and NGS-generated SNPs helped to uncover the species status of *Toona ciliate* (McPherson et al., 2013). SNPs of 149 candidate genes revealed population and inbreeding structure among *Pinus radiata* mainland California populations of conservation concern (Dillon et al., 2013). Whole-genome sequencing of *P. persica* have provided valuable insights into demographic history and local adaptation (Verde et al., 2013).

6. FUTURE PERSPECTIVES

Genomics present a novel insight for the structure, function and evolution of genes and genomes. Soon, it will be routine to 'paste on' the genomes of related genera and even families (Walbot, 2013). It will become feasible to sequence hundreds or even thousands of related genomes for thousands of genetic variants, in order to assess intraspecific and interspecific genomic diversity. With diminishing sequencing costs and increasing sequence capacity at an unprecedented rate, whole-genome resequencing will become routine. It will also be feasible to identify all the sequence variants that occur at moderate to high frequencies in any breeding population. A stage would be reached where a diagnostic marker could predict the phenotype of interest in a breeding (or even in a natural) population.

To date, the process of plant domestication has focused on securing specific traits that occurred at random, in nature or in the laboratory. Important traits that have been selected for are, e.g. easy harvest, high yield and low toxicity (Palmgren et al., 2014). By contrast, variants that compromise the hereditary basis of crop (and in many cases fruit tree or a highly bred and planted forest tree such as poplar) survival during environmental stresses, biotic and abiotic are rarely selected against. Thus, many survival traits may have been weakened or completely lost in cultivated species selected by humans. NGS technologies and data algorithms can now be used to detecting natural population or wild relative variants, important for survival. Once such genes have been identified, the next step would be re-wilding that would allow trees and plants not only to better utilize available resources in the environment and have higher nutritional value, but also to better resist diseases, pests, and weeds (Palmgren et al., 2014).

Our ability to understand natural selection, adaptation and speciation will be limited more by conceptual, analytical and technical weaknesses (such as the availability of appropriate material, sufficiently replicated across field sites and precise phenotyping; Grattapaglia & Kirst, 2008), than by the amount of molecular data produced. Tree genomics, slowly but surely are moving towards a paradigm shift and the genomization of tree genetics and breeding will lead to novel systems biology approach (Aravanopoulos, 2014).

REFERENCES

Abbott, A. G. (2014). Comparative genomics of key forest and fruit tree species: a strategic path to forest health improvement. In *Plant and animal genome XXII conference. Plant and animal genome, January 11–15, San Diego CA.*

Abril, N., Gion, J. M., Kerner, R., Müller-Starck, G., Cerrillo, R. M. N., Plomion, C., et al. (2011). Proteomics research on forest trees, the most recalcitrant and orphan plant species. *Phytochemistry, 72*, 1219–1242.

Abrouk, M., Murat, F., Pont, C., Messing, J., Jackson, S., Faraut, T., et al. (2010). Paleogenomics of plants: modern species synteny-based modelling of extinct ancestors. *Trends in Plant Science, 15*, 479–487.

Allendorf, F. W., Hohenlohe, P. A., & Luikart, G. (2010). Genomics and the future of conservation genetics. *Nature Reviews Genetics, 11*, 697–709.

Aravanopoulos, F. A. (2001). Genetic linkage analysis in *Salix eriocephala* suggests linkage conservation above the species level. *Forest Genetics, 8*(2), 145–149.

Aravanopoulos, F. A. (2010). The importance of the biodiversity of Hellenic forest and land-based ecosystems for Europe and the Mediterranean. In M. Ananiadou-Tzimopoulou (Ed.), *Proc. Conf. Aristotle University Environmental Council* (pp. 1–11). Thessaloniki: Aristotle University Press.

Aravanopoulos, F. A. (2011). Genetic monitoring in natural perennial plant populations. *Botany, 89*, 75–81.

Aravanopoulos, F. A. (2014). Genomics of trees. In K. G. Ramawat, J. M. Merillon, & M. R. Ahuja (Eds.), *Tree biotechnology* (pp. 514–557). Boca Raton: CRC Press.

Austerlitz, F., Mariette, S., Machon, N., Gouyon, P. H., & Godelle, B. (2000). Effects of colonization processes on genetic diversity: differences between annual plants and tree species. *Genetics, 154*, 1309–1321.

Avise, J. C. (2010). Perspective: conservation genetics enters the genomics era. *Conservation Genetics, 11*, 665–669.

Biswas, S., & Akey, J. M. (2006). Genomic insights into positive selection. *Trends in Genetics, 22*, 437–446.

Brown, G. R., Gill, G. P., Kuntz, R. J., Langley, C. H., & Neale, D. B. (2004). Nucleotide diversity and linkage disequilibrium in loblolly pine. *Proceedings of National Academy of Sciences of the United States of America, 101*, 15255–15260.

Buckler, E. S., Holland, J. B., Bradbury, P. J., Acharya, C. B., Brown, P. J., Browne, C., et al. (2009). The genetic architecture of maize flowering time. *Science, 325*, 714–718.

Cabrera, A., Rosyara, U. R., van der Knaap, E., De Franceschi, P., Sebolt, A., Sooriyapathirana, S. S., et al. (2012). Rosaceae conserved orthologous sequences marker polymorphism in sweet cherry germplasm and construction of a SNP-based Map. *Tree Genetics & Genomes, 8*, 237–247.

Cao, K., Zheng, Z., Wang, L., Liu, X., Zhu, G., Fang, W., et al. (2014). Comparative population genomics reveals the domestication history of the peach, *Prunus persica*, and human influences on perennial fruit crops. *Genome Biology, 15*, 415.

Cevik, V., Ryder, C. D., Popovich, A., Manning, K., King, G. J., & Seymour, G. B. (2010). A FRUITFULL-like gene is associated with genetic variation for fruit flesh firmness in apple (*Malus domestica* Borkh.). *Tree Genetics & Genomes, 6*, 271–279.

Chen, J., Källman, T., Ma, X., Gyllenstrand, N., Zaina, G., Morgante, M., et al. (2012). Disentangling the roles of history and local selection in shaping clinal variation of allele frequencies and gene expression in Norway spruce (*Picea abies*). *Genetics, 191*, 865–881.

Clauss, M. J., & Mitchell-Olds, T. (2006). Population genetic structure of *Arabidopsis lyrata* in Europe. *Molecular Ecology, 15*, 2753–2766.

De Franceschi, P., Stegmeir, T., Cabrera, A., van der Knaap, E., Rosyara, U. R., Sebolt, A. M., et al. (2013). Cell number regulator genes in *Prunus* provide candidate genes for the control of fruit size in sweet and sour cherry. *Molecular Breeding, 32*, 311–326.

Dillon, S. K., Nolan, M. F., Matter, P., Gapare, W. J., Bragg, J. G., & Southerton, S. G. (2013). Signatures of adaptation and genetic structure among the mainland populations of *Pinus radiata* (D. Don) inferred from SNP loci. *Tree Genetics & Genomes, 9*, 1447–1463.

Dvornyk, V., Sirviö, A., Mikkonen, M., & Savolainen, O. (2002). Low nucleotide diversity at the pal1 locus in the widely distributed *Pinus sylvestris*. *Molecular Biology and Evolution, 19*, 179–188.

Eckert, A. (2011). Population and conservation genomics of forest trees: seeing the forest for the trees. *BMC Proceedings, 5*(Suppl. 7), I2. http://www.biomedcentral.com/1753-6561/5/S7/I2.

Eckert, A. J., Bower, A. D., Wegrzyn, J. L., Pande, B., Jermstad, K. D., Krutovsky, K. V., et al. (2009). Association genetics of coastal Douglas fir (*Pseudotsuga menziesii* var. *menziesii*, Pinaceae). I. Cold-hardiness related traits. *Genetics, 182*, 1289–1302.

Ekblom, R., & Galindo, J. (2010). Applications of next generation sequencing in molecular ecology of non-model organisms. *Heredity, 107*, 1–15.

Ekblom, R., & Galindo, J. (2011). Applications of next generation sequencing in molecular ecology of non-model organisms. *Heredity, 107*, 1–15.

Emanuelli, F., Battilana, J., Costantini, L., Le Cunff, L., Boursiquot, J. M., This, P., et al. (2010). A candidate gene association study on muscat flavor in grapevine (*Vitis vinifera* L.). *BMC Plant Biology, 10*, 241.

Evans, L. M., Slavov, G. T., Rodgers-Melnick, E., Martin, J., Ranjan, P., Muchero, W., et al. (2014). Population genomics of *Populus trichocarpa* identifies signatures of selection and adaptive trait associations. *Nature Genetics, 46*, 1089–1096.

Eveno, E., Collada, C., Guevara, M. A., Léger, V., Soto, A., Díaz, L., et al. (2008). Contrasting patterns of selection at *Pinus pinaster* Ait. drought stress candidate genes as revealed by genetic differentiation analyses. *Molecular Biology and Evolution, 25*, 417–437.

Farjon, A. (2008). *A natural history of conifers*. Portland: Timber Press.

Fournier-Level, A., Le Cunff, L., Gomez, C., Doligez, A., Ageorges, A., Roux, C., et al. (2009). Quantitative genetic bases of anthocyanin variation in grape (*Vitis vinifera* L. ssp. *sativa*) berry: a quantitative trait locus to quantitative trait nucleotide integrated study. *Genetics, 183*, 1127–1139.

Funk, W. C., McKay, J. K., Hohenlohe, P. A., & Allendorf, F. W. (2012). Harnessing genomics for delineating conservation units. *Trends in Ecology and Evolution, 27*, 489–496.

Gailing, O., Vornam, B., Leinemann, L., & Finkeldey, R. (2009). Genetic and genomic approaches to assess adaptive genetic variation in plants: forest trees as a model. *Physiologia Plantarum, 137*, 509–519.

Ganopoulos, I., Aravanopoulos, F. A., Argiriou, A., Kalivas, A., & Tsaftaris, A. (2011). Is the genetic diversity of small scattered forest tree populations at the southern limits of their range more prone to stochastic events? A wild cherry case study by microsatellite-based markers. *Tree Genetics & Genomes, 7*, 1299–1313.

Ganopoulos, I., Aravanopoulos, F., Argiriou, A., & Tsaftaris, A. (2012). Genome and population dynamics under selection and neutrality: an example of *S*-allele diversity in wild cherry (*Prunus avium* L.). *Tree Genetics & Genomes, 8*, 1181–1190.

Ganopoulos, I. V., Aravanopoulos, F. A., & Tsaftaris, A. (2013). Genetic differentiation and gene flow between wild and cultivated *Prunus avium*: an analysis of molecular genetic evidence at a regional scale. *Plant Biosystems, 147*, 678–685.

Garcia-Lor, A., Curk, F., Snoussi-Trifa, H., Morillon, R., Ancillo, G., Luro, F., et al. (2013). A nuclear phylogenetic analysis: SNPs, indels and SSRs deliver new insights into the relationships in the 'true citrus fruit trees' group (Citrinae, Rutaceae) and the origin of cultivated species. *Annals of Botany, 111*, 1–19.

Goddard, M. E., & Hayes, B. J. (2009). Mapping genes for complex traits in domestic animals and their use in breeding programmes. *Nature Reviews Genetics, 10*, 381–391.

González-Martínez, S. C., Ersoz, E., Brown, G. R., Wheeler, N. C., & Neale, D. (2006). DNA sequence variation and selection of tag single-nucleotide polymorphisms at candidate genes for drought-stress response in *Pinus taeda* L. *Genetics, 172*, 1915–1926.

González-Martinez, S. C., Huber, D., Ersoz, E., Davis, J. M., & Neale, D. B. (2008). Association genetics in *Pinus taeda* L. II. Carbon isotope discrimination. *Heredity, 101*, 19–26.

González-Martínez, S. C., Krutovsky, K. V., & Neale, D. B. (2006). Forest tree population genomics and adaptive evolution. *New Phytologist, 170*, 227–238.

González-Martínez, S. C., Wheeler, N. C., Ersoz, E., Nelson, C. D., & Neale, D. B. (2007). Association genetics in *Pinus taeda* L. I. Wood property traits. *Genetics, 175*, 399–409.

Grattapaglia, D., & Kirst, M. (2008). *Eucalyptus* applied genomics: from gene sequences to breeding tools. *New Phytologist, 179*, 911–929.

Grattapaglia, D., Plomion, C., Kirst, M., & Sederoff, R. R. (2009). Genomics of growth traits in forest trees. *Current Opinion in Plant Biology, 12*, 148–156.

Grattapaglia, D., & Resende, M. D. V. (2011). Genomic selection in forest tree breeding. *Tree Genetics & Genomes, 7*, 241–255.

Graudal, L., Aravanopoulos, F., Bennadji, Z., Changtragoon, S., Fady, B., Kjær, E. D., et al. (2014). Global to local genetic diversity indicators of evolutionary potential in tree species within and outside forests. *Forest Ecology and Management, 333*, 35–51.

Grivet, D., Sebastiani, F., González-Martínez, S. C., & Vendramin, G. G. (2009). Patterns of polymorphism resulting from long-range colonization in the Mediterranean conifer Aleppo pine. *New Phytologist, 184*, 1016–1028.

Hall, D., Luquez, V., Garcia, V. M., St Onge, K. R., Jansson, S., & Ingvarsson, P. K. (2007). Adaptive population differentiation in phenology across a latitudinal gradient in European aspen (*Populus tremula*, L.): a comparison of neutral markers, candidate genes and phenotypic traits. *Evolution, 61*, 2849–2860.

Hamrick, J. L., Godt, M. J. W., & Sherman-Broyles, S. L. (1992). Factors influencing levels of genetic diversity in woody plant species. *New Forests, 6*, 95–124.

Hirschhorn, J. N., & Daly, M. J. (2005). Genome-wide association studies for common diseases and complex traits. *Nature Reviews Genetics, 6*, 95–108.

Hoban, S. M., Hauffe, H. C., Pérez-Espona, S., Arntzen, J. W., Bertorelle, G., Bryja, J., et al. (2013). Bringing genetic diversity to the forefront of conservation policy and management. *Conservation Genetics Resources, 5*, 593–598.

Houel, C., Bounon, R., Chaïb, J., Guichard, C., Péros, J. P., Bacilieri, R., et al. (2010). Patterns of sequence polymorphism in the fleshless berry locus in cultivated and wild *Vitis vinifera* accessions. *BMC Plant Biology, 10*, 284.

Ingvarsson, P. K. (2005). Nucleotide polymorphism and linkage disequilibrium within and among natural populations of European aspen (*Populus tremula* L., Salicaceae). *Genetics, 169*, 945–953.

Ingvarsson, P. K., García, M. V., Hall, D., Luquez, V., & Jansson, S. (2006). Clinal variation in *phyB2*, a candidate gene for day-length-induced growth cessation and bud set, across a latitudinal gradient in European aspen (*Populus tremula*). *Genetics, 172*, 1845–1853.

IPGI. (2011). *International plant genome initiative.* http://www.rosaceae.org/species/prunus/prunus_persica.

Jansson, S., & Douglas, C. J. (2007). *Populus*: a model system for plant biology. *Annual Review of Plant Biology, 58*, 435–458.

Kenrick, P. (1999). The family tree flowers. *Nature, 402*, 358–359.

Khan, M. A., Zhao, Y., & Korban, S. S. (2013). Identification of genetic loci associated with fire blight resistance in *Malus* through combined use of QTL and association mapping. *Physiologia Plantarum, 148*, 344–353.

Klagges, C., Campoy, J. A., Quero-Garcia, J., Guzmán, A., Mansur, L., Gratacós, E., et al. (2013). Construction and comparative analyses of highly dense linkage maps of two sweet cherry intra-specific progenies of commercial cultivars. *PLoS One, 8*, e54743.

Kort, H., Vandepitte, K., Bruun, H. H., Closset-Kopp, D., Honnay, O., & Mergeay, J. (2014). Landscape genomics and a common garden trial reveal adaptive differentiation to temperature across Europe in the tree species *Alnus glutinosa*. *Molecular Ecology, 23*, 4709–4721.

Kremer, A., & Le Corre, V. (2011). Decoupling of differentiation between traits and their underlying genes in response to divergent selection. *Heredity, 108*, 375–385.

Krutovsky, K. V., & Neale, D. B. (2005). Nucleotide diversity and linkage disequilibrium in cold-hardiness and wood quality-related candidate genes in Douglas fir. *Genetics, 171*, 2029–2041.

Kumar, S., Chagné, D., Bink, M. C. A. M., Volz, R. K., Whitworth, C., & Carlisle, C. (2012). Genomic selection for fruit quality traits in apple (*Malus×domestica* Borkh.). *PLoS One*, 7, e36674.

Lawrie, D. S., & Petrov, D. A. (2014). Comparative population genomics: power and principles for the inference of functionality. *Trends in Genetics, 30*, 133–139.

Le Corre, V., & Kremer, A. (2003). Genetic variability at neutral markers, quantitative trait loci and trait in a subdivided population under selection. *Genetics, 164*, 1205–1219.

Le Dantec, L., Cardinet, G., Bonet, J., Fouché, M., Boudehri, K., Monfort, A., et al. (2010). Development and mapping of peach candidate genes involved in fruit quality and their transferability and potential use in other Rosaceae species. *Tree Genetics & Genomes, 6*, 995–1012.

Luikart, G., England, P. R., Tallmon, D., Jordan, S., & Taberlet, P. (2003). The power and promise of population genomics: from genotyping to genome typing. *Nature Reviews Genetics, 4*, 981–994.

Mackay, J., Dean, J. F., Plomion, C., Peterson, D. G., Cánovas, F. M., Pavy, N., et al. (2012). Towards decoding the conifer giga-genome. *Plant Molecular Biology, 80*, 555–569.

Magallo, S. A., & Sanderson, M. J. (2005). Angiosperm divergence times: the effect of genes, codon positions and time constrains. *Evolution, 59*, 1653–1670.

Malliarou, E., & Aravanopoulos, F. A. (2012). A comparative analysis of genetic diversity of tree species in Greece and Europe. In *Proc. 10th Pan-Hellenic conf., Hellenic Sci. Soc. Plant Genet. Breeding* (pp. 413–418). Thessaloniki.

Manel, S., & Holderegger, R. (2013). Ten years of landscape genetics. *Trends in Ecology and Evolution, 28*, 614–621.

Manescu, C., Hamamouch, N., Maios, C., Harfouche, A., Doulis, A. G., & Aravanopoulos, F. A. (2011). Linkage mapping of the Mediterranean cypress, *Cupressus sempervirens*, based on molecular and morphological markers. *Genetics and Molecular Research, 10*, 1891–1909.

Mardis, E. R. (2013). Next-generation sequencing platforms. *Annual Review of Analytical Chemistry, 6*, 287–303.

McKown, A. D., Klápště, J., Guy, R. D., Geraldes, A., Porth, I., Hannemann, J., et al. (2014). Genome-wide association implicates numerous genes underlying ecological trait variation in natural populations of *Populus trichocarpa*. *New Phytologist, 203*, 535–553.

McMahon, B. J., Teeling, E. C., & Höglund, J. (2014). How and why should we implement genomics into conservation? *Evolutionary Applications, 7*, 999–1007.

McPherson, H., van der Merwe, M., Delaney, S. K., Edwards, M. A., Henry, R. J., McIntosh, E., et al. (2013). Capturing chloroplast variation for molecular ecology studies: a simple next generation sequencing approach applied to a rainforest tree. *BMC Ecology, 13*, 8.

Miller, A. J., Matasci, N., Schwaninger, H., Aradhya, M. K., Prins, B., Zhong, G. Y., et al. (2013). *Vitis* phylogenomics: hybridization intensities from a SNP array outperform genotype calls. *PLoS One, 8*, e78680.

Müller, T., Freund, F., Wildhagen, H., & Schmid, K. J. (2015). Targeted re-sequencing of five Douglas-fir provenances reveals population structure and putative target genes of positive selection. *Tree Genetics & Genomes, 11*, 1–17.

Myles, S., Boyko, A. R., Owens, C. L., Brown, P. J., Grassi, F., Aradhya, M. K., et al. (2011). Genetic structure and domestication history of the grape. *Proceedings of National Academy of Sciences of the United States of America, 108*, 3530–3535.

Narum, S. R., Buerkle, C. A., Davey, J. W., Miller, M. R., & Hohenlohe, P. A. (2013). Genotyping-by-sequencing in ecological and conservation genomics. *Molecular Ecology, 22*, 2841–2847.

Neale, D. B. (2007). Genomics to tree breeding and forest health. *Current Opinion in Genetics and Development, 17*, 1–6.

Neale, D. B., & Ingvarsson, P. K. (2008). Population, quantitative and comparative genomics of adaptation in forest trees. *Current Opinion in Plant Biology, 11*, 149–155.

Neale, D. B., & Kremer, A. (2011). Forest tree genomics, growing resources and applications. *Nature Reviews Genetics, 12*, 111–122.

Neale, D. B., & Savolainen, O. (2004). Association genetics of complex traits in conifers. *Trends in Plant Science, 9*, 325–330.

Nordborg, M., Hu, T. T., Ishino, Y., Jhaveri, J., Toomajian, C., Zheng, H., et al. (2005). The pattern of polymorphism in *Arabidopsis thaliana*. *PLoS Biology, 3*, e196.

Ouborg, N., Pertoldi, C., Loeschcke, V., Bijlsma, R. K., & Hedrick, P. W. (2010). Conservation genetics in transition to conservation genomics. *Trends in Genetics, 26*, 177–187.

Palmgren, M. G., Edenbrandt, A. K., Vedel, S. E., Andersen, M. M., Landes, X., Østerberg, J. T., et al. (2014). Are we ready for back-to-nature crop breeding? *Trends in Plant Science, 20*, 155–164.

Parchman, T. L., Gompert, Z., Mudge, J., Schilkey, F. D., Benkman, C. W., & Buerkle, C. (2012). Genome-wide association genetics of an adaptive trait in lodgepole pine. *Molecular Ecology, 21*, 2991–3005.

Plomion, C., Chagné, D., Pot, D., Kumar, S., Wilcox, P. L., Burdon, R. D., et al. (2007). Pines. In *Forest trees* (pp. 29–92). Berlin - Heidelberg: Springer.

Prunier, J., Laroche, J., Beaulieu, J., & Bousquet, J. (2011). Scanning the genome for gene SNPs related to climate adaptation and estimating selection at the molecular level in boreal black spruce. *Molecular Ecology, 20*, 1702–1716.

Rafalski, A., & Morgante, M. (2004). Corn and humans, recombination and linkage disequilibrium in two genomes of similar size. *Trends in Genetics, 20*, 103–111.

Ramawat, K. G., Mérillon, J. M., & Ahuja, M. R. (Eds.). (2014). *Tree biotechnology*. CRC Press.

Salse, J. (2012). In silico archeogenomics unveils modern plant genome organization, regulation and evolution. *Current Opinion in Plant Biology, 15*, 122–130.

Savolainen, O., Lascoux, M., & Merilä, J. (2013). Ecological genomics of local adaptation. *Nature Reviews Genetics, 14*, 807–820.

Savolainen, O., & Pyhajarvi, T. (2007). Genomic diversity in forest trees. *Current Opinion in Plant Biology, 10*, 162–167.

Savolainen, O., Pyhäjärvi, T., & Knürr, T. (2007). Gene flow and local adaptation in trees. *Annual Review of Ecology, Evolution, and Systematics*, 595–619.

Schoville, S. D., Bonin, A., François, O., Lobreaux, S., Melodelima, C., & Manel, S. (2012). Adaptive genetic variation on the landscape: methods and cases. *Annual Review of Ecology, Evolution, and Systematics, 43*, 23–43.

Schueler, S., Falk, W., Koskela, J., Lefèvre, F., Bozzano, M., Hubert, J., et al. (2014). Vulnerability of dynamic genetic conservation units of forest trees in Europe to climate change. *Global Change Biology, 20*, 1498–1511.

Sork, V. L., Aitken, S. N., Dyer, R. J., Eckert, A. J., Legendre, P., & Neale, D. B. (2013). Putting the landscape into the genomics of trees: approaches for understanding local adaptation and population responses to changing climate. *Tree Genetics & Genomes, 9*, 901–911.

Thumma, B. R., Nolan, M. F., Evans, R., & Moran, G. F. (2005). Polymorphisms in cinnamoyl CoA reductase (CCR) are associated with variation in microfibril angle in *Eucalyptus* spp. *Genetics, 171*, 1257–1265.

Tsumura, Y., Kado, T., Takahashi, T., Tani, N., Ujino-Ihara, T., & Iwata, H. (2007). Genome scan to detect genetic structure and adaptive genes of natural populations of *Cryptomeria japonica*. *Genetics, 176*, 2393–2403.

Tsumura, Y., Uchiyama, K., Moriguchi, Y., Kimura, K. M., Ueno, S., & Ujino-Ihara, T. (2014). Genetic differentiation and evolutionary adaptation in *Cryptomeria japonica*. *G3: Genes, Genomes, Genetics, 10*, 1257–1270.

Tuskan, G. A., Difazio, S., Jansson, S., Bohlmann, J., Grigoriev, I., Hellsten, U., et al. (2006). The genome of black cottonwood, *Populus trichocarpa* (Torr. & Gray). *Science, 313*, 1596–1604.

Uchiyama, K., Miyamoto, N., Takahashi, M., Watanabe, A., & Tsumura, Y. (2014). Population genetic structure and the effect of historical human activity on the genetic variability of *Cryptomeria japonica* core collection, in Japan. *Tree Genetics & Genomes, 10*, 1257–1270.

Velasco, R., Zharkikh, A., Affourtit, J., Dhingra, A., Cestaro, A., Kalyanaraman, A., et al. (2010). The genome of the domesticated apple (*Malus x domestica* Borkh.). *Nature Genetics, 42*, 833–839.

Verde, I., Abbott, A. G., Scalabrin, S., Jung, S., Shu, S., Marroni, F., et al. (2013). The high-quality draft genome of peach (*Prunus persica*) identifies unique patterns of genetic diversity, domestication and genome evolution. *Nature Genetics, 45*, 487–494.

Visendi, P., Batley, J., & Edwards, D. (2014). Next generation sequencing and germplasm resources. In *Genomics of plant genetic resources* (pp. 369–390). Netherlands: Springer.

Visscher, P. M. (2008). Sizing up human height variation. *Nature Genetics, 40*, 489–490.

Walbot, V. (2013). Open questions: reflections on plant development and genetics. *BMC Biology, 11*, 25.

Wang, R. L., Stec, A., Hey, J., Lukens, L., & Doebley, J. (1999). The limits of selection during maize domestication. *Nature, 398*, 236–239.

White, T. L., Adams, W. T., & Neale, D. B. (2007). *Forest genetics*. Wallinford: CABI Publ.

Willyard, A., Syring, J., Gernandt, D. S., Liston, A., & Cronn, R. (2007). Fossil calibration of molecular divergence infers a moderate mutation rate and recent radiations for *Pinus*. *Molecular Biology and Evolution, 24*, 90–101.

Wright, S. I., & Gaut, B. S. (2005). Molecular population genetics and the search for adaptive evolution in plants. *Molecular Biology and Evolution, 22*, 506–519.

Wu, G. A., Prochnik, S., Jenkins, J., Salse, J., Hellsten, U., Murat, F., et al. (2014). Sequencing of diverse mandarin, pummelo and orange genomes reveals complex history of admixture during citrus domestication. *Nature Biotechnology, 32*, 656–662.

Yu, Q., Li, B., Nelson, C. D., McKeand, S. E., Batista, V. B., & Mullin, T. J. (2006). Association of the cad-n1 allele with increased stem growth and wood density in full-sib families of loblolly pine. *Tree Genetics & Genomes, 2*, 98–108.

Zohary, D., & Spiegel-Roy, P. (1975). Beginnings of fruit growing in the old world. *Science, 187*(4174), 319–327.

Zoldos, V., Papes, D., Brown, S. C., Panaud, O., & Siljak-Yakovlev, S. (1998). Genome size and base composition of seven *Quercus* species: inter- and intra-population variation. *Genome, 41*, 162–168.

CHAPTER FIVE

Genetics and Genomics of Tree Architecture

Evelyne Costes*[,1], Jean-Marc Gion[§]
*INRA, UMR AGAP, CIRAD-INRA-SupAgro, AFEF Team (Architecture et Fonctionnement des Espèces Fruitières), Montpellier, France
[§]CIRAD, UMR AGAP, UMR BIOGECO, Cestas, France
[1]Corresponding author: E-mail: costes@supagro.inra.fr

Contents

Abstract

Even though the repetitive and decentralized nature of plant organogenesis confers a large plasticity to plants, their architecture is highly organized throughout ontogeny and at different scales. Growth, branching and flowering constitute the main processes leading to plant architectures and are key determinants of plant productivity, whatever the target product (biomass, flowers, seeds or fruits). Their genetic and genomic study is thus of great interest for breeding purpose. However, progresses have been limited by many constraints resulting from the large size and long-live of trees and therefore the difficult to assess phenotypes on large numbers of individuals for exploring both the genetic effect or genetic by environment interaction. Despite these limitations, significant

Advances in Botanical Research, Volume 74
ISSN 0065-2296
http://dx.doi.org/10.1016/bs.abr.2015.05.001

advances have been recently obtained in quantitative genetics studies performed on architectural traits in forest and fruit trees, and on the physiological and molecular comprehension of mechanisms underlying plant architectural edification. Moreover, tools and methods for describing plant architectures and for exploring genetic and molecular controls have been improved, leading to new perspectives in terms of high-throughput genotyping, phenotyping and genomic studies. In this chapter, we review the genetic, physiological and molecular analyses that have been performed in the last years on tree architecture, with a particular focus on advances realized on tree crops.

1. INTRODUCTION

Plant architecture results from the coordinated development of organs during their ontogeny. Architectural plant categories, or 'architectural models', which have been highlighted by Hallé, Oldeman and Tomlinson (1978) result from the behaviours of the meristems and phytomers they produce. Even though the repetitive and decentralized nature of plant organogenesis confers a large plasticity to plants, their architecture is highly organized throughout ontogeny and at different scales: the phytomers (including the leaf, its subtending internode and the attached axillary meristem (AxM)) which appear in succession within axes, the relative positioning of laterals along main axes at successive branching orders, as well as the reiteration process (Barthélémy & Caraglio, 2007). Architectural models can thus be viewed as an abstraction for the deterministic genetic blueprint on which the construction of the plant is based (Sussex & Kerk, 2001). Under optimal growth conditions, the development and architecture of the plant conform strictly to the model, but environmental conditions and interactions with other organisms may result in deviations from the model. Thus, most developmental morphologies of plants represent a balance between deterministic genetic and opportunistic environmental events. For each species, this balance presides to the coordinated development of organs at different levels of organization, phytomers, growth units, axes.

Considering perennial crops such as fruit and forest trees, environmental factors that affect growth and development are supplemented by age-induced morphological changes throughout their life. Even though tree structure results from repetitive processes (White, 1979), the successive growth units are not entirely similar because of morphogenetic gradients during tree ontogeny (Barthélémy & Caraglio, 2007; Costes, Sinoquet, Kelner, & Godin, 2003; Renton, Guédon, Godin, & Costes, 2006). The observed tree phenotype is thus plastic and complex, since it results from genetic, environmental, and ontogenetic factors and their interactions.

Understanding the genetic mechanisms underlying this phenotypic plasticity is of great interest in plant science. Indeed, growth, branching and flowering constitute the main processes leading to plant architectures and are key determinants of plant productivity and adaptation to environmental changes, whatever the target product (biomass, flowers, seeds or fruits). Moreover, in the specific case of fruit trees, it is of major importance to account for interactions with cultural practices when studying tree architecture. Indeed, yield performance results from the 'orchard system puzzle' which integrates a number of tree manipulations carried out at different time steps (Barritt, 2000; Costes, Lauri, & Regnard, 2006). Most of these manipulations which modify tree architecture aim at (1) optimizing light capture at both orchard and within-tree scales; (2) improving biomass partition to fruiting shoots and (3) avoiding competition with vegetative sinks (Fumey, Lauri, Guédon, Godin, & Costes, 2011; Miller, Broom, Thorp, & Barnett, 2001).

The main constraints for studying tree architecture are their great size and long-live. Indeed, in genetics, the assessment of phenotypes for a large number of individuals is mandatory to explore both the genetic effect or genetic by environment interaction. For organisms such as forest and fruit trees, genetic approaches on plant architecture have often been limited by the number of phenotypes, over temporal and spatial scales. Despite these limitations, significant advances have been recently obtained in quantitative genetics studies performed on architectural traits in forest and fruit trees, and on the physiological and molecular comprehension of mechanisms underlying plant architectural edification. In addition, both the molecular tools and the methods for describing plant architectures have been improved, leading to new perspectives in terms of high-throughput genotyping, phenotyping and for the exploration of candidate genes. In this chapter, we will briefly review the genetic, physiological and molecular analyses that have been performed in the last years to study meristem and axis behaviours, with a particular focus on advances realized on tree crops.

2. HIGHLIGHTING THE GENETIC CONTROL OF TREE ARCHITECTURE: FROM PLANT MATERIAL TO NEW PHENOTYPING METHODS

2.1 Evidences of Genetic Determinism of Tree Architecture Through Remarkable Mutants of Tree Shape

The first evidence for a genetic control in tree architecture has been provided by the identification of genetic variants, known as fastigiated forms, or mutants in different forest and fruit species. Fastigiated forms have been identified in *Cupressus sempervirens* (Barthélémy & Caraglio, 2007; Figure 1), in *Quercus* (*Quercus robur* and *Quercus pedunculata*, var. fastigiata) and *Populus* (*Populus nigra* var. italic Muenchh.) which are commonly used in horticulture. A natural mutant of *Hevea brasiliensis* has been evaluated

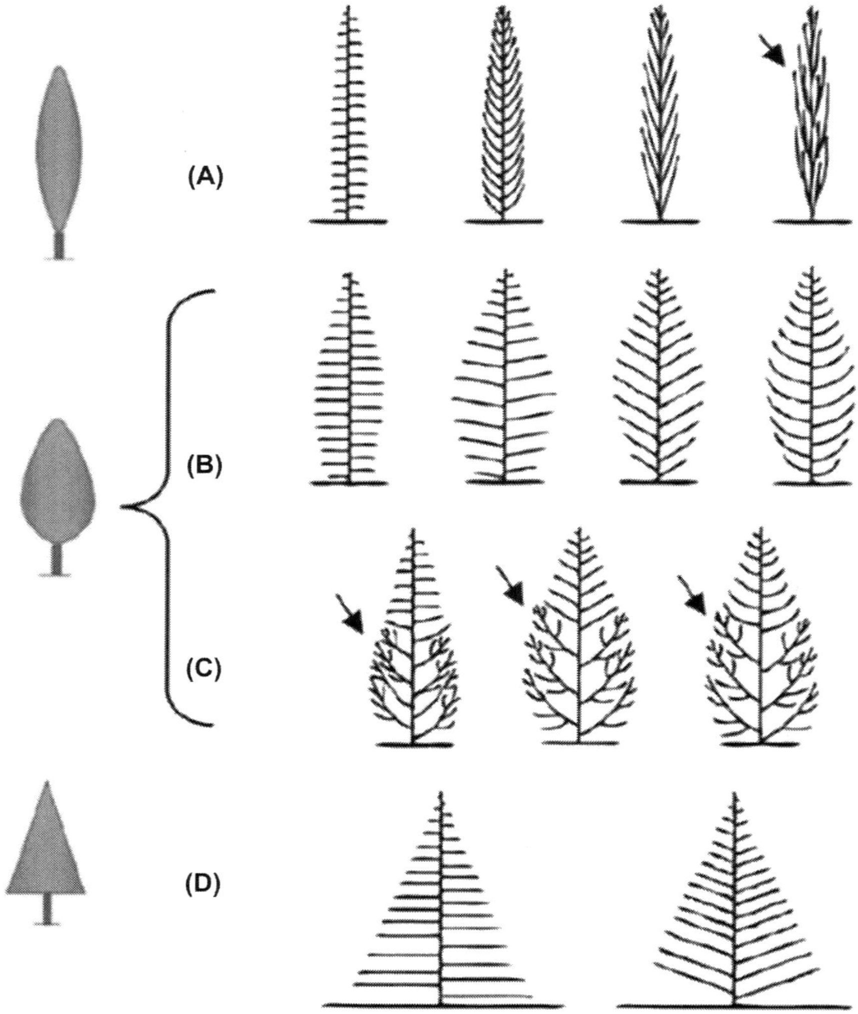

Figure 1 Different architectural types of *Cupressus sempervirens*: (A) fastigiate, (B and C) Intermediate and (D) horizontal crown shape groups. *From Barthélémy and Caraglio (2007)*. (See colour plate)

showing distinct morphological variation within the crown which has been qualified as compact morphotype (Gireesh & Mydin, 2014). In *Picea abies*, the acrocona mutant has been characterized by a dwarf phenotype and early cone formation on main shoots and branches (Fladung, Tusch, Markussen, & Ziegenhagen, 1999). In Scots pine and Norway spruce, the inheritance of crown shape from narrow-crowned parents could be explained by the effect of single dominant gene (Karki & Tigerstedt, 1985). Inter-clonal variation has also been observed in crown structure for loblolly pine and slash pine (Emhart, Martin, White, & Huber, 2007), *Eucalyptus grandis* (Lambeth, Endo, & Wright, 1994) and *Populus* (Ceulemans, Stettler, Hinckley, Isebrands, & Heilman, 1990). Similarly, in fruit trees, natural mutants have been identified such as 'Wijcick McIntosh', a sport of 'McIntosh' exhibiting a columnar compact growth habit (Fisher, 1970). In peach, both weeping (Monet, Bastard, & Gibault, 1988) and fastigiated mutant with pronounced vertical growth of branches (Scorza, Miller, Glenn, & Tworkoski, 2006) have been described. The genetic origin of these mutations has been confirmed through crosses produced for breeding perspective, for instance, by using a columnar parent in the apple tree (Kenis & Keulemans, 2007; Petersen & Krost, 2013). However, the mechanisms involved have been rarely deciphered (see below for the review of most recent results on columnar apple and weeping peach trees).

2.2 Genetic Determinism of Tree Architecture

More generally, a range of tree shapes can be observed in most forest and fruit tree species. In *C. sempervirens*, several forms ranking from fastigiated, horizontal and intermediate are well known (Figure 1; Barthélémy & Caraglio, 2007). In apple, peach and apricot, varieties have been classified depending on their shapes ranging from erected to weeping (Costes et al., 2006; Lespinasse, 1992; Scorza, 1984). Barthélémy and Caraglio (2007) have proposed that in such cases, the observed phenotypic variability could be explained by the different expressions of several basic morphological and geometrical features such as (1) the relative main stem/branch length, (2) the straightness or gradual straightening-up status of the branches, (3) the insertion angle, (4) homogeneity or heterogeneity of branch types within a single tree and (5) the occurrence, early manifestation and importance of the reiteration process.

Other proofs have been brought by more detailed analysis of the genetic determinism of tree architecture done in general for classical growth traits such as height and diameter, easily assessed on a large number of individuals from several progeny tests. Over the last decades, many studies have analyzed the genetic determinism of tree growth and branching in forest (Bradshaw & Stettler, 1995; Plomion, Durel, & O'Malley, 1996; Scotti-Saintagne et al., 2004; Wu & Stettler, 1996, 1998) and fruit species such as apple (Conner, Brown, & Weeden, 1998; Kenis & Keulemans, 2007; Liebhard, Kellerhals, Pfammatter, Jertmini, & Gessler, 2003; Segura, Cilas, Laurens, & Costes, 2006). In forest trees, from gymnosperms to angiosperms, numerous of results have been reported about the variance components of growth traits and their low to medium heritabilities (Hannrup et al., 2004; Kube, Raymond, & Banham, 2001; Lepoittevin et al., 2011). Consistently, in apple, basic morphological traits such as basis diameter and trunk length which are usually measured to characterize the tree 'vigor' have been shown to have medium to low heritability values (Durel, Laurens, Fouillet, & Lespinasse, 1998; Liebhard et al., 2003; Segura et al., 2006; Watkins & Spangelo, 1970).

The relative importance of genetic, genetic by environment interaction and stochastic effects controlling the metamer's development, and finally the form of tree's crown, has been shown to differ depending on sylleptic or proleptic branches (these terms refer to AxM development which may develop into a lateral axis immediately after AxM initiation or may remain dormant; they also refer to as immediate or delayed branching, respectively) (Diao, De Reffye, Lei, Guo, & Letort, 2012; Wu & Hinckley, 2001). In poplar, the study of branching phenotypic plasticity revealed that sylleptic branches played a major role in tree architecture (Marron, Bastien, Sabatti, Taylor, & Ceulemans, 2006; Wu & Hinckley, 2001). Very similar findings have been highlighted in a study on an apple tree F1 segregating population in which phenotypes were decomposed in elementary variables related to primary growth and branching. Based on heritability values and low correlations between traits, a selection of variables has been proposed, combining topological and geometric traits measured on both trunks and long axillary sylleptic shoots (Segura, Denancé, Durel, & Costes, 2007). More recently, based on a similar strategy, medium to high broad sense heritability have also been reported in olive tree form, whereas reproductive traits were highly heritable (Sadok et al., 2012).

All these studies have brought evidences for a genetic control for topological, geometric and reproductive traits, with high within-tree variability. The evolution of the effect of genetic determinants at different developmental stages from juvenile phase to more advanced stages age has also been characterized even though only few studies have reported the genetic determinism of several morphological and topological attributes of the tree architecture over time. In Douglas-Fir, St Clair (1993) showed a strong genetic variation for tree crown attributes indicating that growing performances were linked to investment in photosynthetic machinery (leaf and branch biomass) with a distribution over a great vertical distance. In poplar, Wu and Stettler (1994) reported a genetic determinism for compartments at a thinner scale, showing a strong genetic basis for leaf size, sylleptic branches (length and number) and growth rate that change with tree age. They also reported strong genotype by environment interactions for branch and canopy traits (Wu & Stettler, 1998). In apple tree, tree architectural plasticity has been dissected into genetic, ontogenetic and environmental effects over the first 4 year of growth. Mixed linear models of repeated data were estimated, considering both spatial (i.e. different axis types) and temporal repetitions (i.e. successive ages of trees) on traits related to both growth and branching processes of trees planted in a staggered-start design. Significant genotype effect and interactions between genotype and year and/or age were found (Segura, Cilas, & Costes, 2008). It also revealed that a higher genetic effect was observed during the first year of tree development, resulting likely from a larger variance of most traits in the early years. This result contrasted with that obtained in olive tree in which genetic effect was revealed at the tree periphery only, after 2–3 years of growth (Sadok et al., 2012). Such discrepancy could result from different propagation mode between the species, grafting and cutting respectively. It also suggests that accounting for tree biology and usual orchard practices is highly recommended when studying genetic effects on complex tree structures.

3. GENETICS AND GENOMICS OF TREE GROWTH AND ARCHITECTURAL TRAITS: A FIELD IN FAST EVOLUTION

Heritability studies have been progressively relayed by quantitative trait loci (QTLs) analyses that have allowed the identification of many genomic regions involved in the control of architectural traits. Using simple growth traits, QTLs have been identified controlling the tree shape as, for instance, on *Populus* (Bradshaw & Stettler, 1995; Wu, 1998), *Eucalyptus* (Verhaegen et al., 1997) and apple tree (Segura et al., 2007; Segura, Durel, & Costes, 2009), with several studies dedicated to QTL identification for columnar phenotypes (Kenis & Keulemans, 2007; Tian, Wang, Zhang, James, & Dai, 2005) and dwarfing traits induced by rootstock (see Foster et al. 2015 and references within). Some transgenic experiments in poplar have also reported that particular genes could affect the crown architecture (as reviewed by Dubouzet, Strabala, and Wagner (2013)).

The accessibility to genomics resources has increased rapidly during the last decade thanks to next generation sequencing (NGS) and high-throughput genotyping technologies. In fruit trees, several reference genomes have been published in the past 5 years, for 3 main species in the Rosaceae family, apple (Velasco et al., 2010), peach (The IPGI et al., 2013) and pear (Chagné et al., 2014), but also on tropical or subtropical fruit species such as *Theobroma cacao* (Argout et al., 2011) or *Citrus sinensis* (Xu et al., 2013). Other reference genomes are under construction and should be able soon, in particular for olive or avocado tree. In forest trees, few reference genomes are available in *Populus* (Tuskan et al., 2006), and more recently loblolly pine (Zimin et al., 2014) and *Eucalyptus grandis* (Myburg et al., 2014). In all species, such resource allows a fine description of both the genome structure (genome features, number and nature of genes) and the associated nucleotidic variability (SNP, insertions/deletions) within genes and between genes, along each chromosome.

Moreover, the annotation of thousands of genes in fruit and forest trees allowed their analysis under different environmental conditions, developmental stages and for different levels of plant organization (tissues, organs...) (Table 1). Several gene families potentially involved in tree growth and tree shape variability have been identified. For instance, exhaustive inventories have been performed for genes related to auxin pathways (ARF, AUX/IAA and TIR) in the 'Golden Delicious' genome (Devoghalaere et al., 2012) or for the genes of the GRAS family in *Prunus mume* (Lu, Wang, Xu, Sun, & Zhang, 2015). The expression profiles of many major genes for tree growth have also been described. As an example, expression profiles of secondary cell wall-related genes implicated in cellulose and xylan biosynthesis have been studied in shoot tips, young and mature leaves, floral buds, roots and wood forming tissues of *Eucalyptus grandis* (Myburg et al., 2014). Gene atlas dedicated to forest trees are also available allowing transcriptional and proteomic profiling in several contexts. The molecular plasticity of shoot apices of eucalyptus was studied in response to water deficit using NGS which provided extensive transcriptome coverage (Villar et al., 2011). In oak, a pyrosequencing strategy to study the bud dormancy induction and release generated 6471 contigs of differentially expressed genes (Ueno et al., 2013) and allowed the detection of several expressional candidate genes. Candidate gene approaches have also been conducted in maritime pine to reveal expression variation of cuticular-related genes in needles submitted to drought stress,

Table 1 List of Quantitative Trait Loci Related to Tree Architecture Identified in Either Fruit or Forest Trees

Trait	Species	Material	LG	References
Whole Tree				
Tree height	Apple	Fiesta × Discovery F1	3, 5, 8	Liebhard et al. (2003)
		Starkrimson × Granny S. F1	14	Segura et al. (2009)
Trunk basal diameter	*Id.*	*Id.*	1, 7	*Id.*
Id.	Apricot	Harostar × Rouge de Mauve F1	1	Socquet-Juglard et al. (2013)
Tree height and circumference	*Eucalyptus*	*Eucalyptus urophylla* × *Eucalyptus grandis* full-sibs	5, 6, 8, 10	Bartholomé et al. (2013)
Tree height	Oak	*Quercus robur*, full-sib family	3, 5, 6, 10	Scotti-Saintagne et al. (2004)
Id.	White spruce	2 full-sibs *Picea glauca*	I, III	Pelgas et al. (2011)
Id.	Pine	*Pinus pinaster* F2 seedlings	1, 2, 3, 4, 5, 6, 7, 9, 11, 12	Plomion et al. (1996)
Branching				
Number of laterals	Apple	Starkrimson × Granny S. F1	4, 13	Segura et al. (2009)
	Apricot	Harostar × Rouge de Mauve F1	6	Socquet-Juglard et al. (2013)
Number of latent buds	Apple	Starkrimson × Granny S. F1	17	Segura et al. (2009)
% Branching nodes		*Id.*	4, 13, 17	*Id.*
Sylleptic branching		*Id.*	1, 9, 13, 16	*Id.*

(Continued)

Table 1 List of Quantitative Trait Loci Related to Tree Architecture Identified in Either Fruit or Forest Trees—cont'd

Trait	Species	Material	LG	References
Id.	Poplar	*Populus trichocarpa* × *Populus deltoïdes* F1 hybrids	1, 2	Bradshaw and Stettler (1995)
Columnar growth habit	Apple	F1s from *Malus* × *domestica* Borkh	10	Tian et al. (2005) and Moriya et al. (2012)
Shoot Growth				
Stem growth and form	*Eucalyptus*	*E. urophylla* × *E. grandis* full-sibs	1, 2, 3, 5, 8, 11	Verhaegen et al. (1997)
Internode length	Apple	Starkrimson × Granny S. F1	3, 6, 7, 10, 14, 15, 16	Segura et al. (2007, 2009)
		Wijcik × NY75441-58 F1		Conner et al. (1998)
		Spur Fuji × Telamon F1	10	Tian et al. (2005)
		F1 progenies		Moriya et al. (2012)
		Telamon × Braeburn F1	3, 14	Kenis and Keulemans (2007)
Phenology				
Bud burst (vegetative and floral)	Apple	Golden D. × Anna F1	9	Van Dyck et al. (2010)
		Sharpe's early × Anna F1		*Id.*
		Starkrimson × Granny S. F1	2, 3, 5, 6	Celton et al. (2011)
		Id.	6, 8	Segura et al. (2007)
		Wijcik × NY75441-58 F1	8	Conner et al. (1998)
		X3263 × Belrène F1	1, 9, 10, 15	Celton et al. (2011)
Flowering date	Peach	F2: BC1 × Zephyr	2, 3, 4, 6, 7	Dirlewanger et al. (2012)

Bud burst	*Salix*	*Salix viminalis* × *Salix schwerinii* F1	1, 3, 10, 15, 21	Tsarouhas et al. (2003)
Id.	Oak	*Q. robur*, full-sib family	1, 2, 3, 4, 5, 6, 7, 9, 10, 11, 12	Scotti-Saintagne et al. (2004)
Id.	Poplar	*P. trichocarpa* × *P. deltoïdes* F1 hybrids	5	Bradshaw and Stettler (1995)
Id. & Bud set	Poplar	*P. trichocarpa* × *P. deltoïdes* F1 hybrids	D, Z, H, F, PY, J	Frewen et al. (2000)
Id. & Bud set	White spruce	2 full-sibs picea glauca	I, III	Pelgas et al. (2011)
Id. & leaf senescence	*Salix*	*S. viminalis* × *S. schwerinii* F1		Ghelardini et al. (2014)
Chilling requirements, heat requirements	Almond	R1000 × Desmayo Largueta F1	1, 2, 3, 4, 7	Sánchez-Pérez et al. (2012)
	Apricot			
Flowering date, Chilling and heat requirements	Cherry tree	*Prunus avium* L. F1	4	Castede et al. (2014)
Fruiting Behaviour				
Yield inflorescences fruits	Apple	Starkrimson × Granny S.	1, 8, 15 1, 5, 8, 11	Guitton et al. (2012)
Id.	Olive	Oliviere × Arbequina	12	Sadok et al. (2013)
Fruit set	Apple	X3263 × Belrène	10, 12	Celton et al. (2014)
Maturity date	Peach	F2: BC1 × 'Zephyr'	4	Dirlewanger et al. (2012)

one of the major environmental stresses influencing tree growth (Le Provost et al., 2013). Finally available gene sequences were also useful to identify the nature of protein showing changes in abundance. Bedon et al. (2012) reported the proteomic plasticity of two clones of eucalyptus in different water regimes. Similar examples of gene or miRNA atlas can be found in fruit trees, for their response to water stress (Bassett et al., 2014; Eldem et al., 2012) or during winter dormancy (Falavigna et al., 2014).

These genomic resources combined with high-density genetic maps constructed for many forest tree species as in eucalyptus (Kullan et al., 2011; Neves, Mamani, Alfenas, Kirst, & Grattapaglia, 2011), maritime pine (Chancerelle et al., 2013), *P. abies* (Lind et al., 2014), *Populus* (Berlin, Lagercrantz, von Arnold, Öst, & Rönnberg-Wästljung, 2010; Muchero et al., 2015), *Cryptomeria japonica* (Moriguchi et al., 2012), constitute a major basis to study the genetic determinants involved in the architectural development of trees. A recent study reported two high-resolution genetic maps constructed using a SNPs array genotyped on a large size pedigree (Bartholomé et al., 2014), which has been useful to identify putative candidate genes underlying QTLs (Bartholomé, Mabiala, et al., 2015). Advances in comparative mapping between species within or between genus (Gion, Chaumeil, & Plomion, 2014) should also constitute a major issue to understand the gene network underlying the architectural development of such long-lived organisms. In apple tree, after the sequencing of the 'Golden Delicious' genome (Velasco et al., 2010), whole-genome genotyping (micro-) arrays have been developed for both apple and pear (Bonghi et al., 2011; Chagné et al., 2012). This greatly facilitated the development of high-density linkage maps for segregating apple progenies (e.g. Antanaviciute et al., 2012). More recently, a high-density map has been obtained after the development of a 20,000 SNP array that should allow to performing robust GWAS and pedigree-based analyses (Bianco et al., 2014; Bink et al., 2007). A similar evolution towards high-density maps can be observed in *Prunus* species (Klagges et al., 2013; Martinez-Garcia et al., 2013).

The ability to genotype large size populations with thousands of molecular markers allowed new approaches in forest genetics from the dissection of QTL effects to whole-genome effects to predict phenotypes (Grattapaglia, 2014), the so-called genomic selection (GS) approach. GS simulation studies are very promising to increase genetic gain per unit time through improved estimation of breeding (parent selection) and genotypic (clone selection) values (Denis & Bouvet, 2012; Grattapaglia & Resende, 2010). Some proof-of-concept experiments are already in progress in maritime pine (Bartholomé, Vidal, et al., 2015). These perspectives in genomic prediction characterizing with a high accuracy the genetic part involved the phenotypic variation should be very helpful in the investigation of the genetic determinants of multi-trait and multi-scale characterizations needed for a fine description of tree architecture. Indeed, the genomic prediction approaches should provide interesting methodological tools to integrate a genetic effect in the architectural models, especially in forest and fruit trees for which phenotyping effort is difficult and time consuming.

4. METHODOLOGICAL IMPROVEMENTS: FROM LOW AND MEDIUM TOWARDS HIGH-THROUGHPUT PHENOTYPING

Whatever the scale of investigation, a single individual or a population, the main obstacles to assess the tree architecture are linked to the number of measurements necessary to characterize the structural complexity of trees. Another major limitation is intrinsically linked to the long-lived organism for which changes in architecture occur according to the developmental stage (from juvenile phase to mature stages). Finally, the tallness of the forest tree species makes it difficult to characterize the architecture at a mature stage. Some indirect methods can serve for prediction of tree architecture using allometric equations or models, which can be available in editing and simulation software, such as AMAPstudio (Griffon & de Coligny, 2014). Such methods require to be calibrated on large amounts of data collected by reliable phenotyping methods, and could benefit from the technological and methodological improvements that are already available or under development for tree high-throughput phenotyping.

Because of the difficulty to directly characterize tree architectures with numerous internodes or axes, many studies have concerned trees with simple structure. This is also well illustrated by the estimation of below-ground biomass and its distribution in the soil. Different sampling strategies have been proposed to facilitate such characterization of rooting systems as volumetric soil—root sample methods ranging from traditional auger cores and monoliths to Voronoi polygons (Saint-André et al., 2005; Snowdon et al., 2002), with different advantages according to coarse or fine root biomass (Levillain, M'Bou, Deleporte, Saint-André, & Jourdan, 2011). Other sampling strategies have been proposed reducing the experimental work for aerial biomass characterization. For instance, protocols have been established to facilitate the topological description of the trees dividing the architectural into levels of organization and combining this decomposition with an adequate within-tree sampling. Such phenotyping strategy has been proposed to determine the genetic determinism of vegetative and reproductive traits in apple tree (Segura et al., 2006) and further applied to olive tree (Sadok et al., 2012). Another sampling strategy has been proposed for 3- and 2-years-old eucalyptus (Diao et al., 2012), dividing the trunk into zones of 1 m in length, within two average branches are selected and characterized for their internode number, length and diameter.

The temporal evolution of the tree architecture, and phenotypes, is an important issue for perennial species. Several studies have been carried out during several years of growth, in which QTL analyses have been performed independently in each year of growth. This strategy consisted of looking for co-localization of the QTLs detected in each analysis which could also be statistically tested by multi-trait QTL and pleiotropic effects analyses (Jiang & Zeng, 1995; Korol, Ronin, & Kirzhner, 1995; Mangin, Thoquet, & Grimsley, 1998). This approach was meant to detect 'stable' QTLs with regard to environmental and ontogenetic effects. However, many QTLs were not common to the locations or years studied, and the challenge was therefore to identify among them those specific to an ontogenetic and/or an environmental effect. A first attempt was reached by introducing an interaction term between genotype and age in the QTL mapping model (Verhaegen et al., 1997). However, neither autocorrelations between measurements performed throughout development nor environmental effects were accounted for in such an approach. Wu and Lin (2006) proposed a methodology, called functional mapping, to detect QTLs involved in growth trajectories. This approach based on growth models integrates autocorrelations between repeated measurements but does not allow a clear distinction between stable and unstable QTLs throughout tree life nor distinguishing among unstable QTLs those related to plant ontogeny from those related to environmental effects. A specific design, called a staggered-start design, in which repetitions of a same genotype are planted in the same field but at least 1 year apart (Loughin, 2006) appears suitable for perennial crops, since their growth occurs during different climatic years. Therefore, traits related to both growth and branching can be annually assessed on different axes of the trees planted in a staggered-start design, with, for instance, the first year of growth along the trunk being developed in different climatic years. This approach has been implemented in a 4-year-old apple tree population and has allowed to mapping QTLs that were specific to these different effects (Segura et al., 2009).

In the last years, the development of automated systems for both the characterization of the environments and some particular properties constituted one more major step to study such long-lived organism in natural conditions. As an example, automated sensors being able to measure the radial daily fluctuations of the trunk, are commonly used in ecophysiology to study the water status of the tree. Various insights into tree growth and physiology can be obtained using such sensors (as a review Drew & Downes, 2009). Electronic point dendrometers were used to study the daily patterns of stem variation in irrigated and unirrigated *Eucalyptus globulus* and *Eucalyptus nitens* (Downes, Beadle, & Worledge, 1999). Continuous-time data for both the secondary meristem activity (secondary growth) and the water status could be another way to decipher both the genetic control underlying wood architecture but also its response to changing environmental conditions. Recently, such instrumentation was deployed on a segregating population of *Eucalyptus urophylla* × *E. grandis*, and revealed QTLs for radial daily fluctuations (Bartholomé, 2014). One major expected result is to identify the genetic and environmental determinants, at a multi time scale (from daily pattern to exploitation age), for the radial growth, i.e. a major component of tree architecture.

Others methods based on imagery/laser scanning allow the phenotyping of particular architectural traits on standing trees at different accuracy scales, over several hectares. Airborne laser scanning enables a three-dimensional characterization of vertical forest structure and is currently used for operational forest inventory (White et al., 2013, p. 50). Ground-based laser imaging is also usable for assessing three-dimensional forest canopy structures (Henning & Radtke, 2006). At a thinner scale, individual tree volume can be obtained by individual tree crown approach based on aerial images and laser point clouds (Hyyppä et al., 2005). The combination of laser scanning (LIDAR data) and high-resolution map (Digital) allows the extraction of classical architectural traits, e.g. tree height, crow size, basal area and stem volume (Chen, Baldocchi, Gong, & Kelly, 2006; Chen, Gong, Baldocchi, & Tian, 2007). The accuracy of individual phenotyping remains difficult to attempt for genetics studies, as it requires the combination of several data sources. Recently, genetic parameters in *C. japonica* for tree height, DBH, stem bending, stem curve, and crown depth were estimated using terrestrial LIDAR (Hiraoka, Takahashi, Nakamura, & Watanabe, 2014), showing a high correlation with measured values. In apple tree, the combination of airborne thermal imagery with calculation of vegetation indexes was applied for field phenotyping of a segregating population response to soil water restriction (Virlet et al., 2014) and QTL detection was performed on a range of indexes (Virlet, Costes, Martinez, Kelner, & Regnard, under submission). Multisensor (optical, thermal, infrared, LIDAR...) platforms for plant imaging have been developed for field-based phenotyping in crop breeding (Busemeyer et al., 2013). Similarly, under greenhouses, i.e. in controlled environments, automated high-throughput monitoring and quantification of plant architecture and performance with a high precision are already used in crops (Sirault et al., 2013). Such technologies could be used in forest/fruit tree experiments but only at the young stages (first years of plantation), and could become essential tools to get measurements of classical architectural traits on a large number of individuals.

Moreover, the measurements of developmental traits can be assessed on harvested material which can require the felling of the tree. In forest trees, the first high-throughput phenotyping technologies have been used principally to assess the wood properties on thousands of individuals. Near-infrared spectrometry (NIRs) technology is now currently used to predict chemical properties of wood (such as lignin content or cellulose content). Such spectral predictions of wood properties need the construction of calibration curves developed in general at a species level using a training sample with both spectral profiles and real measures (Hein, Lima, & Chaix, 2010). So far, calibration curves have been developed and successfully used to estimate genetic parameters of wood chemistry in eucalyptus (Mandrou et al., 2012) and maritime pine (Lepoittevin et al., 2011). Robots improving the preparation of powder samples from solid or core wood on a large number of individuals are also available in some phenotyping facilities (e.g. GenoBois; http://mic.xyloforest.org/en/Research-team/Phenotyping-wood-quality-GenoBois/r161.html). Finally, tomography using the RX technology allows characterizing the 3D structure of tree logs. This technology was, for instance, used to identify rameal traces in sessile oak (Colin et al., 2010) and to study epicormic ontogeny (Morisset, Mothe, Bock, Breda, & Colin, 2012).

5. UNDERSTANDING THE BIOLOGICAL PROCESSES UNDERLYING TREE ARCHITECTURE

5.1 The Control of Primary Growth and Shoot Elongation

Shoot elongation is a major trait for controlling tree height and vegetative development. As leaves separators, shoot internodes play an important role in optimizing water conductance and light interception (Cochard, Coste, Chanson, Guehl, & Nicolini, 2005; Da Silva, Han, Faivre, & Costes, 2014). Internode length thus appears as a key variable for preserving a balance between an optimized hydraulic efficiency and a mechanical stability (McCulloh & Sperry, 2005). In fruit trees, variables linked to internode length have been shown to be genetically determined (Hammami, León, Rapoport, & De la Rosa, 2011; Segura et al., 2007) and, in olive tree, to be stable across environments (Sadok et al., in press). This strong genetic control strengthens the relevance of such traits for breeding programs which has been underlined in various woody species – Coffea (Leroy, Montagnon, Charrier, & Eskes, 1993) or radiata pine (Shelbourne, Apialoza, Jayawickrama, & Sorensson, 1997).

In apple tree, columnar mutants which exhibit short internodes, reduced branching and increased number of short laterals, are considered useful to establish high-density planting because columnar trees require minimal pruning and facilitate mechanical harvesting (Janick, Cummins, Brown, & Hemmat, 1996) and despite major drawbacks have been revealed, in particular regarding their irregular bearing behaviour (Looney & Lane, 1984). Since numerous crosses between a columnar and a non-columnar parent have been performed, the dominant versus recessive nature of the gene involved in this mutation is still under debate. A major point regards the large fluctuations in the proportions of columnar phenotypes observed among progenies, depending on the study. Since the columnar phenotype includes multiple traits and is assessed visually, this situation could result from differences regarding 'columnar' trait phenotyping. However, several groups have successfully identified the genomic region controlling the columnar phenotype, using molecular markers and recombinants. The Co region has been identified on the chromosome 10, in two presumably adjacent intervals: one between SSR markers CH02c11 and CH03d11 (Fernández-Fernández et al., 2008), the other between CH03d11 and Hi01a03 (Moriya, Okada, Haji, Yamamoto, & Abe, 2012). It has been gradually narrowed down to 18.76–18.92 Mb (Bai et al., 2012; Baldi et al., 2013; Morimoto & Banno, 2015; Moriya et al., 2012). Recently, comparing the genomic sequence of the Co region of 'Wijcik McIntosh' and its wild type 'McIntosh', two different groups identified the presence of insertion element in the Co region of 'Wijcik McIntosh' (Otto, Petersen, Brauksiepe, Braun, & Schmidt, 2013; Wolters, Schouten, Velasco, Si-Ammour, & Baldi, 2013). However, the results are still contradictory because Otto et al. (2013) detected a gypsy-like retrotransposon into the

Co region, whereas Wolters et al. (2013) found a novel type of mobile elements lacking terminal inverted repeats and did not find the retrotransposon. Moreover, transcriptomic studies have revealed a high number of differentially expressed genes between roots or apices of columnar versus wild-type apple trees (Krost, Petersen, & Schmidt, 2012; Zhang, Zhu, & Dai, 2012). Among them, several genes mapped on LG10, including gibberellic acid-insensitive repressor of gibberellic acid-insensitive and scarecrow (GRAS) transcription factors like DELLAs, which play a role in the gibberellin signal transduction (Zhang et al., 2012). Also, Krost et al. (2012) found that four genes differentially expressed are associated with phytohormonal regulations, in particular auxin, cytokinines and gibberellins metabolism, signal transduction and perception. These authors underlined that further time course gene expression experiments will be necessary to increase the knowledge about the function of the Co gene product and to yield a comprehensive picture of the physiological state of the plant.

Similarly, the molecular mechanisms by which dwarfing rootstocks reduce tree stature and enhance yield efficiency are still poorly understood, despite numerous studies on their morphology, histology or physiology (e.g. in apple tree: Prassinos, Ko, Lang, Iezzoni, & Han, 2009; Tombesi, Johnson, Day, & DeJong, 2010). Mutations in DELLA proteins have been shown to induce dwarfism in various species and are therefore strong candidates. Because DELLA proteins function similarly in many species, various plants expressing mutant DELLA repressors have been shown to become gibberellic acid (GA)-insensitive, dark-green, late-flowering dwarfs (Dill, Thomas, Hu, Steber, & Sun, 2004; Fleck & Harberd, 2002). The most famous cases are represented by the semi-dominant dwarf wheat mutants (Rht-B1/Rht-D1), well known as 'Green Revolution' cultivars (Peng et al., 1999), and the grape VvGAI highly productive and semi-dwarfed grape cultivar, which is the first and only DELLA-related dwarf mutant found so far in a woody perennial plant (Boss & Thomas, 2002). Arabidopsis plants ectopically expressing GA-insensitive alleles from other plant species, exhibit dwarf stature (Silverstone et al., 2001; Willige et al., 2007). Similarly, the ectopic expression of GA-sensitive alleles of Arabidopsis DELLA genes have been shown to be sufficient to induce dwarf phenotypes in tomato, apple and other plant species (Martí et al., 2007; Zhu, Li, & Welander, 2007). Therefore, the ability of GA-insensitive DELLA proteins to cause dwarf phenotypes appears well conserved in various plant species. Furthermore, a series of grafting experiments showed that DELLA gene mRNAs could be trafficked over long-distances via the phloem (Iqbal, Nazar, Khan, Masood,

& Khan, 2011; Kim, Canio, Kessler, & Sinha, 2001; Wang et al., 2012), opening the possibility of rootstock containing GA-insensitive alleles of DELLA to dwarf the scion cultivars grafted on to them. Recently, among six DELLA genes isolated from the 'Royal Gala' apple cultivar (MdRGL1a/b, MdRGL2a/b and MdRGL3a/b), the ectopic expression of MdRGL2a has been shown to cause GA-insensitive dwarfism in Arabidopsis (Foster et al., 2007). MhGAI1 has been cloned from the apomictic tea crabapple (*Malus hupehensis* Redh. var. pingyiensis), and the ectopic expression of GA-insensitive allele Mhgai1, which contains a point mutation in the DELLA domain, has been shown to reduce plant stature (Wang et al., 2012). These authors suggest Mhgai1, could turn out to be useful targets for the genetic improvement of dwarfing rootstocks in apples. With the recent detection of two major markers for dwarfing rootstock (Foster, Celton, Chagné, Tustin, & Gardiner, 2015), rapid progress can be expected in the next future.

5.2 The Control of Growth Cessation and Resumption

Trees, as other polycarpic plants, can be viewed as a population of meristems, each of them having different stages of differentiation and passing from one stage to the next one through remarkable transitions during ontogeny. However, these different meristems are linked together by the plant age, resource availability and environmental conditions in which the plant develops. In trees growing in temperate climatic zones, growth in the different meristems is rhythmed and thus synchronized by the alternation of favourable and unfavourable periods. They are characterized by the formation of buds that protect the meristematic tissues and make them able to survive winter periods and resume growth at spring. These buds go through a 'dormancy' stage, this generic term including three different physiological states of the bud — ecodormancy, paradormancy and endodormancy (Lang, Early, Martin, & Darnell, 1987; Shim et al., 2014) — depending on whether bud inhibition is conditioned on environmental causes or on a correlation with other tree parts, or lies within the bud itself, respectively. Flowering process is also tightly linked to these annual climatic fluctuations, especially in fruit trees in which flower induction and differentiation occurs in the year preceding bloom and, as vegetative buds, floral buds go through a dormancy period during winter. Because in all species, cultivars or clones may be characterized as precocious or late for bud burst, the genetic control of flowering and vegetative phenology has been widely studied in both fruit and forest trees. Many QTLs zones have been detected for the time of vegetative and floral bud break in apple (Celton et al., 2011; van Dyk, Soeker, Labuschagne, & Rees, 2010), almond (Sánchez-Pérez, Dicenta, & Martínez-Gómez, 2012), apricot (Campoy, Ruiz, & Egea, 2011), peach (Dirlewanger et al., 2012) and cherry tree (Castede et al., 2014). In forest trees, QTLs have also been reported for bud set and bud flush in *Populus* (Bradshaw & Stettler, 1995; Frewen et al., 2000), white spruce (Pelgas, Bousquet, Meirmans, Ritland, & Isabel, 2011), *Salix* (Ghelardini et al., 2014; Tsarouhas, Gullberg, & Lagercrantz, 2003) and bud burst in *Quercus* (Scotti-Saintagne et al., 2004). In *Salix*, genomic regions harbouring hotspots have been identified for spring and autumn phenology where QTL for different traits co-localized (Ghelardini et al., 2014). Moreover, the anchoring of the SNP genetic map to the *Populus trichocarpa* genome allowed (1) the identification of co-localized QTLs with poplar indicating putative common determinants for phenology trait in Salicaceae, (2) the identification of candidate genes among the gene models (PtFT2, LHY1, LHY2 and SVP homologue) which have been shown to be involved in growth cessation and bud burst.

Recent studies have outlined close relationships between molecular control of shoot apical meristem transitions, especially floral transitions, entrance/release of dormancy and the control of AxM branching (Paul, Rinne, & van der Schoot, 2014; van der Schoot, Paul, & Rinne, 2014). As suggested by van der Schoot et al. (2014), growth cessation and constitution of the embryonic shoot within the dormant bud is a key step for perennial species with a dormancy period. However, among the population of meristems, this step may occur at different times depending, and the different meristems within a given tree are not necessarily synchronized for growth cessation. Indeed, short shoots grow for a shorter period and stop growing earlier than long shoots (which often have the capability to develop neoformed organs), this leading to axis polymorphism, well known in many fruit and forest species (Champagnat, 1965; Costes et al., 2006). As a consequence, it is very likely that growth cessation may have different origins, resulting from nutritional starvation and correlative inhibition in the case of early cessation of short shoots whereas a climatic control is more likely to explain growth cessation of long shoots at fall.

In reviews of the basic genetic factors governing bud and dormancy induction (e.g. Horvath, Anderson, Chao, & Foley, 2003; Rohde & Bhalerao, 2007), the involvement of genes controlling perceptive events (e.g. phytochrome and the perception of photoperiod) as well as the downstream regulation of the cell cycle that must occur for growth arrest have been pointed out. Moreover, the analysis of the *Populus* CO/FT regulatory genes have demonstrated a role for these genes in seasonal growth dynamics in addition to the role they have in flowering regulation (Bohlenius, 2006). Hormones (GA, abscisic acid and ethylene) and metabolic regulation are also clearly involved in the regulatory network(s) connecting signal and response, as demonstrated by reverse genetic approaches (Horvath et al., 2003; Ruonala et al., 2006). In an evergreen, nondormant, peach mutant, the nondormant trait has been described as a recessive nuclear gene and the exploration of the corresponding zone has revealed a cluster of six MADS-box transcription factors, designated as dormancy-associated MADS-box (DAM) genes, which were not expressed in the evg mutant (Bielenberg et al., 2004). The peach DAM genes are members of the SVP/StMADS11 lineage (Jiménez, Lawton-Rauh, Reighard, Abbott, & Bielenberg, 2009). Among them, DAM1, DAM2 and DAM4 have expression patterns more closely associated with growth cessation and bud set whereas DAM3, DAM5 and DAM6 are winter expressed reaching a minimum just prior to bud break (Li, Reighard, Abbott, & Bielenberg, 2009). Moreover, the expression patterns of DAM5 and DAM6 have been shown to be consistent with a role as quantitative repressors of bud break, with different expression levels between high and low chilling requirement peach cultivars (Jiménez, Reighard, & Bielenberg, 2010). These DAM genes have also been associated with endodormancy in Japanese pear (*Pyrus pyrifolia*) where low expression is associated with lack of dormancy (Ubi et al., 2010). Two apple DAM genes have been identified that have high similarity to *Pyrus* DAM genes (Wisniewski, Norelli, Bassett, Artlip, & Macarisin, 2011). These authors have also revealed that the ectopic expression of a C-repeat binding factor (CBF/DREB) from *Prunus persica* (PpCBF1), which belongs to a family of transcription factors known to induce genes associated with increased cold tolerance, resulted in strong sensitivity to short day length in apple, with growth cessation and leaf senescence induced in transgenic lines. However, whether the overexpression of PpCBF1 in apple affects the expression of apple homologues of DAM genes remains to be determined. These results indicate that genes regulating dormancy and those that regulate cold hardiness may be more interactive than previously expected, especially in woody plants. Ueno et al. (2013) proposed a functional network predicted from the genes upregulated during ecodormancy (Figure 2). Several pathways known to be regulated during ecodormancy such as response to abscisic acid, response to gibberellin stimulus, response to cold and response to oxidative stress, were differentially regulated in

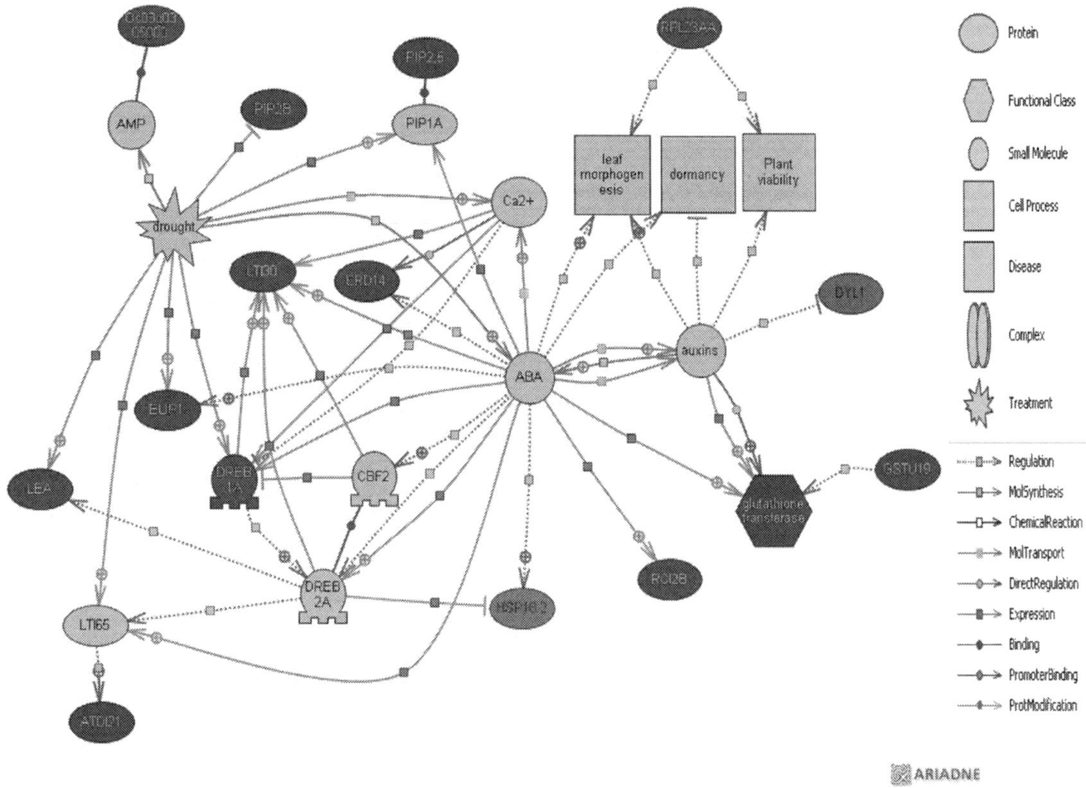

Figure 2 Functional network predicted from the genes upregulated during ecodormancy. *From Ueno et al. (2013).* (See colour plate)

oak buds. Conde, González-Melendi, and Allona (2012) put forward an epigenetic control of winter dormancy in poplar stems. Other results suggest that active growth-regulating genes might be under epigenetic regulation, repressed during dormancy induction through chromatin compaction mechanisms, i.e. histone deacetylation and methylation, histone ubiquitination, DNA methylation and polycomb activity (see Shim et al., 2014 as a review).

5.3 The Control of Aerial Axillary Bud Break: Apical Dominance, Acrotony and Branching

The physiologic and genetic mechanisms underlying meristem organogenetic activity and AxM outgrowth control have been extensively studied. Since a long time, apical dominance has been considered as the result of auxin (IAA) production in the apical bud (Cline, 2000). The screening of mutants, mainly in annual plants such as *Arabidopsis thaliana* (Leyser, 2003), rice (Li et al., 2003), pea (Beveridge, 2000; Foo, Turnbull, & Beveridge, 2001) has led to considerable improvement of knowledge in these domains. Apical dominance has been shown to involve auxin transport down the shoot via active transporter in the parenchyma associated to xylem (Booker, Chatfield, & Leyser, 2003). However, when AxM to be controlled are distant from apex, the speed of IAA transport appears incompatible with its putative role in axillary bud inhibition (Renton, Hanan, Ferguson, & Beveridge, 2012). For genetic mechanisms, the formation of AxM has been shown to require lateral suppressor which is expressed in the boundary region between the leaf primordium and stem (Ls in tomato, Schumacher, Schmitt, Rossberg, Schmitz, & Theres, 1999; or LAS in *A. thaliana*, Greb et al., 2003; rice, Li et al., 2003). Moreover, lateral branching has been shown to be under the control of a complex interaction between plant hormones, cytokinines that are promoters of bud outgrowth, auxin (IAA) as a repressor, but also strigolactones (SL) (Leyser, 2009), which inhibit the axillary bud outgrowth (Gomez-Roldan et al., 2008; Umehara et al., 2008). In the debate on the hormonal control of branching, some studies have suggested that SLs act directly in the bud to inhibit bud outgrowth, thus acting as second messenger to auxin (Brewer, Dun, Ferguson, Rameau, & Beveridge, 2009; Dun, de Saint Germain, Rameau, & Beveridge, 2013) while another hypothesis proposed that SLs impede the ability of buds to export auxin into the main stem, and hence inhibit their outgrowth (Domagalska & Leyser, 2011). The TEOSINTE BRANCHED 1 (BRC1) gene has been proposed as a possible integrator of these multiple bud outgrowth pathways (Braun et al., 2012). However, the shoot tip's strong demand for sugars, rather than auxin supply, could also inhibit axillary bud outgrowth by limiting the amount of sugar translocated to those buds (Mason, Ross, Babst, Wienclaw, & Beveridge, 2014). In a recent review, Rameau et al. (2015) proposed a more complex view in which BRC1, CYCLOIDEA, PCF transcription factor TB1/BRC1 and the polar auxin transport stream in the stem could be possible integrators of different pathways, involving not only the classical hormones, but also other signals with a strong influence on shoot branching such as sugars or molecular actors of plant phase transition, in particular FT (Figure 3). This proposition relies mainly on a corpus of studies on annual plants and sylleptic branching, i.e. to the outgrowth of AxMs during the year of their parent shoot growth. By contrast, the molecular control of AxMs latency during the parent shoot growth and the outgrowth of a proportion of them after dormancy, which leads to proleptic branching, have been far less studied. In *Populus*, Bradshaw and Stettler (1995) detected few QTLs for the number of proleptic branch. Proleptic branching which is widely represented in trees and usually accompanied by acrotony

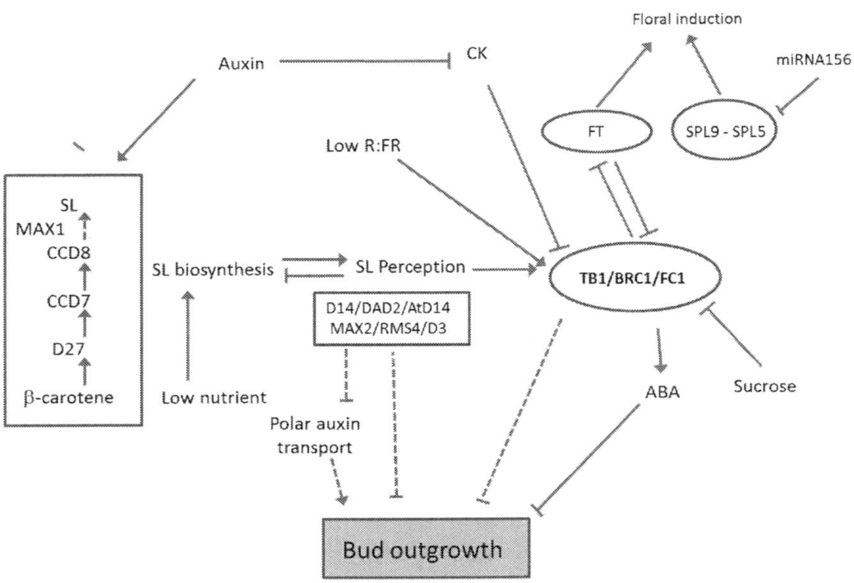

Figure 3 Putative hormone signaling and control of axillary meristem outgrowth and floral initiation, with BRC1 as an integrator. Dotted lines and arrow indicate relationship still with contrasting results depending on the study. *Adapted from Janssen, Drummond, and Snowden (2014) and Rameau et al. (2015).*

(see Barthélémy & Caraglio, 2007; Costes et al., 2014 for definitions) still require molecular and physiological investigations, using the knowledge acquired on annual plants as a reference.

It must be also mentioned that, because dwarf mutants often have a modified branching habit (see above the apple columnar mutant case), the pathways involved in the control of AxM outgrowth could be partly common or could interact with those controlling shoot elongation (Rameau et al., 2015). Similarly, several genes have been recently shown to be involved in both floral induction and growth cessation events, especially in species with terminal flowering. FT-like genes, which are known to promoting flowering, are likely to be also involved in seasonal growth and dormancy in trees (Bohlenius, 2006; Foster, Watson, van Hooijdonk, & Schaffer, 2013; Hsu et al., 2011). In some cases, these genes have also been shown to regulate branching (Srinivasan, Dardick, Callahan, & Scorza, 2012). These studies have been relayed by more recent ones confirming the coordination of hormonal and nutritional controls of shoot elongation and growth cessation (Wingler, 2015) with genes previously known as being involved in flowering control, in particular FT2, as found in Norway spruce (Karlgren, Gyllenstrand, Clapham, & Lagercrantz, 2013), Scot pine (Avia, Kärkkäinen, Lagercrantz, & Savolainen, 2014) or hybrid aspen (Ericksonn et al., 2015). This is consistent with an environmental control of meristem activity, especially during the period of growth cessation prior dormancy as emphasized by van der Schoot et al. (2014), but also suggests a combination with unfavourable nutritional status of buds within the plant during this period as emphasized by physiological studies on the role of carbohydrates in floral induction (Bangerth, 2009; Wahl et al., 2013) or axillary shoot control (Mason et al., 2014).

5.4 The Control of the Transition between Developmental Phases: Juvenility versus Mature Phase

Contrary to monocarpic plants where floral differentiation occurs in all aerial meristems and ends the plant's life cycle, floral differentiation in perennial polycarpic species occurs recurrently throughout ontogeny but in a proportion of meristems only (Bangerth, 2009). The occurrence of the first flowering ends up the juvenile phase and the plant then enters its mature phase (Hackett, 1985). The location of flowering buds, from lateral to terminal, is an important criterion for defining the tree architecture, with, for instance, terminal flowering linked to sympodial branching. Moreover, this location is a key feature is structuring the annual shoots, as reviewed in the Rosaceae family (Costes et al., 2014).

In higher plants, the transition towards flowering occurs under the control of environmental and endogenous stimuli, which have been categorized into four genetic flowering pathways: photoperiodic, autonomous, vernalization and gibberellin (Boss, Bastow, Mylne, & Dean, 2004; Mouradov, Cremer, & Coupland, 2002). In *A. thaliana*, this transition has been shown to require the accumulation of flower-promoting proteins in the meristem, especially two of them encoded by the genes FLOWERING LOCUS T (FT) and SUPPRESSOR OF OVEREXPRESSION OF CONSTANCE 1 (SOC1) (Kobayashi, Kaya, Goto, Iwabuchi, & Araki, 1999). These proteins activate floral meristem identity genes such as LEAFY (LFY) and APETALA1 (AP1), which in turn activate floral homeotic genes responsible for floral organ formation (Lee & Lee, 2010; Robles & Pelaz, 2005). However, flowering transition can be delayed by floral repressors such as TERMINAL FLOWER1 (TFL1) and BROTHER of FT and TFL1 (BFT), which repress LFY and AP1 (Bradley, Ratcliffe, Vincent, Carpenter, & Coen, 1997; Yoo et al., 2010).

Numerous studies have contributed to decipher the flowering process in tree species, particularly in the model tree species poplar and in apple (for reviews, see Brunner & Nilsson, 2004; Hanke, Flachowsky, Peil, & Hättasch, 2007; Wilkie, Sedgley, & Olesen, 2008). In these species, genes known to be involved in flowering have been isolated from apple based on their sequence similarity with *Arabidopsis* genes, and a number of them have been shown to have similar function. In particular, FT genes involvement in flowering transition appears conserved in many plants, including fruit and forest trees. In poplar, the overexpression of two genes encoding FT-like proteins induces flowers within several months, whereas nontransgenic trees generally produce first flowers after 5–10 years (Bohlenius, 2006; Hsu, Liu, Luthe, & Yuceer, 2006). In *E. globulus*, the expression of a FLOWERING LOCUS T homologue was also associated with the annual transition from vegetative to reproductive growth, i.e. flower bud initiation (Jones, Hecht, Potts, Vaillancourt, & Weller, 2012). Similarly, in *Citrus* (trifoliate orange) and apple, overexpression of FT-homologous genes led to a precocious flowering (Endo et al., 2005; Tränkner et al., 2010).

However many other actors could be involved for triggering floral induction. Two highly conserved microRNAs (miRNAs), miR156 and miR172, and their targets have been identified as key components of the genetic control mechanisms involved in transition changes (Wu et al., 2009).

Indeed, miR156 targets SQUAMOSA promoter-binding protein-like (SPL) family transcription factors, which promote the transition from the juvenile to the adult phase in *Populus* (Wang et al., 2011). Antagonistically to miR156, miR172 promotes adult transition, targeting key floral repressors belonging to the AP2-like transcription factor family genes (Zhu & Helliwell, 2011). Other age-dependent pathways have been revealed that involve carbohydrates directly (Wahl et al., 2013), or in conjunction with photoperiod (Lu et al., 2014). Moreover, a number of antiflorigens, especially SHORT VEGETATIVE PHASE (SVP), FLOWERING LOCUS C (FLC) and TEMPRANILLO (TEM1 and TEM2) have been shown to interact with the CONSTANS (CO)-FT complex to determine the optimal timing of floral transition depending on day length and temperature (Sawa & Kay, 2011; Song, Ito, & Imaizumi, 2013), or in a decreasing way with increasing age of the plant with TEM1 and TEM2 (Johansson & Staiger, 2014; Sgamma, Jackson, Muleo, Thomas, & Massiah, 2014). In addition, the redox status of the plant, which is linked to its perception of environment and is involved in the control of plant growth and development through the reprogramming of metabolism, might be involved in the control of floral transition (Blanvillain, Wei, Wei, Kim, & Ow, 2011; Kocsy et al., 2013).

Once the tree is mature, the floral initiation and differentiation of shoot meristems are not 'automatic', but have often been reported as irregular, in particular in fruit trees (e.g. Costes et al., 2014; Wilkie et al., 2008). At the tree level, the timing of floral induction may vary depending on the type of axis and its position (Costes et al., 2003; Costes et al., 2014), and complex patterns of synchronization versus resynchronization may be observed between shoots and branches within the tree (Durand et al., 2013; Monselise & Goldschmidt, 1982). Superabundance and synchronized fruit/seeds productions have also been reported in some noncultivated perennial species, and is known as 'masting' (Samach & Smith, 2013). Masting often occurs between intervals longer than 2 years thus it is not biennial. However, this phenomenon is likely caused by similar physiological constraints than in fruit trees, and similarly during a mast year there is little vegetative growth whereas the following year there is little flowering. This phenomenon is also referred as biennial bearing which is a problem observed in many fruiting trees (Monselise & Goldschmidt, 1982). Genetic control of biennial bearing has been proved in apple tree and first QTLs have been detected (Guitton et al., 2012). On a molecular point of view, the comparison of genes and proteins differentially expressed between organs of fruited and de-fruited

Citrus trees has demonstrated that the pathways involved in the control of FI in fluctuating temperature and photoperiod conditions were affected by the presence of fruit. Proteins involved in primary metabolism and redox state have been shown to be differentially expressed in mandarin leaves, depending on fruit load (Muñoz-Fambuena et al., 2013). In addition, photosynthetic genes and calcium-dependent IAA transport were induced rapidly after de-fructification and in 'OFF' trees (Shalom et al., 2014).

6. PERSPECTIVES: INTEGRATING ARCHITECTURAL TRAITS IN BREEDING PROGRAMS AND ASSISTING IDEOTYPE DEFINITION BY COMPUTER-BASED MODELLING

Despite the difficulty to collect precise and numerous measurements or to assess estimations, several traits related to plant architecture have been included in breeding scheme. Diametral growth is indeed a crucial trait for forest tree production. But, other advantageous traits such as efficient physiological metabolism and adequate tree architecture could be gathered for a forest tree ideotype definition (Wu, 1998). In fruit trees, the focus has been often put on a general tree shape such as compact or weeping shapes, likely because of their observation simplicity. In apple tree, several propositions have been formulated for promoting either dwarfed spur types (Dickman, Gold, & Flore, 1994) or trees characterized by long and weeping branching (Type IV) which were considered prone to regular bearing (Laurens et al., 2000; Lespinasse, 1992). Recently, new indices able to estimate alternate bearing behaviour have been proposed which could be applicable on young trees during the early production period, when production per tree is still increasing over years (Durand et al., 2013). In all species, it must be underlined that breeding these ideotypes may be seriously limited by unfavourable correlations due to negative pleiotropic effects of the genes involved (Da Silva et al., 2014; Han et al., 2012; Wu, 1998).

An alternative to laborious, costly, and time-consuming phenotyping and in complement to high-throughput phenotyping strategies, computer-based virtual experiments have been initiated to explore the impact of traits variations (Da Silva et al., 2014; Martre et al., 2015). The modelling of topological development using stochastic models has been proved in several species (Costes et al., 2008; Diao et al., 2012). Among different traits that can be considered to define ideotype depending on the target organs, traits or cultivation systems (Andrivon et al., 2012), a particular attention has been paid to the relationship between tree architecture and light interception because of its strong impact on tree transpiration, as well as carbon acquisition and allocation. Different combinations of organ geometrical and branching traits have been explored in silico as well as their consequences on light interception, here considered as an output variable (Da Silva et al., 2014; Han et al., 2012). This strategy required the use of several models, methods and information already available from previous studies and combined them into three steps: (1) a number of geometrical and topological traits whose range of variation was previously observed within a segregating population of apple hybrids were considered as input parameters in an apple tree architectural model; (2) the 3D representations resulting from this model were used by an environmental simulation tool to calculate light interception; and (3) a sensitivity analysis was performed on an output variable considered representative of LIE. Despite some limitations that must be overcome through further investigations, this strategy appeared very promising for identifying traits that could control light interception optimization, from the breeders point of view. Among the main limitations, we can cite (1) the necessity to combine several output variables each one being potentially a trait to be selected, (2) the possible antagonistic genetic determinism of these traits (3) the computational complexity resulting from such combinations and from the changes in scales between input and output variables. For instance, in Da Silva et al. (2014) other output variables should be combined with light interception, considering physiological (e.g. photosynthesis) or agronomic (e.g. production) traits simultaneously in a multi-objective approach as recently performed by Quilot-Turion et al. (2012). In the next future, the use of in silico scenarios and optimization procedures should allow researchers, growers and breeders to better account for antagonistic effects of input parameters, possibly at several scales of tree organization, and their combined effects on multiple objectives.

REFERENCES

Andrivon, D., Giorgetti, C., Baranger, A., Calonnec, A., Cartolaro, P., Faivre, R., et al. (2012). Defining and designing plant architectural ideotypes to control epidemics? *European Journal of Plant Pathology, 135*(3), 611–617. http://doi.org/10.1007/s10658-012-0126-y.

Antanaviciute, L., Fernández-Fernández, F., Jansen, J., Banchi, E., Evans, K. M., Viola, R., et al. (2012). Development of a dense SNP-based linkage map of an apple rootstock progeny using the *Malus Infinium* whole genome genotyping array. *BMC Genomics, 13*(1), 203. http://doi.org/10.1186/1471-2164-13-203.

Argout, X., Salse, J., Aury, J.-M., Guiltinan, M. J., Droc, G., Gouzy, J., et al. (2011). The genome of *Theobroma cacao*. *Nature Genetics, 43*(2), 101–108. http://doi.org/10.1038/ng.736.

Avia, K., Kärkkäinen, K., Lagercrantz, U., & Savolainen, O. (2014). Association of *FLOWERING LOCUS T/TERMINAL FLOWER 1*-like gene *FTL2* expression with growth rhythm in Scots pine (*Pinus sylvestris*). *New Phytologist, 204*(1), 159–170. http://doi.org/10.1111/nph.12901.

Bai, T., Zhu, Y., Fernández-Fernández, F., Keulemans, J., Brown, S., & Xu, K. (2012). Fine genetic mapping of the Co locus controlling columnar growth habit in apple. *Molecular Genetics and Genomics, 287*(5), 437–450. http://doi.org/10.1007/s00438-012-0689-5.

Baldi, P., Wolters, P. J., Komjanc, M., Viola, R., Velasco, R., & Salvi, S. (2013). Genetic and physical characterisation of the locus controlling columnar habit in apple (*Malus × domestica* Borkh.). *Molecular Breeding, 31*(2), 429–440. http://doi.org/10.1007/s11032-012-9800-1.

Bangerth, K. F. (2009). Floral induction in mature, perennial angiosperm fruit trees: similarities and discrepancies with annual/biennial plants and the involvement of plant hormones. *Scientia Horticulturae, 122*(2), 153–163. http://doi.org/10.1016/j.scienta.2009.06.014.

Barritt, B. H. (2000). Selecting an orchard system for apples. *Compact Fruit Tree, 33*(3), 89–92.

Barthélémy, D., & Caraglio, Y. (2007). Plant architecture: a dynamic, multilevel and comprehensive approach to plant form, structure and ontogeny. *Annals of Botany, 99*(3), 375–407. http://doi.org/10.1093/aob/mcl260.

Bartholomé, J. (2014). Déterminisme génétique de la dynamique de croissance et de la composition isotopique du carbone chez l'Eucalyptus en réponse aux variations environnementales. Montpellier SupAgro, 296 p. Thèse de doctorat : Systèmes intégrés en biologie, agronomie, géosciences, hydrosciences et environnement (SIBAGHE). Biologie intégrative des plantes (BIP).

Bartholomé, J., Mabiala, A., Savelli, B., Bert, D., Brendel, O., Plomion, C., et al. (2015). Genetic architecture of carbon isotope composition and growth in Eucalyptus across multiple environments. *New Phytologist.* n/a–n/a http://doi.org/10.1111/nph.13301.

Bartholomé, J., Mandrou, E., Mabiala, A., Jenkins, J., Nabihoudine, I., Klopp, C., et al. (2014). High-resolution genetic maps of Eucalyptus improve *Eucalyptus grandis* genome assembly. *New Phytologist.* n/a–n/a http://doi.org/10.1111/nph.13150.

Bartholomé, J., Vidal, M., van Heerwaarden, J., Isik, F., Plomion, C., & Bouffier, L. (2015). Genomic selection in a multi-generation breeding population of maritime pine. In *Plant and animal genome XXIII conference, San Diego, USA, January 2015.*

Bartholomé, J., Salmon, F., Vigneron, P., Bouvet, J.-M., Plomion, C., & Gion, J.-M. (2013). Plasticity of primary and secondary growth dynamics in *Eucalyptus* hybrids: a quantitative genetics and QTL mapping perspective. *BMC Plant Biology, 13*(1), 120.

Bassett, C. L., Baldo, A. M., Moore, J. T., Jenkins, R. M., Soffe, D. S., Wisniewski, M. E., et al. (2014). Genes responding to water deficit in apple (*Malus × domestica* Borkh.) roots. *BMC Plant Biology, 14*(1), 182.

Bedon, F., Villar, E., Vincent, D., Dupuy, J.-W., Lomenech, A.-M., Mabialangoma, A., et al. (2012). Proteomic plasticity of two Eucalyptus genotypes under contrasted water regimes in the field. *Plant, Cell & Environment, 35*(4), 790–805. http://doi.org/10.1111/j.1365-3040.2011.02452.x.

Berlin, S., Lagercrantz, U., von Arnold, S., Öst, T., & Rönnberg-Wästljung, A. C. (2010). High-density linkage mapping and evolution of paralogs and orthologs in Salix and Populus. *BMC Genomics, 11*(1), 129. http://doi.org/10.1186/1471-2164-11-129.

Beveridge, C. A. (2000). Long-distance signalling and a mutational analysis of branching in pea. *Plant Growth Regulation, 32*(2–3), 193–203. http://doi.org/10.1023/A:1010718020095.

Bianco, L., Cestaro, A., Sargent, D. J., Banchi, E., Derdak, S., Di Guardo, M., et al. (2014). Development and validation of a 20K single nucleotide polymorphism (SNP) whole genome genotyping array for apple (*Malus × domestica* Borkh). *PLoS One, 9*(10), e110377. http://doi.org/10.1371/journal.pone.0110377.

Bielenberg, D. G., Wang, Y., Fan, S., Reighard, G. L., Scorza, R., & Abbott, A. G. (2004). A deletion affecting several gene candidates is present in the evergrowing peach mutant. *Journal of Heredity, 95*(5), 436–444. http://doi.org/10.1093/jhered/esh057.

Bink, M. C. A. M., Boer, M. P., ter Braak, C. J. F., Jansen, J., Voorrips, R. E., & van de Weg, W. E. (2007). Bayesian analysis of complex traits in pedigreed plant populations. *Euphytica, 161*(1–2), 85–96. http://doi.org/10.1007/s10681-007-9516-1.

Blanvillain, R., Wei, S., Wei, P., Kim, J. H., & Ow, D. W. (2011). Stress tolerance to stress escape in plants: role of the OXS2 zinc-finger transcription factor family. *The EMBO Journal, 30*(18), 3812–3822. http://doi.org/10.1038/emboj.2011.270.

Bohlenius, H. (2006). CO/FT regulatory module controls timing of flowering and seasonal growth cessation in trees. *Science, 312*(5776), 1040–1043. http://doi.org/10.1126/science.1126038.

Bonghi, C., Trainotti, L., Botton, A., Tadiello, A., Rasori, A., Ziliotto, F., et al. (2011). A microarray approach to identify genes involved in seed-pericarp cross-talk and development in peach. *BMC Plant Biology, 11*(1), 107. http://doi.org/10.1186/1471-2229-11-107.

Booker, J., Chatfield, S., & Leyser, O. (2003). Auxin acts in xylem-associated or medullary cells to mediate apical dominance. *Plant Cell, 15*, 495–507. http://dx.doi.org/10.1105/tpc.007542.

Boss, P. K., Bastow, R. M., Mylne, J. S., & Dean, C. (2004). Multiple pathways in the decision to flower: enabling, promoting, and resetting. *The Plant Cell, 16*(Suppl.), S18–S31. http://doi.org/10.1105/tpc.015958.

Boss, P. K., & Thomas, M. R. (2002). Association of dwarfism and floral induction with a grape 'green revolution' mutation. *Nature, 416*(6883), 847–850. http://doi.org/10.1038/416847a.

Bradley, D., Ratcliffe, O., Vincent, C., Carpenter, R., & Coen, E. (1997). Inflorescence commitment and architecture in Arabidopsis. *Science, 275*(5296), 80–83. http://doi.org/10.1126/science.275.5296.80.

Bradshaw, H. D., & Stettler, R. F. (1995). Molecular genetics of growth and development in populus. IV. Mapping QTLs with large effects on growth, form, and phenology traits in a forest tree. *Genetics, 139*(2), 963–973.

Braun, N., de Saint Germain, A., Pillot, J.-P., Boutet-Mercey, S., Dalmais, M., Antoniadi, I., et al. (2012). The pea TCP transcription factor PsBRC1 acts downstream of Strigolactones to control shoot branching. *Plant Physiology, 158*(1), 225–238. http://doi.org/10.1104/pp.111.182725.

Brewer, P. B., Dun, E. A., Ferguson, B. J., Rameau, C., & Beveridge, C. A. (2009). Strigolactone acts downstream of auxin to regulate bud outgrowth in pea and Arabidopsis. *Plant Physiology, 150*(1), 482–493. http://doi.org/10.1104/pp.108.134783.

Brunner, A. M., & Nilsson, O. (2004). Revisiting tree maturation and floral initiation in the poplar functional genomics era. *New Phytologist, 164*(1), 43–51. http://doi.org/10.1111/j.1469-8137.2004.01165.x.

Busemeyer, L., Mentrup, D., Möller, K., Wunder, E., Alheit, K., Hahn, V., et al. (2013). BreedVision – a multi-sensor platform for non-destructive field-based phenotyping in plant breeding. *Sensors, 13*(3), 2830–2847. http://dx.doi.org/10.3390/s130302830.

Campoy, J. A., Ruiz, D., & Egea, J. (2011). Dormancy in temperate fruit trees in a global warming context: a review. *Scientia Horticulturae, 130*(2), 357–372. http://doi.org/10.1016/j.scienta.2011.07.011.

Castède, S., Campoy, J. A., García, J. Q., Le Dantec, L., Lafargue, M., Barreneche, T., et al. (2014). Genetic determinism of phenological traits highly affected by climate change in *Prunus avium*: flowering date dissected into chilling and heat requirements. *New Phytologist, 202*(2), 703–715. http://doi.org/10.1111/nph.12658.

Celton, J.-M., Martinez, S., Jammes, M.-J., Bechti, A., Salvi, S., Legave, J.-M., et al. (2011). Deciphering the genetic determinism of bud phenology in apple progenies: a new insight into chilling and heat requirement effects on flowering dates and positional candidate genes. *New Phytologist, 192*(2), 378–392. http://doi.org/10.1111/j.1469-8137.2011.03823.x.

Celton, J. M., Kelner, J. J., Martinez, S., Bechti, A., Khelifi Touhami, A., Jammes, M. J., et al. (2014). Fruit Self-Thinning: A Trait to Consider for Genetic Improvement of Apple Tree. *PLoS One, 9*(3). e91016.

Ceulemans, R., Stettler, R. F., Hinckley, T. M., Isebrands, J. G., & Heilman, P. E. (1990). Crown architecture of Populus clones as determined by branch orientation and branch characteristics. *Tree Physiology,* 7(1_2_3_4), 157–167. http://doi.org/10.1093/treephys/7.1-2-3-4.157.

Chagné, D., Crowhurst, R. N., Pindo, M., Thrimawithana, A., Deng, C., Ireland, H., et al. (2014). The draft genome sequence of European pear (*Pyrus communis* L. 'Bartlett'). *PLoS One, 9*(4), e92644. http://doi.org/10.1371/journal.pone.0092644.

Chagné, D., Crowhurst, R. N., Troggio, M., Davey, M. W., Gilmore, B., Lawley, C., et al. (2012). Genome-wide SNP detection, validation, and development of an 8K SNP array for apple. *PLoS One,* 7(2), e31745. http://doi.org/10.1371/journal.pone.0031745.

Champagnat, P. (1965). Physiologie de la croissance et de l'inhibition des bourgeons: Dominance apicale et phénomènes analogues. In *Rameaux courts et rameaux longs: Problèmes physiologiques. Encyclopedia of Plant Physiology* (Vol. 15) (pp. 1106–1171).

Chancerel, E., Lamy, J.-B., Lesur, I., Noirot, C., Klopp, C., Ehrenmann, F., et al. (2013). High-density linkage mapping in a pine tree reveals a genomic region associated with inbreeding depression and provides clues to the extent and distribution of meiotic recombination. *BMC Biology, 11*(1), 50. http://doi.org/10.1186/1741-7007-11-50.

Chen, Q., Baldocchi, D., Gong, P., & Kelly, M. (2006). Isolating individual trees in a savanna woodland using small footprint lidar data. *Photogrammetric Engineering & Remote Sensing, 72*(8), 923–932.

Chen, Q., Gong, P., Baldocchi, D., & Tian, Y. Q. (2007). Estimating basal area and stem volume for individual trees from lidar data. *Photogrammetric Engineering & Remote Sensing, 73*(12), 1355–1365.

Cline, M. G. (2000). The role of hormones and apical dominance. New approaches to an old problem in plant development. *Physiologia Plantarum, 90*, 230–237. http://dx.doi.org/10.1111/j.1399-3054.1994.tb02216.x.

Cochard, H., Coste, S., Chanson, B., Guehl, J. M., & Nicolini, E. (2005). Hydraulic architecture correlates with bud organogenesis and primary shoot growth in beech (*Fagus sylvatica*). *Tree Physiology, 25*(12), 1545–1552.

Colin, F., Mothe, F., Freyburger, C., Morisset, J. B., Leban, J. M., & Fontaine, F. (2010). Tracking rameal traces in sessile oak trunks with X-ray computer tomography: biological bases, preliminary results and perspectives. *Trees Structure and Function, 24*, 953–967.

Conde, D., González-Melendi, P., & Allona, I. (2012). Poplar stems show opposite epigenetic patterns during winter dormancy and vegetative growth. *Trees, 27*(1), 311–320. http://doi.org/10.1007/s00468-012-0800-x.

Conner, P. J., Brown, S. K., & Weeden, N. F. (1998). Molecular-marker analysis of quantitative traits for growth and development in juvenile apple trees. *Theoretical and Applied Genetics, 96*(8), 1027–1035. http://doi.org/10.1007/s001220050835.

Costes, E., Crespel, L., Denoyes, B., Morel, P., Demene, M.-N., Lauri, P.-E., et al. (2014). Bud structure, position and fate generate various branching patterns along shoots of closely related Rosaceae species: a review. *Frontiers in Plant Science, 5.* http://doi.org/10.3389/fpls.2014.00666.

Costes, E., Lauri, P. E., & Regnard, J. L. (2006). Tree architecture and production. *Horticultural Reviews, 32*, 1–60. John Wiley [distributor]. Janick, J., & Wiley InterScience. http://site.ebrary.com/id/10383565

Costes, E., Sinoquet, H., Kelner, J.-J., & Godin, C. (2003). Exploring within-tree architectural development of two apple tree cultivars over 6 years. *Annals of Botany, 91*(1), 91−104.

Costes, E., Smith, C., Renton, M., Guédon, Y., Prusinkiewicz, P., & Godin, C. (2008). MAppleT: simulation of apple tree development using mixed stochastic and biomechanical models. *Functional Plant Biology, 35*(10), 936−950.

Da Silva, D., Han, L., Faivre, R., & Costes, E. (2014). Influence of the variation of geometrical and topological traits on light interception efficiency of apple trees: sensitivity analysis and metamodelling for ideotype definition. *Annals of Botany, 114*(4), 739−752. http://doi.org/10.1093/aob/mcu034.

Denis, M., & Bouvet, J.-M. (2012). Efficiency of genomic selection with models including dominance effect in the context of Eucalyptus breeding. *Tree Genetics & Genomes, 9*(1), 37−51. http://doi.org/10.1007/s11295-012-0528-1.

Devoghalaere, F., Doucen, T., Guitton, B., Keeling, J., Payne, W., Ling, T. J., et al. (2012). A genomics approach to understanding the role of auxin in apple (*Malus × domestica*) fruit size control. *BMC Plant Biology, 12*(1), 7.

Diao, J., De Reffye, P., Lei, X., Guo, H., & Letort, V. (2012). Simulation of the topological development of young eucalyptus using a stochastic model and sampling measurement strategy. *Computers and Electronics in Agriculture, 80*, 105−114. http://doi.org/10.1016/j.compag.2011.10.019.

Dickman, D. I., Gold, M. A., & Flore, J. A. (1994). The ideotype concept and the genetic improvement of tree crops. *Plant Breeding Review, 12*, 163−193.

Dill, A., Thomas, S. G., Hu, J., Steber, C. M., & Sun, T. (2004). The Arabidopsis F-box protein SLEEPY1 targets gibberellin signaling repressors for gibberellin-induced degradation. *The Plant Cell, 16*(6), 1392−1405. http://doi.org/10.1105/tpc.020958.

Dirlewanger, E., Quero-García, J., Le Dantec, L., Lambert, P., Ruiz, D., Dondini, L., et al. (2012). Comparison of the genetic determinism of two key phenological traits, flowering and maturity dates, in three Prunus species: peach, apricot and sweet cherry. *Heredity, 109*(5), 280−292.

Domagalska, M. A., & Leyser, O. (2011). Signal integration in the control of shoot branching. *Nature Reviews Molecular Cell Biology, 12*(4), 211−221. http://doi.org/10.1038/nrm3088.

Downes, G. M., Beadle, C., & Worledge, D. (1999). Daily stem growth patterns in irrigated *Eucalyptus globules* and *E. nitens* in relation to climate. *Trees, 14*, 102−111.

Drew, D. M., & Downes, G. M. (2009). The use of precision dendrometers in research on daily stem size and wood property variation: a review. *Dendrochronologia, 27*, 159−172.

Dubouzet, J. G., Strabala, T. J., & Wagner, A. (2013). Potential transgenic routes to increase tree biomass. *Plant Science, 212*, 72−101. http://doi.org/10.1016/j.plantsci.2013.08.006.

Dun, E. A., de Saint Germain, A., Rameau, C., & Beveridge, C. A. (2013). Dynamics of strigolactone function and shoot branching responses in *Pisum sativum*. *Molecular Plant, 6*(1), 128−140. http://doi.org/10.1093/mp/sss131.

Durand, J.-B., Guitton, B., Peyhardi, J., Holtz, Y., Guédon, Y., Trottier, C., et al. (2013). New insights for estimating the genetic value of segregating apple progenies for irregular bearing during the first years of tree production. *Journal of Experimental Botany*, 5099−5113. http://doi.org/10.1093/jxb/ert297.

Durel, C. E., Laurens, F., Fouillet, A., & Lespinasse, Y. (1998). Utilization of pedigree information to estimate genetic parameters from large unbalanced data sets in apple. *Theoretical and Applied Genetics, 96*(8), 1077−1085. http://doi.org/10.1007/s001220050842.

van Dyk, M. M., Soeker, M. K., Labuschagne, I. F., & Rees, D. J. G. (2010). Identification of a major QTL for time of initial vegetative budbreak in apple (*Malus × domestica* Borkh.). *Tree Genetics & Genomes, 6*(3), 489−502. http://doi.org/10.1007/s11295-009-0266-1.

Eldem, V., Çelikkol Akçay, U., Ozhuner, E., Bakır, Y., Uranbey, S., & Unver, T. (2012). Genome-wide identification of miRNAs responsive to drought in peach (*Prunus persica*) by high-throughput deep sequencing. *PLoS One, 7*(12), e50298. http://doi.org/10.1371/journal.pone.0050298.

Emhart, V. I., Martin, T. A., White, T. L., & Huber, D. A. (2007). Clonal variation in crown structure, absorbed photosynthetically active radiation and growth of loblolly pine and slash pine. *Tree Physiology, 27*(3), 421–430. http://doi.org/10.1093/treephys/27.3.421.

Endo, T., Shimada, T., Fujii, H., Kobayashi, Y., Araki, T., & Omura, M. (2005). Ectopic expression of an FT homolog from citrus confers an early flowering phenotype on trifoliate orange (*Poncirus trifoliata* L. Raf.). *Transgenic Research, 14*(5), 703–712. http://doi.org/10.1007/s11248-005-6632-3.

Eriksson, M. E., Hoffman, D., Kaduk, M., Mauriat, M., & Moritz, T. (2015). Transgenic hybrid aspen trees with increased gibberellin (GA) concentrations suggest that GA acts in parallel with FLOWERING LOCUS T2 to control shoot elongation. *New Phytologist, 205*(3), 1288–1295. http://doi.org/10.1111/nph.13144.

Falavigna, V. da S., Porto, D. D., Buffon, V., Margis-Pinheiro, M., Pasquali, G., & Revers, L. F. (2014). Differential transcriptional profiles of dormancy-related genes in apple buds. *Plant Molecular Biology Reporter, 32*(4), 796–813. http://doi.org/10.1007/s11105-013-0690-0.

Fernández-Fernández, F., Evans, K. M., Clarke, J. B., Govan, C. L., James, C. M., Marič, S., et al. (2008). Development of an STS map of an interspecific progeny of Malus. *Tree Genetics & Genomes, 4*(3), 469–479. http://doi.org/10.1007/s11295-007-0124-y.

Fisher, D. V. (1970). Spur strains of McIntosh discovered in British Columbia, Canada. *Fruit Varieties & Horticulture Digest, 24*, 27–32.

Fladung, M., Tusch, A., Markussen, T., & Ziegenhagen, B. (1999). Analysis of morphological mutants in Picea. In *Proceedings of the international congress on the applications of biotechnology to forest genetics, Vitoria-Gasteiz, September 22–25, 1999* (pp. 167–170).

Fleck, B., & Harberd, N. P. (2002). Evidence that the Arabidopsis nuclear gibberellin signalling protein GAI is not destabilised by gibberellin. *The Plant Journal, 32*(6), 935–947. http://doi.org/10.1046/j.1365-313X.2002.01478.x.

Foo, E., Turnbull, C. G., & Beveridge, C. A. (2001). Long-distance signaling and the control of branching in therms1 mutant of pea. *Plant Physiology, 126*(1), 203–209.

Foster, T., Kirk, C., Jones, W. T., Allan, A. C., Espley, R., Karunairetnam, S., et al. (2007). Characterisation of the DELLA subfamily in apple (*Malus* × *domestica* Borkh.). *Tree Genetics & Genomes, 3*(3), 187–197. http://doi.org/10.1007/s11295-006-0047-z.

Foster, T. M., Celton, J.-M., Chagné, D., Tustin, D. S., & Gardiner, S. E. (2015). Two quantitative trait loci, Dw1 and Dw2, are primarily responsible for rootstock-induced dwarfing in apple. *Horticulture Research, 2*, 15001. http://doi.org/10.1038/hortres.2015.1.

Foster, T. M., Watson, A. E., van Hooijdonk, B. M., & Schaffer, R. J. (2013). Key flowering genes including FT-like genes are upregulated in the vasculature of apple dwarfing rootstocks. *Tree Genetics & Genomes, 10*(1), 189–202. http://doi.org/10.1007/s11295-013-0675-z.

Frewen, B. E., Chen, T. H. H., Howe, G. T., Davis, J., Rohde, A., Boerjan, W., et al. (2000). Quantitative trait loci and candidate gene mapping of bud set and bud flush in populus. *Genetics, 154*(2), 837–845.

Fumey, D., Lauri, P.-É., Guédon, Y., Godin, C., & Costes, E. (2011). How young trees cope with removal of whole or parts of shoots: an analysis of local and distant responses to pruning in 1-year-old apple (*Malus* × *domestica*; Rosaceae) trees. *American Journal of Botany, 98*(11), 1737–1751.

Ghelardini, L., Berlin, S., Weih, M., Lagercrantz, U., Gyllenstrand, N., & Rönnberg-Wästljung, A. C. (2014). Genetic architecture of spring and autumn phenology in Salix. *BMC Plant Biology, 14*(1), 31. http://doi.org/10.1186/1471-2229-14-31.

Gion, J.-M., Chaumeil, P., & Plomion, C. (2014). EucaMaps: linking genetic maps and associated QTLs to the *Eucalyptus grandis* genome. *Tree Genetics & Genomes, 11*(1), 1–7. http://doi.org/10.1007/s11295-014-0795-0.

Gireesh, T., & Mydin, K. K. (2014). Field evaluation of progenies of a canopy mutant of *Hevea brasiliensis*. *Rubber Science, 27*(1), 61–68.

Gomez-Roldan, V., Fermas, S., Brewer, P. B., Puech-Pagès, V., Dun, E. A., Pillot, J.-P., et al. (2008). Strigolactone inhibition of shoot branching. *Nature, 455*(7210), 189–194. http://doi.org/10.1038/nature07271.

Grattapaglia, D. (2014). Breeding Forest trees by genomic selection: current progress and the way forward. In R. Tuberosa, A. Graner, & E. Frison (Eds.), *Genomics of plant genetic resources* (pp. 651–682). Dordrecht: Springer. Netherlands. Consulté à l'adresse http://link.springer.com/10.1007/978-94-007-7572-5_26.

Grattapaglia, D., & Resende, M. D. V. (2010). Genomic selection in forest tree breeding. *Tree Genetics & Genomes, 7*(2), 241–255. http://doi.org/10.1007/s11295-010-0328-4.

Greb, T., Clarenz, O., Schäfer, E., Müller, D., Herrero, R., Schmitz, G., et al. (2003). Molecular analysis of the LATERAL SUPPRESSOR gene in Arabidopsis reveals a conserved control mechanism for axillary meristem formation. *Genes & Development, 17*(9), 1175–1187.

Griffon, S., & de Coligny, F. (2014). AMAPstudio: an editing and simulation software suite for plants architecture modelling. *Ecological Modelling, 290*, 3–10. http://doi.org/10.1016/j.ecolmodel.2013.10.037.

Guitton, B., Kelner, J.-J., Velasco, R., Gardiner, S. E., Chagné, D., & Costes, E. (2012). Genetic control of biennial bearing in apple. *Journal of experimental botany, 63*(1), 131–149.

Hackett, W. P. (1985). Juvenility, maturation, and rejuvenation in woody plants. In J. Janick (Ed.), *Horticultural reviews* (pp. 109–155). John Wiley & Sons, Inc.

Hallé, F., Oldeman, R. A. A., & Tomlinson, P. B. (1978). *Tropical trees and forests*. Berlin: Springer-Verlag.

Hammami, S. B. M., León, L., Rapoport, H. F., & De la Rosa, R. (2011). Early growth habit and vigour parameters in olive seedlings. *Scientia Horticulturae, 129*(4), 761–768. http://doi.org/10.1016/j.scienta.2011.05.038.

Han, L., Costes, E., Boudon, F., Cokelaer, T., Pradal, C., Da Silva, D., et al. (2012). Investigating the influence of geometrical traits on light interception efficiency of apple trees: a modelling study with MAppleT. In *Plant growth modeling, simulation, visualization and applications (PMA), 2012 IEEE fourth international symposium on* (pp. 152–159). IEEE. Consulté à l'adresse http://ieeexplore.ieee.org/xpls/abs_all.jsp?arnumber=6524827.

Hanke, M.-V., Flachowsky, H., Peil, A., & Hättasch, C. (2007). No flower no fruit—genetic potentials to trigger flowering in fruit trees. *Genes Genomes Genomics, 1*(1), 1–20.

Hannrup, B., Cahalan, C., Chantre, G., Grabner, M., Karlsson, B., Le Bayon, I., et al. (2004). Genetic parameters of growth and wood quality traits in *Picea abies*. *Scandinavian Journal of Forest Research, 19*, 14–29.

Hein, P. R. G., Lima, J. T., & Chaix, G. (2010). Effects of sample preparation on NIR spectroscopic estimation of chemical properties of *Eucalyptus urophylla* ST Blake wood. *Holzforschung, 64*(1), 45–54.

Henning, J. G., & Radtke, P. J. (2006). Ground-based laser imaging for assessing three-dimensional forest canopy structure. *Photogrammetric Engineering & Remote Sensing, 72*, 1349–1358.

Hiraoka, Y., Takahashi, M., Nakamura, H., & Watanabe, A. (2014). Above-ground phenotyping of sugi using terrestrial lidar. In *Plant and animal genome XXII conference, San Diego, USA, January 2014.*

Horvath, D. P., Anderson, J. V., Chao, W. S., & Foley, M. E. (2003). Knowing when to grow: signals regulating bud dormancy. *Trends in Plant Science, 8*(11), 534–540. http://doi.org/10.1016/j.tplants.2003.09.013.

Hsu, C.-Y., Adams, J. P., Kim, H., No, K., Ma, C., Strauss, S. H., et al. (2011). FLOWERING LOCUS T duplication coordinates reproductive and vegetative growth in perennial poplar. *Proceedings of the National Academy of Sciences of the United States of America, 108*(26), 10756–10761. http://doi.org/10.1073/pnas.1104713108.

Hsu, C.-Y., Liu, Y., Luthe, D. S., & Yuceer, C. (2006). Poplar FT2 shortens the juvenile phase and promotes seasonal flowering. *The Plant Cell, 18*(8), 1846–1861. http://doi.org/10.1105/tpc.106.041038.

Hyyppä, J., Mielonen, T., Hyyppä, H., Maltamo, M., Yu, X., Honkavaara, E., et al. (2005). Using individual tree crown approach for forest volume extraction with aerial images and laser point clouds. In *International archives of photogrammetry, remote sensing and spatial information sciences, 12–14 September, Enschede, Netherlands, ISPRS WG III/3, III/4, V/3 Workshop laser scanning 2005* (Vol. XXXVI, pp. 144–149). Part 3/W19.

The International Peach Genome Initiative, Verde, I., Abbott, A. G., Scalabrin, S., Jung, S., Shu, S., & Rokhsar, D. S. (2013). The high-quality draft genome of peach (*Prunus persica*) identifies unique patterns of genetic diversity, domestication and genome evolution. *Nature Genetics, 45*(5), 487–494. http://doi.org/10.1038/ng.2586.

Iqbal, N., Nazar, R., Khan, M. I. R., Masood, A., & Khan, N. A. (2011). Role of gibberellins in regulation of source-sink relations under optimal and limiting environmental conditions. *Current Science (Bangalore), 100*(7), 998–1007.

Janick, J., Cummins, J. N., Brown, S. K., & Hemmat, M. (1996). In Janick, & Moore (Eds.), *Apples in fruit breeding, tree and tropical fruits*. John Wiley & Sons.

Janssen, B. J., Drummond, R. S., & Snowden, K. C. (2014). Regulation of axillary shoot development. *Current Opinion in Plant Biology, 17*, 28–35. http://doi.org/10.1016/j.pbi.2013.11.004.

Jiang, C., & Zeng, Z. B. (1995). Multiple trait analysis of genetic mapping for quantitative trait loci. *Genetics, 140*(3), 1111–1127.

Jiménez, S., Lawton-Rauh, A. L., Reighard, G. L., Abbott, A. G., & Bielenberg, D. G. (2009). Phylogenetic analysis and molecular evolution of the dormancy associated MADS-box genes from peach. *BMC Plant Biology, 9*(1), 81. http://doi.org/10.1186/1471-2229-9-81.

Jiménez, S., Reighard, G. L., & Bielenberg, D. G. (2010). Gene expression of DAM5 and DAM6 is suppressed by chilling temperatures and inversely correlated with bud break rate. *Plant Molecular Biology, 73*(1–2), 157–167. http://doi.org/10.1007/s11103-010-9608-5.

Johansson, M., & Staiger, D. (2014). SRR1 is essential to repress flowering in non-inductive conditions in *Arabidopsis thaliana*. *Journal of Experimental Botany, 65*(20), 5811–5822. http://doi.org/10.1093/jxb/eru317.

Jones, R. C., Hecht, V. F. G., Potts, B. M., Vaillancourt, R. E., & Weller, J. L. (2012). Expression of a FLOWERING LOCUS T homologue is temporally associated with annual flower bud initiation in *Eucalyptus globulus* subsp globulus (Myrtaceae). *Australian Journal of Botany, 59*(8), 756–769. http://doi.org/10.1071/bt11251.

Karki, L., & Tigerstedt, P. M. A. (1985). Definition and exploitation of forest tree ideotypes in Finland. In M. G. R. Cannell, & J. E. Jackson (Eds.), *Attributes of trees as crop plants* (pp. 102–109). UK: Inst. of Terrestrial Ecology.

Karlgren, A., Gyllenstrand, N., Clapham, D., & Lagercrantz, U. (2013). *FLOWERING LOCUS T/TERMINAL FLOWER1*-like genes affect growth rhythm and bud set in Norway Spruce. *Plant Physiology, 163*(2), 792–803. http://doi.org/10.1104/pp.113.224139.

Kenis, K., & Keulemans, J. (2007). Study of tree architecture of apple (*Malus* × *domestica* Borkh.) by QTL analysis of growth traits. *Molecular Breeding, 19*(3), 193–208. http://doi.org/10.1007/s11032-006-9022-5.

Kim, M., Canio, W., Kessler, S., & Sinha, N. (2001). Developmental changes due to long-distance movement of a homeobox fusion transcript in tomato. *Science, 293*(5528), 287–289. http://doi.org/10.1126/science.1059805.

Klagges, C., Campoy, J. A., Quero-García, J., Guzmán, A., Mansur, L., Gratacós, E., et al. (2013). Construction and comparative analyses of highly dense linkage maps of two sweet cherry intra-specific progenies of commercial cultivars. *PLoS One, 8*(1), e54743. http://doi.org/10.1371/journal.pone.0054743.

Kobayashi, Y., Kaya, H., Goto, K., Iwabuchi, M., & Araki, T. (1999). A pair of related genes with antagonistic roles in mediating flowering signals. *Science (New York, N.Y.), 286*(5446), 1960–1962.

Kocsy, G., Tari, I., Vanková, R., Zechmann, B., Gulyás, Z., Poór, P., et al. (2013). Redox control of plant growth and development. *Plant Science, 211*, 77–91. http://doi.org/10.1016/j.plantsci.2013.07.004.

Korol, A. B., Ronin, Y. I., & Kirzhner, V. M. (1995). Interval mapping of quantitative trait loci employing correlated trait complexes. *Genetics, 140*, 1137–1147.

Krost, C., Petersen, R., & Schmidt, E. R. (2012). The transcriptomes of columnar and standard type apple trees (Malus × domestica) – a comparative study. *Gene, 498*(2), 223–230. http://doi.org/10.1016/j.gene.2012.01.078.

Kube, P. D., Raymond, C. A., & Banham, P. W. (2001). Genetic parameters for diameter, basic density, cellulose content and fibre properties for Eucalyptus. *Forest Genetics, 8*, 285–294.

Kullan, A. R. K., van Dyk, M. M., Jones, N., Kanzler, A., Bayley, A., & Myburg, A. A. (2011). High-density genetic linkage maps with over 2,400 sequence-anchored DArT markers for genetic dissection in an F2 pseudo-backcross of *Eucalyptus grandis* × *E. urophylla*. *Tree Genetics & Genomes, 8*(1), 163–175. http://doi.org/10.1007/s11295-011-0430-2.

Lambeth, C. C., Endo, M., & Wright, J. (1994). Genetic analysis of 16 clonal trials of *Eucalyptus grandis* and comparisons with seedlings checks. *Forest Science, 40*, 397–411.

Lang, G. A., Early, J. D., Martin, G. C., & Darnell, R. L. (1987). Endo-, para-, and ecodormancy: physiological terminology and classification for dormancy research. *HortScience, 22*, 371–377.

Laurens, F., Audergon, J., Claverie, J., Duval, H., Germain, E., Kervella, J., et al. (2000). Integration of architectural types in French programmes of ligneous fruit species genetic improvement. *Fruits, 55*, 141–152.

Le Provost, G., Domergue, F., Lalanne, C., Ramos Campos, P., Grosbois, A., Bert, D., et al. (2013). Soil water stress affects both cuticular wax content and cuticle-related gene expression in young saplings of maritime pine (*Pinus pinaster* Ait). *BMC Plant Biology, 13*, 95.

Lee, J., & Lee, I. (2010). Regulation and function of SOC1, a flowering pathway integrator. *Journal of Experimental Botany, 61*(9), 2247–2254. http://doi.org/10.1093/jxb/erq098.

Lepoittevin, C., Rousseau, J. P., Guillemin, A., Gauvrit, C., Besson, F., Hubert, F., et al. (2011). Genetic parameters of growth, straightness and wood chemistry traits in *Pinus pinaster*. *Annals for Forest Science, 68*, 873–884.

Leroy, T., Montagnon, C., Charrier, A., & Eskes, A. B. (1993). Reciprocal recurrent selection applied to *Coffea canephora* Pierre. I. Characterization and evaluation of breeding populations and value of intergroup hybrids. *Euphytica, 67*(1–2), 113–125. http://doi.org/10.1007/BF00022734.

Lespinasse, Y. (1992). Breeding apple tree: aims and methods. In *Proc. joint conf. of the EAPR breeding and varietal assessment section and the Eucarpia potato section* (pp. 103–110).

Levillain, J., M'Bou, A. T., Deleporte, P., Saint-André, L., & Jourdan, C. (2011). Is the simple auger coring method reliable for below-ground standing biomass estimation in Eucalyptus forest plantations? *Annals of Botany*. mcr102 http://doi.org/10.1093/aob/mcr102.

Leyser, O. (2003). Regulation of shoot branching by auxin. *Trends in Plant Science, 8*(11), 541–545. http://doi.org/10.1016/j.tplants.2003.09.008.

Leyser, O. (2009). The control of shoot branching: an example of plant information processing. *Plant, Cell & Environment, 32*(6), 694–703. http://doi.org/10.1111/j.1365-3040.2009.01930.x.

Li, X., Qian, Q., Fu, Z., Wang, Y., Xiong, G., Zeng, D., et al. (2003). Control of tillering in rice. *Nature, 422*(6932), 618–621.

Li, Z., Reighard, G. L., Abbott, A. G., & Bielenberg, D. G. (2009). Dormancy-associated MADS genes from the EVG locus of peach [*Prunus persica* (L.) Batsch] have distinct seasonal and photoperiodic expression patterns. *Journal of Experimental Botany, 60*(12), 3521–3530. http://doi.org/10.1093/jxb/erp195.

Liebhard, R., Kellerhals, M., Pfammatter, W., Jertmini, M., & Gessler, C. (2003). Mapping quantitative physiological traits in apple (*Malus* × *domestica* Borkh.). *Plant Molecular Biology, 52*(3), 511–526. http://doi.org/10.1023/A:1024886500979.

Lind, M., Källman, T., Chen, J., Ma, X.-F., Bousquet, J., Morgante, M., et al. (2014). A *Picea abies* linkage map based on SNP markers identifies QTLs for four aspects of resistance to *Heterobasidion parviporum* infection. *PLoS One, 9*(7), e101049. http://doi.org/10.1371/journal.pone.0101049.

Looney, N. E., & Lane, W. D. (1984). Spur-type growth mutants of McIntosh apple: a review of their genetics, physiology and field performance. *Acta Horticultura, 146*, 31–46.

Loughin, T. M. (2006). Improved experimental design and analysis for long-term experiments. *Crop Science, 46*(6), 2492. http://doi.org/10.2135/cropsci2006.04.0271.

Lu, J., Wang, T., Xu, Z., Sun, L., & Zhang, Q. (2015). Genome-wide analysis of the GRAS gene family in *Prunus mume*. *Molecular Genetics and Genomics, 290*(1), 303–317. http://doi.org/10.1007/s00438-014-0918-1.

Lu, Q., Zhao, L., Li, D., Hao, D., Zhan, Y., & Li, W. (2014). A GmRAV ortholog is involved in photoperiod and sucrose control of flowering time in soybean. *PLoS One, 9*(2), e89145. http://doi.org/10.1371/journal.pone.0089145.

Mandrou, E., Villar, E., Hein, P., Vigneron, P., Plomion, C., & Gion, J. M. (2012). A candidate gene for lignin composition in Eucalyptus: cinnamoyl CoA reductase (CCR). *Trees Genetics & Genomes, 8*, 353–364.

Mangin, B., Thoquet, P., & Grimsley, N. (1998). Pleiotropic QTL analysis. *Biometrics, 54*(1), 88–99. http://doi.org/10.2307/2533998.

Marron, N., Bastien, C., Sabatti, M., Taylor, G., & Ceulemans, R. (2006). Plasticity of growth and sylleptic branchiness in two poplar families grown at three sites across Europe. *Tree Physiology, 26*(7), 935–946. http://doi.org/10.1093/treephys/26.7.935.

Martí, C., Orzáez, D., Ellul, P., Moreno, V., Carbonell, J., & Granell, A. (2007). Silencing of DELLA induces facultative parthenocarpy in tomato fruits. *The Plant Journal, 52*(5), 865–876. http://doi.org/10.1111/j.1365-313X.2007.03282.x.

Martínez-García, P. J., Parfitt, D. E., Ogundiwin, E. A., Fass, J., Chan, H. M., Ahmad, R., et al. (2013). High density SNP mapping and QTL analysis for fruit quality characteristics in peach (*Prunus persica* L.). *Tree Genetics & Genomes, 9*(1), 19–36. http://doi.org/10.1007/s11295-012-0522-7.

Martre, P., Quilot-Turion, B., Luquet, D., Memmah, M.-M. O.-S., Chenu, K., & Debaeke, P. (2015). Model-assisted phenotyping and ideotype design. In *Crop physiology* (pp. 349–373). Elsevier. Consulté à l'adresse http://linkinghub.elsevier.com/retrieve/pii/B9780124171046000145.

Mason, M. G., Ross, J. J., Babst, B. A., Wienclaw, B. N., & Beveridge, C. A. (2014). Sugar demand, not auxin, is the initial regulator of apical dominance. *Proceedings of the National Academy of Sciences of the United States of America, 111*(16), 6092–6097. http://doi.org/10.1073/pnas.1322045111.

McCulloh, K. A., & Sperry, J. S. (2005). Patterns in hydraulic architecture and their implications for transport efficiency. *Tree Physiology, 25*(3), 257–267. http://doi.org/10.1093/treephys/25.3.257.

Miller, S. A., Broom, F. D., Thorp, T. G., & Barnett, A. M. (2001). Effects of leader pruning on vine architecture, productivity and fruit quality in kiwifruit (*Actinidia deliciosa* cv. Hayward). *Scientia Horticulturae, 91*(3), 189–199.

Monet, R., Bastard, Y., & Gibault, B. (1988). Etude génétique du caractère 'port pleureur' chez le pêcher. *Agronomie, 8*(2), 127–132.

Monselise, S. P., & Goldschmidt, E. E. (1982). Alternate bearing in fruit trees. *Horticultural Reviews, 4*, 128–173.

Moriguchi, Y., Ujino-Ihara, T., Uchiyama, K., Futamura, N., Saito, M., Ueno, S., et al. (2012). The construction of a high-density linkage map for identifying SNP markers that are tightly linked to a nuclear-recessive major gene for male sterility in *Cryptomeria japonica* D. Don. *BMC Genomics, 13*(1), 95. http://doi.org/10.1186/1471-2164-13-95.

Morimoto, T., & Banno, K. (2015). Genetic and physical mapping of Co, a gene controlling the columnar trait of apple. *Tree Genetics & Genomes, 11*(1). http://doi.org/10.1007/s11295-014-0807-0.

Morisset, J. B., Mothe, F., Bock, J., Breda, N., & Colin, F. (2012). Epicormic ontogeny in *Quercus petraea* Liebl. constrains the highly plausible control of epicormic sprouting by water and carbohydrates. *Annals of Botany, 109*, 365–377.

Moriya, S., Okada, K., Haji, T., Yamamoto, T., & Abe, K. (2012). Fine mapping of *Co*, a gene controlling columnar growth habit located on apple (*Malus* × *domestica* Borkh.) linkage group 10: Fine mapping of *Co* in apple (*Malus* × *domestica* Borkh. *Plant Breeding, 131*(5), 641–647. http://doi.org/10.1111/j.1439-0523.2012.01985.x.

Mouradov, A., Cremer, F., & Coupland, G. (2002). Control of flowering time interacting pathways as a basis for diversity. *The Plant Cell, 14*(Suppl. 1), S111–S130. http://doi.org/10.1105/tpc.001362.

Muchero, W., Guo, J., DiFazio, S. P., Chen, J.-G., Ranjan, P., Slavov, G. T., et al. (2015). High-resolution genetic mapping of allelic variants associated with cell wall chemistry in *Populus*. *BMC Genomics, 16*(1), 24. http://doi.org/10.1186/s12864-015-1215-z.

Muñoz-Fambuena, N., Mesejo, C., Agustí, M., Tárraga, S., Iglesias, D. J., Primo-Millo, E., et al. (2013). Proteomic analysis of "Moncada" mandarin leaves with contrasting fruit load. *Plant Physiology and Biochemistry, 62*, 95–106. http://doi.org/10.1016/j.plaphy.2012.10.020.

Myburg, A. A., Grattapaglia, D., Tuskan, G. A., Hellsten, U., Hayes, R. D., Grimwood, J., et al. (2014). The genome of *Eucalyptus grandis*. *Nature, 510*(7505), 356–362. http://doi.org/10.1038/nature13308.

Neves, L. G., Mamani, E. M., Alfenas, A. C., Kirst, M., & Grattapaglia, D. (2011). A high-density transcript linkage map with 1,845 expressed genes positioned by microarray-based Single Feature Polymorphisms (SFP) in Eucalyptus. *BMC Genomics, 12*(1), 189. http://doi.org/10.1186/1471-2164-12-189.

Otto, D., Petersen, R., Brauksiepe, B., Braun, P., & Schmidt, E. R. (2013). The columnar mutation ("Co gene") of apple (*Malus* × *domestica*) is associated with an integration of a Gypsy-like retrotransposon. *Molecular Breeding, 33*(4), 863–880. http://doi.org/10.1007/s11032-013-0001-3.

Paul, L. K., Rinne, P. L., & van der Schoot, C. (2014). Shoot meristems of deciduous woody perennials: self-organization and morphogenetic transitions. *Current Opinion in Plant Biology, 17*, 86–95. http://doi.org/10.1016/j.pbi.2013.11.009.

Pelgas, B., Bousquet, J., Meirmans, P. G., Ritland, K., & Isabel, N. (2011). QTL mapping in white spruce: gene maps and genomic regions underlying adaptive traits across pedigrees, years and environments. *BMC Genomics, 12*(1), 145. http://doi.org/10.1186/1471-2164-12-145.

Peng, J., Richards, D. E., Hartley, N. M., Murphy, G. P., Devos, K. M., Flintham, J. E., et al. (1999). 'Green revolution' genes encode mutant gibberellin response modulators. *Nature, 400*(6741), 256–261. http://doi.org/10.1038/22307.

Petersen, R., & Krost, C. (2013). Tracing a key player in the regulation of plant architecture: the columnar growth habit of apple trees (*Malus* × *domestica*). *Planta, 238*(1), 1–22. http://doi.org/10.1007/s00425-013-1898-9.

Plomion, C., Durel, C.-E., & O'Malley, D. M. (1996). Genetic dissection of height in maritime pine seedlings raised under accelerated growth conditions. *Theoretical and Applied Genetics, 93*(5–6), 849–858. http://doi.org/10.1007/BF00224085.

Prassinos, C., Ko, J.-H., Lang, G., Iezzoni, A. F., & Han, K.-H. (2009). Rootstock-induced dwarfing in cherries is caused by differential cessation of terminal meristem growth and is triggered by rootstock-specific gene regulation. *Tree Physiology, 29*(7), 927–936. http://doi.org/10.1093/treephys/tpp027.

Quilot-Turion, B., Ould-Sidi, M.-M., Kadrani, A., Hilgert, N., Génard, M., & Lescourret, F. (2012). Optimization of parameters of the 'Virtual Fruit' model to design peach genotype for sustainable production systems. *European Journal of Agronomy, 42*, 34–48. http://doi.org/10.1016/j.eja.2011.11.008.

Rameau, C., Bertheloot, J., Leduc, N., Andrieu, B., Foucher, F., & Sakr, S. (2015). Multiple pathways regulate shoot branching. *Frontiers in Plant Science, 5*. http://doi.org/10.3389/fpls.2014.00741.

Renton, M., Guédon, Y., Godin, C., & Costes, E. (2006). Similarities and gradients in growth unit branching patterns during ontogeny in 'Fuji'apple trees: a stochastic approach. *Journal of Experimental Botany, 57*(12), 3131–3143.

Renton, M., Hanan, J., Ferguson, B. J., & Beveridge, C. A. (2012). Models of long-distance transport: how is carrier-dependent auxin transport regulated in the stem? *New Phytologist, 194*(3), 704–715. http://doi.org/10.1111/j.1469-8137.2012.04093.x.

Robles, P., & Pelaz, S. (2005). Flower and fruit development in *Arabidopsis thaliana*. *The International Journal of Developmental Biology, 49*(5–6), 633–643. http://doi.org/10.1387/ijdb.052020pr.

Rohde, A., & Bhalerao, R. P. (2007). Plant dormancy in the perennial context. *Trends in Plant Science, 12*(5), 217–223. http://doi.org/10.1016/j.tplants.2007.03.012.

Ruonala, R., Rinne, P. L. H., Baghour, M., Moritz, T., Tuominen, H., & Kangasjärvi, J. (2006). Transitions in the functioning of the shoot apical meristem in birch (*Betula pendula*) involve ethylene. *The Plant Journal, 46*(4), 628–640. http://doi.org/10.1111/j.1365-313X.2006.02722.x.

Sadok, I. B., Celton, J. M., Essalouh, L., Zine El Abidine, A., Garcia, G., Martinez, S., et al. (2013). QTL mapping of flowering and fruiting traits in olive. *Plos One, 8*(5), e62831. http://dx.doi.org/10.1371/journal.pone.0062831.

Sadok, I. B., Garcia, G., Martinez, S., Moutier, N., Leon, L., Belaj, A., et al. Plasticity in vegetative growth over contrasted growing sites of an F1 olive tree progeny during its juvenile phase. *Plos One*, in press.

Sadok, I. B., Moutier, N., Garcia, G., Dosba, F., Grati-Kamoun, N., Rebai, A., et al. (2012). Genetic determinism of the vegetative and reproductive traits in an F1 olive tree progeny. *Tree Genetics & Genomes, 9*(1), 205–221. http://doi.org/10.1007/s11295-012-0548-x.

Saint-André, L., M'Bou, A. T., Mabiala, A., Mouvondy, W., Jourdan, C., Roupsard, O., et al. (2005). Age-related equations for above- and below-ground biomass of a Eucalyptus hybrid in Congo. *Forest Ecology and Management, 205*(1–3), 199–214. http://doi.org/10.1016/j.foreco.2004.10.006.

Samach, A., & Smith, H. M. (2013). Constraints to obtaining consistent annual yields in perennials. II: environment and fruit load affect induction of flowering. *Plant Science, 207*, 168–176.

Sánchez-Pérez, R., Dicenta, F., & Martínez-Gómez, P. (2012). Inheritance of chilling and heat requirements for flowering in almond and QTL analysis. *Tree Genetics & Genomes, 8*(2), 379–389. http://doi.org/10.1007/s11295-011-0448-5.

Sawa, M., & Kay, S. A. (2011). GIGANTEA directly activates Flowering Locus T in *Arabidopsis thaliana*. *Proceedings of the National Academy of Sciences of the United States of America, 108*(28), 11698–11703. http://doi.org/10.1073/pnas.1106771108.

van der Schoot, C., Paul, L. K., & Rinne, P. L. H. (2014). The embryonic shoot: a lifeline through winter. *Journal of Experimental Botany, 65*(7), 1699–1712. http://doi.org/10.1093/jxb/ert413.

Schumacher, K., Schmitt, T., Rossberg, M., Schmitz, G., & Theres, K. (1999). The Lateral suppressor (Ls) gene of tomato encodes a new member of the VHIID protein family. *Proceedings of the National Academy of Sciences of the United States of America, 96*(1), 290–295.

Scorza, R. (1984). Characterization of four distinct peach tree growth types. *Journal of the American Society for Horticultural Science, 109*, 455–457.

Scorza, R., Miller, S., Glenn, D. M., & Tworkoski, W. O. T. (2006). USDA-ARS USDA-ARS. In *Proceedings of the VIth international peach symposium: Santiago, Chile, January 9–14, 2005* (p. 61). International Society for Horticultural Science.

Scotti-Saintagne, C., Bodénès, C., Barreneche, T., Bertocchi, E., Plomion, C., & Kremer, A. (2004). Detection of quantitative trait loci controlling bud burst and height growth in *Quercus robur* L. *Theoretical and Applied Genetics, 109*(8), 1648–1659. http://doi.org/10.1007/s00122-004-1789-3.

Segura, V., Cilas, C., & Costes, E. (2008). Dissecting apple tree architecture into genetic, ontogenetic and environmental effects: mixed linear modelling of repeated spatial and temporal measures. *New Phytologist, 178*(2), 302–314. http://doi.org/10.1111/j.1469-8137.2007.02374.x.

Segura, V., Cilas, C., Laurens, F., & Costes, E. (2006). Phenotyping progenies for complex architectural traits: a strategy for 1-year-old apple trees (*Malus × domestica* Borkh.). *Tree Genetics & Genomes, 2*(3), 140–151. http://doi.org/10.1007/s11295-006-0037-1.

Segura, V., Denancé, C., Durel, C.-E., & Costes, E. (2007). Wide range QTL analysis for complex architectural traits in a 1-year-old apple progeny. *Genome, 50*(2), 159–171. http://doi.org/10.1139/G07-002.

Segura, V., Durel, C.-E., & Costes, E. (2009). Dissecting apple tree architecture into genetic, ontogenetic and environmental effects: QTL mapping. *Tree Genetics & Genomes, 5*(1), 165–179.

Sgamma, T., Jackson, A., Muleo, R., Thomas, B., & Massiah, A. (2014). TEMPRANILLO is a regulator of juvenility in plants. *Scientific Reports, 4*. http://doi.org/10.1038/srep03704.

Shalom, L., Samuels, S., Zur, N., Shlizerman, L., Doron-Faigenboim, A., Blumwald, E., et al. (2014). Fruit load induces changes in global gene expression and in abscisic acid (ABA) and indole acetic acid (IAA) homeostasis in citrus buds. *Journal of Experimental Botany*. eru148. http://doi.org/10.1093/jxb/eru148

Shelbourne, C. J. A., Apialoza, L. A., Jayawickrama, K. J. S., & Sorensson, C. T. (1997). Developing breeding objectives for radiata pine in New Zealand. In R. D. Burdon, & J. M. Moore (Eds.), *"IUFRO '97 genetics of radiata Pine". Proceedings of NZ FRI-IUFRO conference 1–4 December and Workshop 5 December, Rotorua, New Zealand. FRI Bull.* (Vol. 203, pp. 129–141).

Shim, D., Ko, J.-H., Kim, W.-C., Wang, Q., Keathley, D. E., & Han, K.-H. (2014). A molecular framework for seasonal growth-dormancy regulation in perennial plants. *Horticulture Research, 1*, 14059. http://doi.org/10.1038/hortres.2014.59.

Silverstone, A. L., Jung, H.-S., Dill, A., Kawaide, H., Kamiya, Y., & Sun, T. (2001). Repressing a repressor gibberellin-induced rapid reduction of the RGA protein in Arabidopsis. *The Plant Cell, 13*(7), 1555–1566. http://doi.org/10.1105/TPC.010047.

Sirault, X., Fripp, J., Paproki, A., Kuffner, P., Nguyen, C., Li, R., et al. (2013). PlantScan: a three-dimensional phenotyping platform for capturing the structural dynamic of plant development and growth. In *Proceedings of the 7th international conference on functional structural plant models* (pp. 45–48). Saariselka, Finland.

Snowdon, P., Raison, J., Keith, H., Ritson, P., Grierson, P., Adams, M., et al. (2002). *Protocol for sampling tree and stand biomass*. Canberra: Australian Greenhouse Office, 2002. NCAS Technical Report No. 31.

Socquet-Juglard, D., Christen, D., Devènes, G., Gessler, C., Duffy, B., & Patocchi, A. (2013). Mapping architectural, phenological, and fruit quality QTLs in apricot. *Plant Molecular Biology Reporter, 31*(2), 387–397.

Song, Y. H., Ito, S., & Imaizumi, T. (2013). Flowering time regulation: photoperiod- and temperature-sensing in leaves. *Trends in Plant Science, 18*(10), 575–583. http://doi.org/10.1016/j.tplants.2013.05.003.

Srinivasan, C., Dardick, C., Callahan, A., & Scorza, R. (2012). Plum (*Prunus domestica*) trees transformed with poplar FT1 result in altered architecture, dormancy requirement, and continuous flowering. *PLoS One, 7*(7), e40715. http://doi.org/10.1371/journal.pone.0040715.

St Clair, J. B. (1993). Family differences in equations for predicting biomass and leaf area in Douglas-fir (*Pseudotsuga menziesii* var. menziesii). *Forest Sciences, 39*, 743–755.

Sussex, I. M., & Kerk, N. M. (2001). The evolution of plant architecture. *Current Opinion in Plant Biology, 4*(1), 33–37.

Tian, Y.-K., Wang, C.-H., Zhang, J.-S., James, C., & Dai, H.-Y. (2005). Mapping Co, a gene controlling the columnar phenotype of apple, with molecular markers. *Euphytica, 145*(1–2), 181–188. http://doi.org/10.1007/s10681-005-1163-9.

Tombesi, S., Johnson, R. S., Day, K. R., & DeJong, T. M. (2010). Interactions between rootstock, inter-stem and scion xylem vessel characteristics of peach trees growing on rootstocks with contrasting size-controlling characteristics. *AoB Plants, 2010*. http://doi.org/10.1093/aobpla/plq013.

Tränkner, C., Lehmann, S., Hoenicka, H., Hanke, M.-V., Fladung, M., Lenhardt, D., et al. (2010). Over-expression of an FT-homologous gene of apple induces early flowering in annual and perennial plants. *Planta, 232*(6), 1309–1324. http://doi.org/10.1007/s00425-010-1254-2.

Tsarouhas, V., Gullberg, U., & Lagercrantz, U. (2003). Mapping of quantitative trait loci controlling timing of bud flush in Salix. *Hereditas, 138*(3), 172–178. http://doi.org/10.1034/j.1601-5223.2003.01695.x.

Tuskan, G. A., DiFazio, S., Jansson, S., Bohlmann, J., Grigoriev, I., Hellsten, U., et al. (2006). The genome of black cottonwood, *Populus trichocarpa* (Torr. & Gray). *Science, 313*(5793), 1596–1604. http://doi.org/10.1126/science.1128691.

Ubi, B. E., Sakamoto, D., Ban, Y., Shimada, T., Ito, A., Nakajima, I., et al. (2010). Molecular cloning of dormancy-associated MADS-box gene homologs and their characterization during seasonal endodormancy transitional phases of Japanese pear. *Journal of the American Society for Horticultural Science, 135*, 174–182.

Ueno, S., Klopp, C., Leplé, J. C., Derory, J., Noirot, C., Léger, V., et al. (2013). Transcriptional profiling of bud dormancy induction and release in oak by next-generation sequencing. *BMC Genomics, 14*(1), 236. http://doi.org/10.1186/1471-2164-14-236.

Umehara, M., Hanada, A., Yoshida, S., Akiyama, K., Arite, T., Takeda-Kamiya, N., et al. (2008). Inhibition of shoot branching by new terpenoid plant hormones. *Nature, 455*(7210), 195–200. http://doi.org/10.1038/nature07272.

Velasco, R., Zharkikh, A., Affourtit, J., Dhingra, A., Cestaro, A., Kalyanaraman, A., et al. (2010). The genome of the domesticated apple (*Malus* × *domestica* Borkh.). *Nature Genetics, 42*(10), 833–839. http://doi.org/10.1038/ng.654.

Verhaegen, D., Plomion, C., Gion, J.-M., Poitel, M., Costa, P., & Kremer, A. (1997). Quantitative trait dissection analysis in Eucalyptus using RAPD markers: 1. Detection of QTL in interspecific hybrid progeny, stability of QTL expression across different ages. *Theoretical and Applied Genetics, 95*(4), 597–608. http://doi.org/10.1007/s001220050601.

Villar, E., Klopp, C., Noirot, C., Novaes, E., Kirst, M., Plomion, C., et al. (2011). RNA-Seq reveals genotype-specific molecular responses to water deficit in eucalyptus. *BMC Genomics, 12*(1), 538. http://doi.org/10.1186/1471-2164-12-538.

Virlet, N., Lebourgeois, V., Martinez, S., Costes, E., Labbe, S., & Regnard, J.-L. (2014). Stress indicators based on airborne thermal imagery for field phenotyping a heterogeneous tree population for response to water constraints. *Journal of Experimental Botany, 65*(18), 5429–5442. http://doi.org/10.1093/jxb/eru309.

Virlet, N., Costes, E., Martinez, S., Kelner, J. J. & Regnard, J. L. (under submission). Multispectral airborne imagery at field opens the way to genetic determinism of morphological and transpiration traits of an apple tree hybrid population in response to water deficit. *J. Exp. Bot.*

Wahl, V., Ponnu, J., Schlereth, A., Arrivault, S., Langenecker, T., Franke, A., et al. (2013). Regulation of flowering by trehalose-6-phosphate signaling in *Arabidopsis thaliana*. *Science, 339*(6120), 704–707. http://doi.org/10.1126/science.1230406.

Wang, J.-W., Park, M. Y., Wang, L.-J., Koo, Y., Chen, X.-Y., Weigel, D., et al. (2011). MiRNA control of vegetative phase change in trees. *PLoS Genetics,* 7(2), e1002012. http://doi.org/10.1371/journal.pgen.1002012.

Wang, S.-S., Liu, Z.-Z., Sun, C., Shi, Q.-H., Yao, Y.-X., You, C.-X., et al. (2012). Functional characterization of the apple MhGAI1 gene through ectopic expression and grafting experiments in tomatoes. *Journal of Plant Physiology, 169*(3), 303–310. http://doi.org/10.1016/j.jplph.2011.09.012.

Watkins, R., & Spangelo, L. P. S. (1970). Components of genetic variance for plant survival and vigor of apple trees. *Theoretical and Applied Genetics, 40*(5), 195–203. http://doi.org/10.1007/BF00285242.

White, J. (1979). The plant as a metapopulation. *Annual Review of Ecological Systems*, 109–145.

White, J. C., Wulder, M. A., Varhola, A., Vastaranta, M., Coops, N. C., Cook, B. D., et al. (2013). *A best practices guide for generating forest inventory attributes from airborne laser scanning data using the area-based approach.* Information Report FI-X-10. Victoria, BC: Natural Resources Canada, Canadian Forest Service, Canadian Wood Fibre Centre, Pacific Forestry Centre.

Wilkie, J. D., Sedgley, M., & Olesen, T. (2008). Regulation of floral initiation in horticultural trees. *Journal of Experimental Botany, 59*(12), 3215–3228. http://doi.org/10.1093/jxb/ern188.

Willige, B. C., Ghosh, S., Nill, C., Zourelidou, M., Dohmann, E. M. N., Maier, A., et al. (2007). The DELLA domain of GA INSENSITIVE mediates the interaction with the GA INSENSITIVE DWARF1A gibberellin receptor of Arabidopsis. *The Plant Cell, 19*(4), 1209–1220. http://doi.org/10.1105/tpc.107.051441.

Wingler, A. (2015). Comparison of signaling interactions determining annual and perennial plant growth in response to low temperature. *Frontiers in Plant Science, 5.* http://doi.org/10.3389/fpls.2014.00794.

Wisniewski, M., Norelli, J., Bassett, C., Artlip, T., & Macarisin, D. (2011). Ectopic expression of a novel peach (*Prunus persica*) CBF transcription factor in apple (*Malus × domestica*) results in short-day induced dormancy and increased cold hardiness. *Planta, 233*(5), 971–983. http://doi.org/10.1007/s00425-011-1358-3.

Wolters, P. J., Schouten, H. J., Velasco, R., Si-Ammour, A., & Baldi, P. (2013). Evidence for regulation of columnar habit in apple by a putative 2OG-Fe(II) oxygenase. *New Phytologist, 200*(4), 993–999. http://doi.org/10.1111/nph.12580.

Wu, G., Park, M. Y., Conway, S. R., Wang, J.-W., Weigel, D., & Poethig, R. S. (2009). The sequential action of miR156 and miR172 regulates developmental timing in Arabidopsis. *Cell, 138*(4), 750–759. http://doi.org/10.1016/j.cell.2009.06.031.

Wu, R., & Hinckley, T. M. (2001). Phenotypic plasticity of sylleptic branching: genetic design of tree architecture. *Critical Reviews in Plant Sciences, 20*, 467–485. http://doi.org/10.1080/07352689.2001.10131827.

Wu, R., & Lin, M. (2006). Functional mapping – how to map and study the genetic architecture of dynamic complex traits. *Nature Reviews. Genetics,* 7(3), 229–237. http://doi.org/10.1038/nrg1804.

Wu, R., & Stettler, R. F. (1994). Quantitative genetics of growth and development in Populus. I. A three-generation comparison of tree architecture during the first 2 years of growth. *Theoretical and Applied Genetics, 89*(7–8), 1046–1054. http://doi.org/10.1007/BF00224537.

Wu, R., & Stettler, R. F. (1996). The genetic resolution of juvenile canopy structure and function in a three-generation pedigree of *Populus*. *Trees, 11*(2), 99–108. http://doi.org/10.1007/PL00009661.

Wu, R., & Stettler, R. F. (1998). Quantitative genetics of growth and development in Populus. III. Phenotypic plasticity of crown structure and function. *Heredity, 81*(3), 299–310. http://doi.org/10.1046/j.1365-2540.1998.00397.x.

Wu, R. L. (1998). Genetic mapping of QTLs affecting tree growth and architecture in Populus: implication for ideotype breeding. *Theoretical and Applied Genetics, 96*(3–4), 447–457. http://doi.org/10.1007/s001220050761.

Xu, Q., Chen, L.-L., Ruan, X., Chen, D., Zhu, A., Chen, C., et al. (2013). The draft genome of sweet orange (Citrus sinensis). *Nature Genetics, 45*(1), 59–66. http://doi.org/10.1038/ng.2472.

Yoo, S. J., Chung, K. S., Jung, S. H., Yoo, S. Y., Lee, J. S., & Ahn, J. H. (2010). BROTHER OF FT and TFL1 (BFT) has TFL1-like activity and functions redundantly with TFL1 in inflorescence meristem development in Arabidopsis: role of BFT in inflorescence meristem development. *The Plant Journal, 63*(2), 241–253. http://doi.org/10.1111/j.1365-313X.2010.04234.x.

Zhang, Y., Zhu, J., & Dai, H. (2012). Characterization of transcriptional differences between columnar and standard apple trees using RNA-seq. *Plant Molecular Biology Reporter, 30*(4), 957–965. http://doi.org/10.1007/s11105-011-0396-0.

Zhu, L. H., Li, X. Y., & Welander, M. (2007). Overexpression of the Arabidopsis gai gene in apple significantly reduces plant size. *Plant Cell Reports, 27*(2), 289–296. http://doi.org/10.1007/s00299-007-0462-0.

Zhu, Q.-H., & Helliwell, C. A. (2011). Regulation of flowering time and floral patterning by miR172. *Journal of Experimental Botany, 62*(2), 487–495. http://doi.org/10.1093/jxb/erq295.

Zimin, A., Stevens, K. A., Crepeau, M. W., Holtz-Morris, A., Koriabine, M., Marçais, G., et al. (2014). Sequencing and assembly of the 22-Gb loblolly pine genome. *Genetics, 196*(3), 875–890. http://doi.org/10.1534/genetics.113.159715.

CHAPTER SIX

The Genetic Control of Bud-Break in Trees

Albert G. Abbott[*,1], **Tatyana Zhebentyayeva**[§], **Abdelali Barakat**[¶], **Zongrang Liu**[||]

[*]Department of Forestry, University of Kentucky, Lexington, KY, USA
[§]Genomics and Computational Biology Laboratory, Clemson University, Clemson, SC, USA
[¶]Department of Biology, University of South Dakota, Vermillion, SD, USA
[||]Appalachian Fruit Research Station, Kearneysville, WV, USA
[1]Corresponding author: E-mail: albert.abbott@uky.edu

Contents

Abstract

To exploit ecogeographical niches of the temperate zone, trees and other perennial plant species optimize the timing of floral and vegetative bud break within the dynamic nature of the annual environmental cycle. This insures successful reproduction and continued growth over their lifespan. Adapting to the annual environmental cycle as well as more long-term environmental fluctuations, such as, global warming is critical to insuring survival of the species. Thus, the tree must utilize physiological systems to sense and respond in time and space to the appropriate environmental cues. Historically, many of the physiological players such as phytohormones and their actions have been well characterized, however, the underlying regulation of the genes and gene networks controlling these physiological systems remains relatively unknown.

Advances in Botanical Research, Volume 74
ISSN 0065-2296
http://dx.doi.org/10.1016/bs.abr.2015.04.002

With advances in molecular genetics and genomics technologies, we are beginning to understand the genetic basis for a number of important phenological traits. It is the intent of this review to highlight our current state of knowledge for the genetics of the timing of bud-break in trees focusing on both fruit and forest trees to provide a framework to guide future efforts to understand how the tree coordinates annual growth, development and reproduction with the environmental cycle.

1. INTRODUCTION

Perennial tree crops, as individual species, underpin many regional and local economies worldwide. Though often seen as minor crop species, when one considers them together, they represent some of the most important species to the success of the human endeavour. These crops are used for: human nutrition, energy production, construction materials, ornamental applications and many are important dominant species of our natural ecosystems.

While comprised of species from many different families, these species are unified in several key features that are commonly evident in their biology and agriculture: (1) individual plants are long lived thus they serve as long-term genetic repositories; (2) they can be clonally propagated for production meaning that individual varieties once developed can dominate the agricultural landscape for a significant period of time; (3) they often exhibit long juvenility periods significantly limiting the rate of development of new varieties and (4) they often require significant space and resources for breeding and production. The long life and long generation times exhibited by many of the tree species and their ecological dominance, however, places them in an extremely precarious position when environmental change occurs rapidly. This is particularly evident in species (such as fruiting trees) that have reproductive systems adapted to particular environmental zones. Unlike annual plants that complete their life cycle in a single season where planting and harvesting can be adjusted to the changing climatic conditions and breeding new selections is more rapid, production in perennial tree crops (which are not replanted every year) relies on repeated vegetative and floral cycles over many years and thus these species are particularly susceptible to the detrimental effects of rapid long-term climate change.

As a likely consequence of broad adaptation by many perennial species to the different temperate zones, many important phenological traits (e.g. bloom date, dormancy, chilling requirement (CR), seed stratification), exhibit complex inheritance quantative trait locus (QTL) patterns with multiple genes contributing to the final phenotype. These phenological traits are effectively adaptive responses to the impacts imposed by environmental stress of which cold and heat are significant in the annual growth and development cycle of perennial trees species. Since adaptation to these abiotic stresses determines the quality and quantity of the production of these trees in agricultural and silvicultural contexts, uncovering the genetic foundations of these phenological traits is critical for rapidly manipulating them in tree breeding programs. Having markers for alleles of genes that influence the extent and timing of bloom are critical needs for tree breeding programs to meet the challenges imposed by the rapidly changing environmental conditions. In this review, we present a current picture of the genetics of the timing of bud-break in fruit and forest trees with focus on the tree species of the temperate ecogeographical regions.

In recent years a significant amount of information on various stress responses in a myriad of different plant species has been published, however much of this work is difficult to integrate because within each species we only have a fragmentary snapshot of the response in time and space. The literature on dormancy in perennial tree species is a prime example of the inherent complexity of trying to build a comprehensive picture of a process in an annual growth cycle from the fragmentary genetic and genomic analyses in different species. Similarly, flowering in perennial species is not the end of the plant's life but only one cycle of many that will be repeated many times in the life of the plant. Dormancy control thus likely requires a number of cyclic inputs orchestrating specific tissue responses and also tree wide responses, as well as, stress inputs imposed on the tree over the yearly climate periods.

In order to present a cohesive picture of what is known of the genetic control of bud-break, we must consider the key events of the annual cycle of growth and flowering in temperate trees (e.g. peach annual cycle, Figure 1). These events are the set of buds in summer under shortening day conditions; endodormancy induction, progression and release; the ecodormant period and finally the budburst period. As gene action in each period likely can and in some cases is known to influence the timing of budburst, it is relevant to present the work in this conceptual framework.

Simple flowering plant systems have the potential to serve as key model study species to identify genetic determinants of the timing of flowering, and for this reason many genetic studies have focused on Arabidopsis and other annuals. However in this regard, we must be cautious to draw strict parallels where the physiological processes are analogous. So for example, although it is tempting to equate vernalization control of annual plant flowering with CR of the endodormant period perennial trees, the physiological events

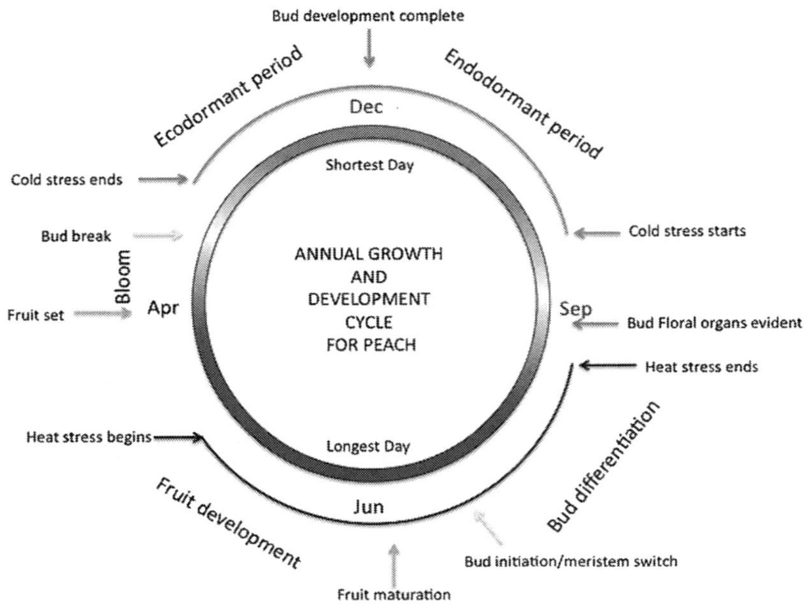

Figure 1 A peach-based conceptual framework that can serve to guide investigations and interpretations of the genetic control of bud-break in fruit and forest tree species. (See colour plate)

of these two cold related phenomena are not strictly comparable. Vernalization is required for winter annuals to initiate the switch from vegetative meristem to a floral meristem. CR is a physiological control of the release of the winter arrest of an already determined floral tissue or vegetative bud tissue. They are analogous only in the fact that cold exposure is the common factor. From a physiological standpoint, it is perhaps more relevant to compare seed stratification requirements of annual plants with CR of trees as these are both cold-related controls on resumption of tissue growth and differentiation, in the case of the seed, the embryo and in the case of the bud the shoot vegetative and floral tissues. However, it is not unreasonable that the cold-related control mechanisms for growth and differentiation of tissues likely utilize common gene networks among the different plant species, thus much can be learned from the simple systems. It is not the purpose of this review to exhaustively cover the genetics of flowering control in annual herbaceous plants and for comprehensive reviews on this topic the reader is directed to a number of excellent recent reviews on this topic (Amasino & Michaels, 2010; Andrés & Coupland, 2012; Finnegan et al., 2010; Greenup, Peacock, Dennis, & Trevaskis, 2009).

What is key to this review is that the temperature-related control of flowering in these annual species is orchestrated by hierarchical networks of genes that are monitoring environmental inputs on several different levels enabling the plant to manage the floral timing as environmental circumstances and growth requirements demand. Similarly in tree species, it is reasonable to assume that bud break in the spring is also orchestrated by hierarchical regulatory networks of genes that integrate environmental signals over the winter and spring periods to control the changes in physiological and developmental state required for buds to emerge at the appropriate time in the plant's annual cycle of spring/summer growth and winter arrest.

Since regulation of the timing of dormancy onset, duration and release strongly correlates with bud flush date in the spring, this review will focus on what is known of the genetic control of the endodormant and ecodormant periods of the annual growth and development cycle of temperate trees.

Day length and temperature are major environmental cues for growth cessation and the induction of dormancy and the timing of dormancy release, however the response to photoperiod and temperature is complex and sometimes contradictory (for review see Tanino, Kalcsits, Silim, Kendall, and Gray (2010)). Rosaceous fruit trees are insensitive to short photoperiods at relatively high temperatures (21 °C). Most *Prunus* species, unlike apple and pear, are sensitive to day length at intermediate temperatures (12–15 °C). At low temperature (9 °C), the reaction of *Prunus* species to shortening day is not uniform (Heide, 2008; Heide & Prestrud, 2005). The physiological basis of these temperature effects remains unclear though changes in water content and mobility have a strong association with induction and release of dormancy in peach cultivars with high and low CR (Yooyongwech et al., 2008).

The same signal that is important in establishing the dormant state in *Prunus* trees is also critical for the breaking of dormant state whereby a period of exposure to suboptimal chilling temperatures (<15 °C) that varies among and within species is absolutely required for dormancy release. For example, peach germplasm has a broad range of variation in CR – from 50 to 1050 h depending on the geographical source of the germplasm (Okie, 1998; Topp, Sherman, & Rasera, 2008). If trees do not fulfill their CR, their blooming is erratic if at all in the spring. Thus, these examples illustrate the complexity of the signaling processes and the tree's responses over time and space implying that there will likely be a complex nature to the genetics that control the timing of bud-break.

2. MOLECULAR GENETIC MAPPING OF FLOWERING/ BUD-BREAK

2.1 Endodormancy

Endodormancy (or 'rest') is the deep or 'true' dormancy seen during winter (Faust, Erez, Rowland, Wang, & Norman, 1997; Fuchigami & Nee, 1987; Lang, 1987) as opposed to the dormancy imposed by apical dominance ('paradormancy') or by unfavourable climatic conditions ('eco-dormancy,' e.g. continued cold temperatures or drought) following release from endodormancy (Lang, 1987).

A primary characteristic of endodormant bud tissues in temperate tree species is the absolute requirement for an accumulated exposure to a prolonged chilling ($0 < T < 7.5\ ^{\circ}C$) period before resumption of active bud break and shoot growth at warm temperatures in early spring (Hesse, 1975). This CR plays a protective role against short intervals of unseasonably warm temperatures followed by resumed cold. The CR varies considerably between and within species; however, CR generally correlates with the climatic origin of the genotype (Bailey & Hough, 1975).

Early genetic studies in perennial tree species, demonstrated that a number of dormancy-related traits are under genetic control, including CR. Most of these traits appear to vary in a quantitative manner, presumably resulting from multigenic control plus environmental effects (Aitken & Adams, 1995; Billington & Pelham, 1991; Ekberg, Eriksson, & Weng, 1985; Eriksson, Ekberg, Dormling, & Matern, 1978; Farmer & Reinholt, 1986; Howe, Saruul, Davis, & Chen, 2000; Pauley & Perry, 1954; Rowland et al., 1999). For example, in *Populus*, multiple QTLs for bud set, bud flush and other dormancy-related traits were reported (Chen, Howe, & Bradshaw, 2002; Frewen et al., 2000).

In fruiting trees of the Rosaceae, qualitative and quantitative inheritance of CR and/or bud break has been described (Table 1). Earlier studies showed that CR was inherited in a semiqualitative fashion in both open-pollinated and control-cross progeny of apple *Malus* x *domestica* Borkh. cultivar 'Anna' (Hauagge & Cummins, 1991), and the authors concluded that the low CR of this cultivar appears to be controlled by at least one major dominant gene. Additionally, a major gene called *Late bloom* (Lb) that delays blooming in almond (*Prunus dulcis* (Mill.) D.A. Webb) was mapped (Ballester, Sociasi Company, Arus, & de Vicente, 2001; Socias i Company, Felipe, & Gomez Aparisi, 1996), but its relationship to CR was not determined. More

Table 1 QTL Number for Endodormant and Ecodormant Traits in Fruit and Forest Tree Species

Species	Trait and Number of QTL				References
	Bud Set	Chilling Req.	Heat Req.	Bud Break[a]	
Prunus persica		8	2	9	Fan et al. (2010)
		4	5	5	Romeu et al. (2014)
Prunus armeniaca		12		3	Olukolu et al. (2009)
					Salazar, Ruiz, Egea, and Martínez-Gómez (2013)
Prunus dulcis		5	2	5	Sánchez-Pérez et al. (2014)
Prunus avium		1		1	Castede et al. (2014)
Malus x *domestica*				11	Celton et al. (2011)
Populus spp.	3			6	Frewen et al. (2000)
Pseudotsuga menziesii				33	Jermstad, Bassoni, Jech, Wheeler, and Neale (2001)
Quercus petraea				19–38	Derory et al. (2010)
Picea glauca	52			33	Pelgas, Bousquet, Meirmans, Ritland, and Isabel (2011)

[a]Data for floral and/or vegetative bud flush are presented as one total.

recently, quantitative genetic studies in two different Prunus species, apricot *P. armeniaca* (Olukolu et al., 2009) and peach (Fan et al., 2010; Romeu et al., 2014) directly dissected the genetics of both of CR and BD identifying overlapping and individual QTL governing the traits (Table 1).

Focusing on BD under the assumption that to some degree it reflects the phenotypic drive of CR and is significantly easier to evaluate, other researchers have subsequently mapped QTLs in almond (*P. dulcis*) (Sánchez-Pérez, Dicenta, & Martínez-Gómez, 2012), apricot (Campoy et al., 2011) and sweet cherry (*P. avium*) (Castede et al., 2014; Dirlewanger et al., 2012). QTL intervals for BD detected in these fruit trees largely overlap with that reported in peach by Fan et al. (2010), although the comparative contribution of individual QTLs to the phenotypic trait variance varies among the *Prunus* species.

Collectively, from QTL mapping data for phenology traits in fruiting and forest trees, bud-break timing is a complex genetic character and depending on the species can vary from a few strongly influential loci to numerous small contributing loci. Also it is apparent from the literature that conservation of QTLs for the various components of the endodormant and ecodormant periods can show conservation among closely related species such as the LG1 *EVG* and the LG4 loci in *Prunus* species (Sánchez-Pérez, Cueto, Dicenta, & Martínez-Gómez, 2014), however in contrast even within different mapping families of the same species, common QTL may not be detected (Celton et al., 2011).

2.2 Ecodormancy

Ecodormancy is defined as the period succeeding endodormancy when CR requirements are met and buds are held or repressed from flowering presumably to surpass the winter freezing period. It is this period where HR requirements a believed to exercise their greatest influence on the timing of bud-break. In contrast to genetic studies the traits that define the endodormant period (e.g. CR) only a few studies have attempted to genetically characterize hallmark traits of the ecodormant period such as HR (Fan et al., 2010; Romeu et al., 2014). The most recently published work of Romeu et al. (2014) is perhaps the most comprehensive study of the genetic nature of ecodormancy characteristic traits (HR, and the timing of the Endo-Eco transition period and Eco-BD period) in peach. Results of these studies illustrate like CR, traits defining the ecodormant period are complex and exhibit multiple QTL (HR-9 loci). Interestingly, in contrast to Fan et al. (2010) and Romeu et al. (2014) identified QTL overlap for CR and HR traits suggesting there may be crosstalk of the gene networks controlling these traits.

3. GENOME-WIDE ASSOCIATION STUDIES ON BUD-BREAK TIMING

One of the major drawbacks of single-family QTL mapping analyses in trees is that the number of progeny in these crosses severely restricts the degree of resolution of identified QTL intervals. In species where sequences are available for the QTL intervals this low resolution makes it difficult to identify candidate genes for the trait since the number of genes in these intervals is usually quite large. Another approach to mapping QTL that has been used successfully in plant and animal species is genetic association mapping which uses natural or breeding populations of limited relatedness among individual genotypes, genome wide markers in high density such as SNPs and detects associations between phenotypic traits and markers based on linkage disequilibrium (LD). Depending on the size of LD interval surrounding a trait locus it is possible to identify a very small number of specific genes in high association with the trait. In recent years, there has been a number of papers utilizing this approach to study the genetics of specific traits in forest tree populations (Alberto et al., 2013; Eckert et al., 2012; Holliday, Ritland, & Aitken, 2010; Ingvarsson, Garcia, Luquez, Hall, & Jansson, 2008; La Mantia et al., 2013; Parchman et al., 2012; Porth et al., 2013; Slavov et al., 2012) and several have investigated bud set and bud break timing (Evans et al., 2014; McKown et al., 2014).

Using this genome-wide association studies (GWAS) approach in poplar, Mckowen et al., 2014 detected significant SNPs associated with phenology traits in numerous genes (up to two orders of magnitude) and these genes encoded proteins with roles in light perception, photoperiod, circadian rhythm and phytohormone regulation though many were associated with the hallmarks of dormancy entry (bud set, leaf drop) and few with bud break. Utilizing 544 poplar trees sampled across a wide latitudinal range, Evans et al. (2014) combined GWAS and adaptive selection analyses to detect 397 genomic regions showing adaptive signatures for specific traits including bud flush and bud set. In respect to the genetics of vegetative bud flush, results of this study support a model strongly implicating the critical importance of genes encoding proteins that control the symplastic mobility of florigens. Thus, the timing of the release of endodormancy may rely on the removal of callous plugs blocking symplastic intercellular connections allowing florigens to mobilize to their targets signaling the resumption of meristem growth and development as suggested by work of Rinne et al. (2011). In this work, it was demonstrated that chilling-based induction of particular gibberellic acid (GA) inducible 1,3-β-glucanases in poplar was associated with the endordormancy completion and with FT hyperinduction and the promotion of vegetative budburst.

Parallel to the ongoing genetic characterization of endodormancy and ecodormancy traits using genetic mapping approaches, identification of potential candidate genes controlling dormancy and bud break employing various genomics and comparative genomics approaches in a number of different fruiting and forest tree species has met with some success (Allona et al., 2008; Arora, Rowland, & Tanino, 2003; Bielenberg et al., 2004, 2008; Rinne et al., 2011; Rohde & Bhalerao, 2007; Shoot & Rinne, 2011; Tanino et al., 2010; Wellinga & Palva, 2006).

4. THE DAM GENES

One of the most significant genetically based studies of genes directly responsible for controlling dormancy was the discovery of the dormancy associated MADS-box (DAM) genes (Bielenberg et al., 2004, 2008; Wang, Georgi, Reighard, Scorza, & Abbott, 2002) and subsequent work that has illustrated the importance of members of this clade to establishment of dormancy and maintenance of dormancy in *Prunus* species (Jiménez, Lawton-Rauh, Reighard, Abbott, Bielenberg, 2009; Li, Reighard, Abbott, & Bielenberg, 2009; Sasaki et al., 2011; Yamane et al., 2011; Zhebentyayeva et al., 2014) and other perennial plant species (Hedley et al., 2010; Horvath, Anderson, Chao, & Foley, 2003; Horvath, Sung, Kim, Chao, & Anderson, 2010; Saito et al., 2013; Ubi et al., 2010; Wu et al., 2012). For a comprehensive review of the potential role of the DAM genes in dormancy initiation and maintenance in perennial species, see Liu, Zhu, and Abbott (in press). For our purposes in this review DAM genes are members of the SVP MADS-box clade and as such likely play central roles in repression of floral development, growth regulator signaling, basic cellular and metabolic processes, protein modification, reproduction, morphology and others (Gregis et al., 2013). In peach it is tantalizing to speculate that the DAM genes are the CR genes and, in an analogous role to flowering locus C (FLC) in Arabidopsis, they regulate the onset and duration of apical bud dormancy process in response to cold and thus impact the timing of flowering in the spring, however currently the link to CR control is correlative and not causative and further work is required to substantiate claims that the DAM genes are the master regulators of CR. In this regard, QTL mapping of CR and BD in some *Prunus* species have not detected the DAM locus as the major control locus in the timing of floral and vegetative bud break (Dirlewanger et al., 2012; Olukolu et al., 2009; Sánchez-Pérez et al., 2014).

5. GENOME SEQUENCES AND CANDIDATE GENES FOR DORMANCY-ASSOCIATED QTLs

The availability of whole genome sequences of particular key species in dormancy and flowering control research enabled the transition from genetically mapped loci to physically characterized genomic regions containing potentially relevant candidate genes. Thus, within recent years numerous papers have proposed candidate genes for control of endodormancy and ecodormancy. The essential problem with these candidate gene lists is that the genomic intervals from which they are derived are often quite large usually in excess 1 cM and thus the reality of many of these candidates as causative in trait manifestation is suspect. Additionally, choices of particular genes from lists of several 100 genes are often driven by analogy to what is known from annual systems such as Arabidopsis, wheat, rice and others and these educated guesses are built on a potentially false premise that molecular physiology of vernalization is equal to that of CR which as explained earlier is likely not the case. However comparative inspection of the QTL gene lists within and among species can provide relevant insight into common recurring types of genes that are found in association within the QTLs. For example, in Rosaceae, where comparative QTL work from different species is possible through a common core *Prunus* map and the whole genome sequence of peach (Verde et al., 2013), several reports have demonstrated that the overlapping and unique dormancy-related QTL loci of different species are populated by genes that are involved in several networks of genetic control: phytohormone regulation, circadian clock, cell cycle control, cold stress and epigenetic regulation (Romeu et al., 2014; Sánchez-Pérez et al., 2014; Zhebentyayeva et al., 2014). Perhaps this is not surprising considering that the timing of bud-break seems to result from the complex interplay of the genetics that define the changing physiological states of the buds during the endodormancy and ecodormancy periods.

6. GENOMICS AND BUD-BREAK CANDIDATE GENE DISCOVERY

Of the genomics technologies available for characterization of genes important to bud break, transcriptomics has provided a wealth of potential gene candidates, however often the shear number of genes responding to the treatments in the studies precludes prioritizing candidates from stand alone data. Additionally in field studies, it is quite difficult to tease apart the effects of multiple changes in environmental inputs from those that are directly driving the response being measured. However, coupling this type of data with QTL mapping and/or Genome Wide Association Studies (GWAS), provides a complementary way to narrow gene lists from transcriptomics analyses to prioritized candidate genes for further study (i.e. those candidate genes that co-localize in QTL controlling the trait under study gain high priority for further investigation).

In poplar, prior research focused on the control of growth cessation and bud development through analysis of bud formation, acclimatization to dehydration and cold, and the acquisition of dormancy (Ruttink et al., 2007). Identification of several candidate genes involved in light perception and growth cessation such as *phytochrome PHYB2* and *CO* (*Constans*) demonstrated the action of photoperiodic control elements (reviewed by Allona et al. (2008)). Transcriptomics studies in sessile oak during the ecodormant period demonstrated the differential expression of numerous genes involving the pathways of cell cycle, metabolism, cell rescue and defense and others showed significant changes in expression over the ecodormant period and more importantly this paper provided direct evidence that genes involved in controlling seed germination (*DAG2*) were potentially also playing an important role in the ecodormant bud suggesting a link in the regulation of these two physiological processes (Derory et al., 2006) Ueno et al. (2013) further demonstrated in sessile oak that during the endodormant period upregulated GO groups included metal ion binding, fatty acid binding and epidermis morphogenesis while the ecodormant period was characterized by GO groups for response to cold, high light intensity and ABA stimulus. Transcriptomics studies in other species of Fagaceae during the ecodormant and budburst periods have highlighted the importance of genes involved in cold/drought stress response and cell division/development, respectively, as hallmarks of the transition from dormancy to active growth (Lesur, Le Provost et al., 2015; Lesur, Bechade et al., 2015). These works further underscore the importance of abiotic stress response pathways to the fundamental nature of the dormant plant state.

Coupling, the QTL mapping approach with whole genome sequences, Rohde et al. (2011) delineated QTLs for bud set and bud flush in poplar and their best-supported genomic QTL contained 394 genes. Other studies (Ibáñez et al., 2010; Ramos et al., 2005) showed that the circadian clock homologs from poplar and chestnut regulate entry and exit of seasonal dormancy and winter hardiness.

Phytohormones seem to play a role in winter dormancy establishment (for review see Arora et al. (2003) and Allona et al. (2008)). For instance, there is evidence of the involvement of GA biosynthesis in growth cessation and bud set and in budburst in poplar though studies of the GA signaling pathway and control of genes responsive to short and long days in poplar (Rinne et al., 2011; Schrader et al., 2004).

The studies of Rinne et al. (2011) coupling proteomics, transcriptomics and intracellular trafficking of proteins convincingly demonstrated that concomitant with endodormancy induction, progression and release there are significant changes in particular GA isomers which are associated with the regulation of particular 1,3-β-glucanases whose roles are linked the deposition and removal of callose plugs in the plasmadesmata (PD). These PD plugs are known to inhibit the movement of florigens and other subcellular signaling molecules in dormant buds and until removed, buds do not proceed to budburst. This is an attractive model for dormancy regulation because it integrates presence of particular environmental signals, the activation or repression of specific phytohormone production and the ability of the phytohormones to access and interact with their targets in time and space. This multilayered gene network system provides opportunities for the plant to fine-tune its growth response over a significant portion of the annual growth and dormancy cycle and to manifest different outcomes related to the different combinations of related players in time and space.

Finally, utilizing activation tagging in transgenic poplar Yordanov, Ma, Strauss, and Busov (2014) identified a gene, *EARLY BUD-BREAK 1* (*EBB1*), which regulates the resumption of growth in poplar buds during dormancy release. Transgene up or down regulation of this AP2/ERF transcription factor advanced or retarded bud burst, respectively. Transcriptomics studies of overexpression of this gene in transgenic poplar identified 2870 differentially expressed genes of which one-third overlap with differentially expressed genes in dormancy induction correlating the presence or absence of this transcription factor and its targets with the actively growing or dormant tissue state, respectively. This gene likely plays a major role in signaling the resumption of the growth to the apical meristem post dormancy and thus potentially plays a significant role in control of bud burst in poplar and likely in other trees as well.

7. EPIGENOMICS AND DORMANCY

Plants have evolved to survive in extremely harsh environments and to defend themselves against biotic and abiotic stresses. Epigenetic modifications seem to play a major role in sensing intracellular and extracellular cues and as such they play a major role in the development and adaptation (see Angers, Castonguay, and Massicotte (2010) and Brautigam et al. (2013) for review). Five major modifications affecting chromatin architecture are known, so far: DNA methylation, posttranslational modification of histones, incorporation of histone iso-forms, incorporation of non-coding RNA into chromatin and chromatin remodelling (see Chen, Lv, and Meng (2010) for review).

7.1 DNA Methylation

DNA methylation plays a dual role in eukaryotic genomes such as genome protection from selfish transposons and regulation of gene expression

(Finnegan, Genger, Peacock, & Dennis, 1998). Important observations on a potential link between decreases in DNA methylation during dormancy release and promotion of vegetative growth were described in over-wintering chestnut buds (Valledor et al., 2007) and changes in DNA methylation patterns in *Castenea sativa* buds associated with bud set and bud burst (Santamaría et al., 2009). A comprehensive study of DNA methylation pattern in multiple tissues was reported for poplar (Vining et al., 2012) where it was demonstrated that methylome is dynamic and that in general promoter and gene body methylation indicated inactivation of genes in particular tissue types.

In plants, DNA methylation occurs on the cytosine nucleotide in CpG, CpHpG or CpHpH sites (for review see Chan, Henderson, and Jacobsen (2015)). Genome-wide mapping of cytosine methylation using bisulphate sequencing was accomplished in three model plant species *Arabidopsis thaliana* (Cokus et al., 2008; Lister et al., 2008), rice (Li et al., 2008) and poplar (Feng et al., 2010). For comprehensive review of methylation in *Arabidopsis* and other species see Teixeira and Colot (2010); Yan et al. (2010) and Feng et al. (2010). CHG and CHH are distributed in non-coding regions including transposable elements (TEs) while CG methylation is distributed in protein coding regions and promoter regions. Three enzymes, methyltransferase 1 (MET1), chromomethylase 3, and domain rearranged methylase 1 and 2 (DRM1/2), maintain the methylation at these three sites. Methylation at CG sites is under the action of MET1 and to a lesser extent DRM1/2. Methylation at TEs is driven by DRM2.

8. SMALL RNA-MEDIATED POSTTRANSCRIPTIONAL EXPRESSION REGULATION AND 'DE NOVO' METHYLATION

During the last few years, microRNAs (miRNAs) and short interfering RNAs (siRNAs) have gained in prominence as regulators of genes in various cellular and developmental processes (Mallory & Vaucheret, 2006). siRNAs and a small subset of miRNAs regulate gene expression at the pre-transcriptional level through DNA methylation while miRNAs regulate it at the posttranscriptional level by cleaving or repressing the expression of target genes (Bao, Lye, & Barton, 2004). RNA-Directed DNA Methylation (RdDM) plays a predominant role in epigenetic modifications by regulating the expression of genes and TEs in plants (Jones-Rhoades, Bartel, & Bartel, 2006; Vazquez, Legrand, & Windels, 2010). RdDM uses siRNAs (usually 24 nucleotides) and to a less extent miRNAs to guide silencing complexes that methylate the genome to specific sequences that are in need of being modified. siRNAs are classified in various categories based on the mechanism of their biogenesis and exist in 1000 copies in plant genomes. It was speculated that siRNAs function in mediating epigenetic adaptation to environmental changes (Dunoyer et al., 2010). While no study was devoted to siRNAs role in dormancy and bud break, there are evidences that these small RNAs might regulate these processes. For instance, siRNAs and long non-coding RNAs were reported to regulate seed dormancy, which is similar to bud dormancy (Nonogaki, 2014). Other studies (Bastow et al., 2004; Sung & Amasino, 2004) show that the repression of the FLC by vernalization was mediated by a 24-nt siRNA.

miRNA (20–21 nucleotides) have emerged as major posttranscriptional regulators of gene expression during plant development and under various stresses (see for review; Jones-Rhoades et al. (2006) and Mallory and Vaucheret (2006) and Jeong, German, Rymarquis, Thatcher, and Green (2010) and Sunkar et al. (2007) and Jay, Renou, Voinnet, and Navarro (2010) and Ruiz-Ferrer and Voinnet (2009)). Studies in *A. thaliana, Brachypodium distachyon* and *Populus trichocarpa* (Ding, Zeng, & He, 2014; Jeyaraj, Chandran, & Gajjeraman, 2014; Liu, Tian, Li, Wu, & Zheng, 2008; Lu, Sun, & Chiang, 2008; Potkar, Recla, & Busov, 2013; Sunkar & Zhu, 2004; Zhang, Xu, Huan, & Chong, 2009; Zhou, Wang, Sutoh, Zhu, & Zhang, 2008) identified several miRNAs with altered expression under cold stress. A subset of these miRNAs was reported being differentially expressed in chilled dormant leaf buds in *Populus tremuloides* and *P. persica* (Barakat et al., 2012, personal communications). Few of these miRNAs or their targets co-localized with QTLs for CR and Bloom date in *P. persica* (Barakat et al., 2012). Among these miRNAs, several (miR164, miR319, miR396, miR156, miR172, miR166 and miR69) are development-related miRNA families functioning in cell proliferation, vascular development, juvenile phase promotion, flowering induction and miRNA biogenesis. These miRNAs exert their functions by regulating the expression of key genes controlling dormancy and bud break at the posttranscriptional or transcriptional levels. Among the genes regulated by miRNAs during dormancy are *VRN1* a known regulator in vernalization (Greb et al., 2007), and genes belonging to the polycomb repressor two silencing complex that regulates the expression of the flowering gene *FLC*. miRNA targets also include genes involved in methylation or RNA biogenesis including *Dicer homolog* (RNA helicase), a *ribonuclease III* that is required for endogenous RDR2-dependent siRNA formation, a *histone 3 lysine 9 specific methyltransferase* involved in the maintenance of DNA methylation, and a *SUVH4/KYP* homolog that is involved in epigenetic control of *FLC* expression.

One of the most significant examples of the importance of miRNA-directed epigenetics to control of phenology in forest trees is that of Norway spruce (Yakovlev, Fossdal, & Johnsen, 2010). In this species, the timing of bud set and bud burst are fixed as a consequence of the temperature at which the zygotic embryo and seed is maintained, high temperatures retarding and low temperatures advancing the timing of these phenological responses (Johnsen Skrøppa, Junttila, & Dæhlen, 1996; Johnsen et al., 2005; Skrøppa & Johnsen, 2000). This environmentally induced phenological trait memory persists as long as 20 years and explains the initial observation that northern Norway spruce trees transferred to southern seed orchards produced progeny exhibiting phenological responses similar to the southern ecotypes (Johnsen et al., 1996; Skrøppa & Johnsen, 2000; Skrøppa, Kohmann, Johnsen, Steffenrem, & Edvardsen, 2007). Comparative transcriptomics studies of ecotypes that do or do not exhibit epialleles for these phenological traits revealed differential responses for genes involved in miRNA-directed epigenetic control of gene expression (e.g. *DICER-LIKE1 (PaDCL1), DICER-LIKE 2 (PaDCL2) and SUPPRESSOR OF GENE SILENCING 3-LIKE (PaSGS3)* and for 16 novel or conserved miRNAs and their targets (e.g. pab-miR159a/*PaGaMYB*, pab-miR 156c/*PaSPB13*) some of which are known from other systems to function in control of flowering (Yakovlev, Asante, Fossdal, Junttila, & Johnsen, 2011; Yaklovev et al., 2010).

Although evidence for such ecogeographically related epigenetic control of flowering in angiosperm trees is lacking, it is likely that components of these systems play a key role in regulating dormancy and flowering.

9. CHROMATIN REMODELLING AND CONTROL OF FLOWERING

Histone methylation and the polycomb group (PcG) proteins: Polycomb (PcG) group proteins are transcriptional repressors that play a central role in epigenetic chromatin-mediated DNA silencing during the vernalization response. PcG proteins are organized into supermolecular complexes whose component peptides vary during development, as a function of cell type, or even from one promoter to another. So far, two PCRC2 and one PCRC1-like complex were identified in plants (Hennig & Derkacheva,

2009). Apparently, the PRC2 mechanism evolved in multicellular eukaryotes and played a significant role in land plant evolution (Mosquna et al., 2009; Shaver, Casa-Mollano, Cerny, & Ceruti, 2010).

All three PCRC2 complexes described in *Arabidopsis* share an FIE subunit and a chromatin remodelling factor, MSI1. The VRN2 complex promotes vernalization-induced flowering by recruiting CLF or SWN methyltransferases to targeted genes including the FLC that is also regulated by deposition of histone H3K27me3 repressive mark (Finnegan et al., 2010). Three gene regulators (*VIN3, VIN5, VEL1*) involved in the recruitment of PRC2 complex are plant-specific homeodomain finger proteins. While PCR1 is involved in histone modification in *drosophila*, there are no typical PCR1 homologs in plants. In *Arabidopsis*, PRC1-like protein – LHP1 (LIKE HETEROCHROMATIN 1) is involved into recognition of H3K27me3 repressive mark and the AtRING1 – E3ubuquitin ligase complex (Hennig & Derkacheva, 2009).

In *C. sativa* Santamaría et al. (2009) demonstrated that changes in acetylation of histone H4 was associated bud set and bud burst suggesting a link with chromatin remodelling and control of phenology. Transcriptomics studies from the same group demonstrated that expression of specific chromatin remodelling enzymes at these stages was correlated with these changes (Santamaría, Rodríguez, Cañal, & Toorop, 2011). These authors also demonstrate that the differentially expressed gene lists for dormant and non-dormant tissues look similar to those of the Arabidopsis seed dormancy studies further suggesting a strong link in the physiology of these two dormant tissue types.

In a study of the regulation of the *DAM* genes in peach, Leida, Conesa, Llácer, Badenes, and Gabino Ríos (2012) demonstrated that *DAM6* shows increases in the H3K27me3 mark with progression in dormancy chilling while the H3K4me3 mark decreases. This is consistent with the known gene repression and activation roles, respectively, for these two chromatin-remodelling marks and the postulated role of *DAM* genes as regulators of dormancy in peach and other species. Intriguingly, Leida, Conejero, et al. (2012) from comparison of gene expression analyses of seeds in stratification and trees in bud dormancy provided evidence that *DAM* genes (e.g. *DAM6*) show similar expression regulation in the chilling period. We currently do not know if the regulation of the *DAM* genes in seed stratification shows similar patterns of chromatin remodelling as evident in over wintering dormant buds, however, this is another example of where the networks of genes that appear to control bud dormancy may also play a role in seed dormancy and adds further support for the importance of DAM proteins as important regulators of these phenological traits. The importance of chromatin remodelling in seed dormancy is well established in Arabidopsis and other plants and the reader is directed to Nonogaki (2014) for a recent review of the key regulatory features.

10. FORWARD GENETIC/TRANSGENIC STUDIES

Juvenility in trees is the time it takes for the tree to develop into a fully mature flowering adult. In different tree species the juvenility period can be years to tens of years. This fact has been one of the primary reasons the genetics of trees has traditionally lagged well behind that of annual plant species. Because trees can have inherently long juvenility periods (3–10 years for most fruit/forest species) and protracted generation times, it takes minimally 20–30 years to complete a varietal selection program or even more, if the desired genetic characters are controlled by multiple genes (quantitative traits or QTLs) and/or if the desired characters come from wild species (e.g. novel traits such as disease resistance or stress tolerance).

Perhaps one of the most important discoveries was the demonstration that the juvenile phase of trees could be eliminated by transgene overexpression of the major flowering control genes such as *FT* (Hsu, Liu, Luthe, & Yuceera, 2006). This has led to the concept of rapid cycle breeding in trees with some initial successes in pilot studies of some transgenic fruit tree systems (plum – Srinivasan, Dardick, Callahan, & Scorza, 2012; citrus – Peña et al., 2001; apple – Flachowsky et al., 2011). These initial successes have demonstrated that introduction and ectopic expression of florigen genes (e.g. *FT*) through transgene technologies can eliminate the juvenility period and produce early flowering genotypes and rapid cycle breeding lines (report in preparation, USDA-ARS sponsored, First International Rapid Cycle Crop Breeding conference, 2014, in Leesburg, VA). Although this technology seems promising, there are major hurdles to be overcome, the greatest of which is the essential cost and lack of transgenic technologies for most tree species of importance. However, recent work by Igarashi et al. (2009) and Yamagishi et al. (2011) demonstrated that early flowering can be induced through introduction of *FT* using the Apple Latent Spherical Virus VIGS viral vector systems in a number of different plants both trees and herbaceous species. This virus approach has the potential to broaden the use of rapid cycle breeding technologies to any tree species and obviates the requirement for transgenic plant production.

11. PERSPECTIVES AND CONCLUSIONS

The search for a genetic understanding of the timing of flowering/bud-break in trees has by necessity leveraged information from annual plant systems where genetic analyses are simpler. However, with the advent of genomics and molecular genetic mapping in trees, it has been possible to advance our understanding of the underlying genetic pathways that appear to control this process. The story is complex and at the moment fragmented, which is not surprising considering that there has not been a single comprehensive analysis of the molecular genetic events of the annual cycle of growth and flowering in any tree species. The current state of our knowledge comes from a number of studies of small windows of the annual cycle (e.g. endodormant period) and the application of a small number of technologies (QTL mapping, transcriptomics and others) in different tree species. This has yielded a complex picture for what seems on surface to be a very specific simple trait, the timing of blooming.

It is perhaps naïve to imagine that bloom date would be governed by a simple genetic system and in fact the research so far indicates that a number of different gene systems appear to exert significant control on this trait. So for example in this review, we have presented arguments that bloom date is likely controlled by a number of gene networks operating at different periods of the annual cycle. These networks are comprised of genes controlling phytohormone balance, stress response, chromatin remodelling and DNA methylation, light signaling and others. They all likely contribute to regulating the annual timing of events leading to bud-break, however, these networks may play significant roles at different periods in the cycle and fragmentary snapshots of gene activity in one period does not lead to a comprehensive understanding of the final outcome. Perturbing any of these systems through genetic mutation or environmental changes will likely impact the date of bud-break. In retrospect, this is probably not surprising since a tree's regulatory systems must choreograph annually a delicate ballet on an ever changing environmental stage.

In consideration of this complex picture, it is certain that a comprehensive genetic/genomic study throughout the annual cycle of growth and development of a few key tree species would likely provide a much more integrated picture of the genetic control of flowering. This would integrate many of the current observations that have been collected for a myriad of systems. In support of this concept there are emerging observations that the genetics of dormancy and flowering control may be more aligned with that of seed dormancy and stratification than vernalization. Again, this is not surprising since comparative examination of these two processes from the phenotypic standpoint reveals common themes (e.g. CR, resumption of arrested meristem growth). We believe that comparisons within a species of these two parts of the annual growth and development cycle will likely lead to a greater understanding of the genetics behind the control of these processes than has previously been obtained by cross comparisons of a number of fragmented studies in different species.

Finally, in conclusion, the genetics of flowering in trees is complex but we are making significant strides in understanding the key pathways that we need to focus on and some of the key players. Manipulation of these key players through genetics and transgenic technologies has already provided exciting possibilities (e.g. rapid cycle breeding technologies) for rapid improvement of our tree resources to meet the challenges of a rapidly evolving environmental landscape. Perhaps the flower of our understanding is beginning to bloom.

REFERENCES

Aitken, S. N., & Adams, W. T. (1995). In B. M. Potts, N. M. G. Borralho, J. B. Reid, R. N. Cromer, & W. N. Tibbits (Eds.), *Screening for cold hardiness in coastal Douglas-fir* (pp. 321–324).

Alberto, F. J., Derory, J., Boury, C., Frigerio, J. M., Zimmermann, N. E., & Kremer, A. (2013). Imprints of natural selection along environmental gradients in phenology-related genes of Quercus petraea. *Genetics, 195*, 495–512.

Allona, I., Ramos, A., Ibáñez, C., Contreras, A., Casado, R., & Aragoncillo, C. (2008). Review. Molecular control of winter dormancy establishment in trees. *Spanish Journal of Agricultural Research, 6*, 201–210.

Amasino, R. M., & Michaels, S. D. (2010). The timing of flowering. *Plant Physiology, 154*, 516–520.

Andrés, F., & Coupland, G. (2012). The genetic basis of flowering responses to seasonal cues. *Nature Reviews Genetics, 13*(9), 627–639. http://dx.doi.org/10.1038/nrg3291.

Angers, B., Castonguay, E., & Massicotte, R. (2010). Environmentally induced phenotypes and DNA methylation: how to deal with unpredictable conditions until the next generation and after. *Molecular Ecology, 19*, 1283–1295.

Arora, R., Rowland, L. J., & Tanino, K. (2003). Induction and release of bud dormancy in woody perennials: a science comes of age. *HortScience, 38*, 911–921.

Bailey, C. H., & Hough, L. F. (1975). Apricots. In J. Janick, & J. N. Moore (Eds.), *Advances in fruit breeding* (pp. 367–383). West Lafayette, Indiana, USA: Purdue University Press.

Ballester, J., Socias i Company, R., Arus, P., & de Vicente, M. C. (2001). Genetic mapping of a major gene delaying blooming time in almond. *Plant Breeding, 120*, 268–270.

Bao, N., Lye, K. W., & Barton, M. K. (2004). MicroRNA binding sites in *Arabidopsis* class IIIHD-ZIP mRNAs are required for methylation of the template chromosome. *Developmental Cell, 7*, 653–662.

Barakat, A., Sriram, A., Park, J., Zhebentyayeva, T., Main, D., & Abbott, A. (2012). Genome wide identification of chilling responsive microRNAs in *Prunus persica*. *BMC Genomics, 13*, 481. http://dx.doi.org/10.1186/1471-2164-13-481.

Bastow, R., Mylne, J. S., Lister, C., Lippman, Z., Martienssen, R. A., & Dean, C. (2004). Vernalization requires epigenetic silencing of FLC by histone methylation. *Nature, 427*, 164–167.

Bielenberg, D. G., Wang, Y., Fan, S., Reighard, G. L., Scorza, R., & Abbott, A. G. (2004). A deletion affecting several gene candidates is present in the *evergrowing* peach mutant. *Journal of Heredity, 95*, 436–444.

Bielenberg, D. G., Wang, Y., Li, Z., Zhebentyayeva, T., Fan, S., Reighard, G. L., et al. (2008). Sequencing and annotation of the *evergrowing* locus in peach [*Prunus persica* (L.) Batsch] reveals a cluster of six MADS-box transcription factors as candidate genes for regulation of terminal bud formation. *Tree Genetics & Genomes, 4*, 495–507.

Billington, H. L., & Pelham, J. (1991). Genetic variation in the date of budburst in Scottish birch populations: Implications for climate change. *Functional Ecology, 5*, 403–409.

Brautigam, K., Vining, K. J., Lafon-Placette, C., Fossda, C. G., Mirouze, M., Marcos, J. G., et al. (2013). Epigenetic regulation of adaptive responses of forest tree species to the environment. *Ecology and Evolution, 3*(2), 399–415.

Campoy, J. A., Ruiz, D., Egea, J., Rees, J., Celton, J. M., & Martínez-Gómez, P. (2011). Inheritance of flowering time in apricot (*Prunus armeniaca* L.) and analysis of linked quantitative trait loci (QTLs) using simple sequence repeat markers. *Plant Molecular Biology Reporter, 29*, 404–410.

Castede, S., Campoy, J. A., Quero-García, J., Le Dantec, L., Lafargue, M., Barreneche, T., et al. (2014). Genetic determinism of phenological traits highly affected by climate change in *Prunus avium*: flowering date dissected into chilling and heat requirements. *New Phytologist, 202*, 703–715. http://dx.doi.org/10.1111/nph. 12658.

Celton, J.-M., Martinez, S., Jammes, M.-J., Bechti, A., Salvi, S., Legave, J.-M., et al. (2011). Deciphering the genetic determinism of bud phenology in apple progenies: a new insight into chilling and heat requirement effects on flowering dates and positional candidate genes. *New Phytologist, 192*, 378–392.

Chan, S. W., Henderson, I. R., & Jacobsen, S. E. (2015). Gardening the genome: DNA methylation in *Arabidopsis thaliana*. *Nature Reviews Genetics, 6*, 351–360.

Chen, T. H. H., Howe, G. T., & Bradshaw, H. D. (2002). Molecular genetic analysis of dormancy-related traits in poplars. *Weed Science, 50*, 232–240.

Chen, M., Lv, S., & Meng, Y. (2010). Epigenetic performers in plants. *Development Growth and Differentiation, 52*, 555–566.

Cokus, S. J., Feng, S. H., Zhang, X. Y., Chen, Z. G., Merriman, B., Haudenshild, C. D., et al. (2008). Shotgun bisulphate sequencing of the *Arabidopsis* genome reveals DNA methylation patterning. *Nature, 452*, 215–219.

Derory, J., Leger, P., Garcia, V., Schaeffer, J., Hauser, M. T., Salin, F., et al. (2006). Transcriptome analysis of bud burst in sessile oak (*Quercus petraea*). *New Phytologist, 170*, 723–738.

Derory, J., Scotti-Saintagne, C., Bertocchi, E., Le Dantec, L., Graignic, N., Jauffres, A., et al. (2010). Contrasting relationships between the diversity of candidate genes and variation of bud burst in natural and segregating populations of European oaks. *Heredity, 104*, 438–448.

Ding, Q., Zeng, J., & He, X. (2014). Deep sequencing on a genome-wide scale reveals diverse stage-specific microRNAs in cambium during dormancy-release induced by chilling in poplar. *BMC Plant Biology, 14*, 267.

Dirlewanger, E., Quero-García, J., Le Dantec, L., Lambert, P., Ruiz, D., Dondini, L., et al. (2012). Comparison of the genetic determinism of two key phenological traits, flowering and maturity dates, in three *Prunus* species: peach, apricot and sweet cherry. *Heredity, 109*, 280–292.

Dunoyer, P., Brosnan, C. A., Schott, G., Wang, Y., Jay, F., Alioua, A., et al. (2010). An endogenous, systemic RNAi pathway in plants. *EMBO Journal, 29*, 1699–1712.

Eckert, A. J., Wegrzyn, J. L., Cumbie, W. P., Goldfarb, B., Huber, D. A., Tolstikov, V., et al. (2012). Association genetics of the loblolly pine (*Pinus taeda*, Pinaceae) metabolome. *New Phytologist, 193*, 890–902.

Ekberg, I., Eriksson, G., & Weng, Y. (1985). Between- and within-population variation in growth rhythm and plant height in four *Picea abies* populations. *Studia Forestalia Suecica*. No. 167.

Eriksson, G., Ekberg, I., Dormling, I., & Matern, B. (1978). Inheritance of bud-set and bud-flushing in *Picea abies* (L.) Karst. *Theoretical and Applied Genetics, 52*, 3–19.

Evans, L. M., Slavov, G. T., Rodgers-Melnick, E., Martin, J., Ranjan, P., Muchero, W., et al. (2014). Population genomics of *Populus trichocarpa* identifies signatures of selection and adaptive trait associations. *Nature Genetics*. http://dx.doi.org/10.1038/ng.3075.

Fan, S., Bielenberg, D. G., Zhebentyayeva, T. N., Reighard, G. L., Okie, W. R., Holland, D., et al. (2010). Mapping quantitative trait loci associated with chilling requirement, heat requirement and bloom date in peach (*Prunus persica*). *New Phytologist, 185*, 917–930.

Farmer, R. E., Jr., & Reinholt, R. W. (1986). Genetic variation in dormancy relations of balsam poplar along a latitudinal transect in northwestern Ontario. *Silvae Genetica, 35*, 38–42.

Faust, M., Erez, A., Rowland, L. J., Wang, S. Y., & Norman, H. A. (1997). Bud dormancy inperennial fruit trees: physiological basis for dormancy induction, maintenance, and release. *HortScience, 32*, 623–629.

Feng, S., Cokus, S., Zhang, X., Chen, P.-Y., Bostick, M., Goll, M. G., et al. (2010). Conservation and divergence of methylation pattening in plants and animals. *Proceedings of the National Academy of Sciences of the United States of America, 107*, 8689–8694.

Finnegan, E. J., Genger, R. K., Peacock, W. J., & Dennis, E. S. (1998). DNA methylation in plants. *Annual Review of Plant Physiology and Plant Molecular Biology, 49*, 223–247.

Finnegan, E. J., Helliwell, C., Sheldon, C., Peacock, J. W., Dennis, E. S., Bagnall, D., et al. (2010). Vernalization. In *Encyclopedia of life sciences (ELS)*. Chichester: John Wiley & Sons, Ltd. http://dx.doi.org/10.1002/9780470015902.a0002048.pub3.

Flachowsky, H., Le Roux, P. M., Peil, A., Patocchi, A., Richter, K., & Hanke, M. V. (2011). Application of a high-speed breeding technology to apple (*Malus × domestica*) based on transgenic early flowering plants and marker-assisted selection. *New Phytologist, 192*(2), 364–377. http://dx.doi.org/10.1111/j.1469-8137.2011.03813.x. Epub Jul 8.

Frewen, B. E., Chen, T. H. H., Howe, G. T., Davis, J., Rohde, A., Boerjan, W., et al. (2000). Quantitative trait loci and candidate gene mapping of bud set and bud flush in *Populus*. *Genetics, 154*, 837–845.

Fuchigami, L. H., & Nee, C.-C. (1987). Degree growth stage model and rest-breaking mechanisms in temperate woody perennials. *HortScience, 22*, 836–845.

Greenup, A., Peacock, W. J., Dennis, E. S., & Trevaskis, B. (2009). The molecular biology of seasonal flowering-responses in *Arabidopsis* and the cereals. *Annals of Botany, 103*, 1165–1172.

Greb, T., Mylne, J. S., Crevillen, P., Geraldo, N., An, H. L., Gendall, A. R., et al. (2007). The PHD finger protein VRN5 functions in the epigenetic silencing of *Arabidopsis FLC*. *Current Biology, 17*, 73–78.

Gregis, V., Andrés, F., Sessa, A., Guerra, R. F., Simonini, S., Mateos, J. L., et al. (2013). Identification of pathways directly regulated by *SHORT VEGETATIVE PHASE* during vegetative and reproductive development in *Arabidopsis*. *Genome Biology, 14*, R56.

Hauagge, R., & Cummins, J. N. (1991). Genetics of length of dormancy period in *Malus* vegetative buds. *Journal of the American Society for Horticultural Science, 116*, 121–126.

Hedley, P. E., Russell, J. R., Jorgensen, L., Gordon, S., Morris, J. A., Hackett, C. A., et al. (2010). Candidate genes associated with bud dormancy release in blackcurrant (*Ribesnigrum* L.). *BMC Plant Biology, 10*, 202. http://dx.doi.org/10.1186/1471-2229-10-202.

Heide, O. M. (2008). Interaction of photoperiod and temperature in the control of growth and dormancy of *Prunus* species. *Scientia Horticulturae, 115*, 309–314.

Heide, O. M., & Prestrud, A. K. (2005). Low temperature, but not photoperiod, controls growth cessation and dormancy induction and release in apple and pear. *Tree Physiology, 25*, 109–114.

Hennig, L., & Derkacheva, M. (2009). Diversity of polycomb group complexes in plants: same rules, different players? *Trends in Genetics, 25*, 414–423.

Hesse, C. O. (1975). Peaches. In J. Janick, & J. N. Moore (Eds.), *Advances in fruit breeding* (pp. 285–335). West Lafayette, Indiana, USA: Purdue University Press.

Holliday, J. A., Ritland, K., & Aitken, S. N. (2010). Widespread, ecologically relevant genetic markers developed from association mapping of climate-related traits in Sitka spruce (*Picea sitchensis*). *New Phytologist, 188*, 501–514.

Horvath, D. P., Anderson, J. V., Chao, W. S., & Foley, M. E. (2003). Knowing when to grow: signals regulating bud dormancy. *Trends in Plant Science, 8*, 534–540.

Horvath, D. P., Sung, S., Kim, D., Chao, W., & Anderson, J. (2010). Characterization, expression and function of *DORMANCY ASSOCIATED MADS-BOX* genes from leafy spurge. *Plant Molecular Biology, 73*, 169–179.

Howe, G. T., Saruul, P., Davis, J., & Chen, T. H. H. (2000). Quantitative genetics of bud phenology, frost damage, and winter survival in an F2 family of hybrid poplars. *Theoretical and Applied Genetics, 101*, 632–642.

Hsu, C., Liu, Y., Luthe, D. S., & Yuceera, C. (2006). Poplar FT2 shortens the juvenile phase and promotes seasonal flowering. *Plant Cell, 18*, 1846–1861.

Ibáñez, C., Kozarewa, I., Johansson, M., Ogren, E., Rohde, A., & Eriksson, M. E. (2010). Circadian clock components regulate entry and affect exit of seasonal dormancy as well as winter hardiness in populus trees. *Plant Physiology, 153*, 1823–1833.

Igarashi, A., Yamagata, K., Sugai, T., Takahashi, Y., Sugawara, E., Tamura, A., et al. (2009). Apple latent spherical virus vectors for reliable and effective virus-induced gene silencing among a broad range of plants including tobacco, tomato, *Arabidopsis thaliana*, cucurbits, and legumes. *Virology, 386*, 407–416.

Ingvarsson, P. K., Garcia, M. V., Luquez, V., Hall, D., & Jansson, S. (2008). Nucleotide polymorphism and phenotypic associations within and around the phytochrome B2 locus in European aspen (*Populus tremula*, Salicaceae). *Genetics, 178*, 2217–2226.

Jay, F., Renou, J. P., Voinnet, O., & Navarro, L. (2010). Biotic stress-associated microRNAs: identification, detection, regulation, and functional analysis. *Methods in Molecular Biology, 592*, 183–202.

Jeong, D. H., German, M. A., Rymarquis, L. A., Thatcher, S. R., & Green, P. J. (2010). Abiotic stress-associated miRNAs: detection and functional analysis. *Methods in Molecular Biology, 592*, 203–230.

Jermstad, K. D., Bassoni, D. L., Jech, K. S., Wheeler, N. C., & Neale, D. B. (2001). Mapping of quantitative trait loci controlling adaptive traits in coastal Douglas-fir. I. Timing of vegetative bud flush. *Theoretical and Applied Genetics, 102*, 1142–1151.

Jeyaraj, A., Chandran, V., & Gajjeraman, P. (2014). Differential expression of microRNAs in dormant bud of tea [*Camellia sinensis* (L.) O. Kuntze]. *Plant Cell Reports, 33*(7), 1053–1069.

Jiménez, S., Lawton-Rauh, A. L., Reighard, G. L., Abbott, A. G., & Bielenberg, D. G. (2009). Phylogenetic analysis and molecular evolution of the dormancy associated *MADS-box* genes from peach. *BMC Plant Biology, 9*, 81.

Johnsen, Ø., Fossdal, C. G., Nagy, N., Molmann, J., Dælen, O. G., & Skrøppa, T. (2005). Climatic adaptation in *Picea abies* progenies is affected by the temperature during zygotic embryogenesis and seed maturation. *Plant Cell Environment, 28*, 1090–1102.

Johnsen, Ø., Skrøppa, T., Junttila, O., & Dæhlen, O. G. (1996). Influence of the female flowering environment on autumn frost-hardiness of *Picea abies* progenies. *Theoretical and Applied Genetics, 92*, 797–802.

Jones-Rhoades, M. W., Bartel, D. P., & Bartel, B. (2006). MicroRNAS and their regulatory roles in plants. *Annual Review of Plant Biology, 57*, 19–53.

La Mantia, J., Klápště, J., El-Kassaby, Y. A., Azam, S., Guy, R. D., Douglas, C. J., et al. (2013). Association analysis identifies Melampsora 9 columbiana poplar leaf rust resistance SNPs. *PLoS One, 8*, e78423.

Lang, G. A. (1987). Dormancy: a new universal terminology. *HortScience, 22*, 817–820.

Leida, C., Conejero, A., Arbona, V., Gómez-Cadenas, A., Llácer, G., Badenes, M. L., et al. (2012). Chilling-dependent release of seed and bud dormancy in peach associates to common changes in gene expression. *PLoS One, 7*, e35777. http://dx.doi.org/10.1371/journal.pone.0035777.

Leida, C., Conesa, A., Llácer, G., Badenes, M. L., & Gabino Ríos, G. (2012). Histone modifications and expression of DAM6 gene in peach are modulated during bud dormancy release in a cultivar-dependent manner. *New Phytologist, 193*, 67–80.

Lesur, I., Le Provost, G., Bento, P., Da Silva, C., Leplé, J. C., Murat, F., et al. (2015). The oak gene expression atlas: insights into Fagaceae genome evolution and the discovery of genes regulated during bud dormancy release. *BMC Genomics, 16*, 112.

Lesur, I., Bechade, A., Lalanne, C., Klopp, C., Noirot, C., Leplé, J. C., et al. (2015). A unigene set for European beech (Fagus sylvatica L.) and its use to decipher the molecular mechanisms involved in dormancy regulation. *Molecular Ecology Resources*. http://dx.doi.org/10.1111/1755-0998.12373.

Li, Z. G., Reighard, G. L., Abbott, A. G., & Bielenberg, D. G. (2009). Dormancy-associated MADS genes from the EVG locus of peach [*Prunus persica* (L.) Batsch] have distinct seasonal and photoperiodic expression patterns. *Journal of Experimental Botany, 60*, 3521–3530.

Lister, R., O'Malley, R. C., Tonti-Filippini, J., Gregory, B. D., Berry, C. C., Millar, A. H., et al. (2008). Highly integrated single-base resolution maps of the epigenome in *Arabidopsis*. *Cell, 133*, 523–536.

Liu, H. H., Tian, X., Li, Y. J., Wu, C. A., & Zheng, C. C. (2008). Microarray-based analysis of stress-regulated microRNAs in *Arabidopsis thaliana*. *RNA, 14*, 836.

Liu, Z., Zhu, H., and Abbott, A. Dormancy behaviors and underlying regulatory mechanisms: from perspective of pathways to epigenetic regulation. In J. V. Anderson (Ed.), *Advances in plant dormancy*, Springer International Publishing AG, Cham, in press.

Li, X., Wang, X., He, K., Ma, Y., Su, N., He, H., et al. (2008). High-resolution mapping of epigenetic modifications of the rice genome uncovers interplay between DNA methylation, histone methylation, and gene expression. *Plant Cell, 20*, 259–276.

Lu, S., Sun, Y., & Chiang, V. (2008). Stress-responsive microRNAs in populus. *Plant Journal, 55*, 131–151.

Mallory, A. C., & Vaucheret, H. (2006). Functions of microRNAs and related small RNAs in plants. *Nature Genetics, 38*(Suppl.), S31–S36.

McKown, A. D., Klápště, J., Guy, R. D., Geraldes, A., Porth, I., Hannemann, J., et al. (2014). Genome-wide association implicates numerous genes underlying ecological trait variation in natural populations of *Populus trichocarpa*. *New Phytologist*. http://dx.doi.org/10.1111/nph.12815.

Mosquna, A., Katz, A., Decker, E. L., Rensing, S. A., Reski, R., & Ohad, N. (2009). Regulation of steam maintenance by the polycomb protein FIE has been conserved during land plant evolution. *Development, 136*, 2433–2444.

Nonogaki, H. (May 2014). Seed dormancy and germination—emerging mechanisms and new hypotheses. *Frontiers in Plant Science, 28*. http://dx.doi.org/10.3389/fpls.2014.00233.

Okie, W. R. (1998). *Handbook of peach and nectarine varieties*. Byron, GA: USDA-ARS, 808 pp.

Olukolu, B. A., Trainin, T., Fan, S., Kole, C., Bielenberg, G., Reighard, G. L., et al. (2009). Genetic linkage mapping for molecular dissection of chilling requirement and bud break in apricot (*Prunus armeniaca* L.). *Genome, 52*, 819–828.

Parchman, T. L., Gompert, Z., Mudge, J., Schilkey, F. D., Benkman, C. W., & Buerkle, C. A. (2012). Genome-wide association genetics of an adaptive trait in lodgepole pine. *Molecular Ecology, 21*(12), 2991–3005.

Pauley, S. S., & Perry, T. O. (1954). Ecotypic variation of the photoperiodic response in *Populus*. *Journal of Arnold Arboretum, 35*, 167–188.

Pelgas, B., Bousquet, J., Meirmans, P. G., Ritland, K., & Isabel, N. (2011). QTL mapping in white spruce: gene maps and genomic regions underlying adaptive traits across pedigrees, years and environments. *BMC Genomics, 12*, 145–168.

Peña, L., Martín-Trillo, M., Juárez, J., Pina, J., Navarro, L., & Martínez-Zapater, J. M. (2001). Constitutive expression of *Arabidopsis* LEAFY or APETALA1 genes in citrus reduces their generation time. *Nature Biotechnology, 19*, 263–267.

Porth, I., Klápště, J., Skyba, O., Hannemann, J., McKown, A. D., Guy, R. D., et al. (2013). Genome-wide association mapping for wood characteristics in *Populus* identifies an array of candidate single nucleotide polymorphisms. *New Phytologist, 200*, 710–726.

Potkar, R., Recla, J., & Busov, V. (2013). ptr-MIR169 is a posttranscriptional repressor of PtrHAP2 during vegetative bud dormancy period of aspen (*Populus tremuloides*) trees. *Biochemical and Biophysical Research Communications, 431*, 512–518.

Ramos, A., Pérez-Solís, E., Ibáñez, C., Casado, R., Collada, C., Gómez, L., et al. (2005). Winter disruption of the circadian clock in chestnut. *Proceedings of the National Academy of Sciences of the United States of America, 102*, 7037–7042.

Rinne, P. L. H., Welling, A., Vahala, J., Ripel, L., Ruonala, R., Kangasjärvi, J., et al. (2011). Chilling of dormant buds hyperinduces *FLOWERING LOCUS T* and recruits GA-inducible 1,3-β-glucanases to reopen signal conduits and release dormancy in *Populus*. *Plant Cell, 23*, 130–146.

Rohde, A., & Bhalerao, R. (2007). Plant dormancy in the perennial context. *Trends in Plant Science, 12*, 217–223.

Rohde, A., Storme, V., Jorge, V., Gaudet, M., Vitacolonna, N., Fabbrini, F., et al. (2011). Bud set in poplar–genetic dissection of a complex trait in natural and hybrid populations. *New Phytologist, 189*, 106–121.

Romeu, J. F., Monforte, A. J., Sánchez, G., Granell, A., García-Brunton, J., Badenes, M. L., et al. (2014). Quantitative trait loci affecting reproductive phenology in peach. *BMC Plant Biology, 14*, 52–68.

Rowland, L. J., Ogden, E. L., Arora, R., Lim, C.-C., Lehman, J. S., Levi, A., et al. (1999). Use of blueberry to study genetic control of chilling requirement and cold hardiness in woody perennials. *Hortscience, 34*(7), 1185–1191.

Ruiz-Ferrer, V., & Voinnet, O. (2009). Roles of plant small RNAs in biotic stress responses. *Annual Review of Plant Biology, 60*, 485–510.

Ruttink, T., Arend, M., Morreel, K., Storme, V., Rombauts, S., Fromm, J., et al. (2007). A molecular timetable for apical bud formation and dormancy induction in poplar. *Plant Cell, 19*, 2370–2390.

Saito, T., Bai, S., Ito, A., Sakamoto, D., Saito, T., Ubi, B. E., et al. (2013). Expression and genomic structure of the *dormancy-associated MADS box* genes *MADS13* in Japanese pears (*Pyrus pyrifolia* Nakai) that differ in their chilling requirement for endodormancy release. *Tree Physiologist, 33*, 654–667.

Salazar, J. A., Ruiz, D., Egea, J., & Martínez-Gómez, P. (2013). Inheritance of fruit quality traits in apricot (*Prunus armeniaca* L.) and analysis of linked quantitative trait loci (QTLs) using simple sequence repeat (SSR) markers. *Plant Molecular Biology Reporter, 31*, 1506–1517. http://dx.doi.org/10.1007/s11105-013-0625-9.

Sánchez-Pérez, R., Cueto, J. D., Dicenta, F., & Martínez-Gómez, P. (2014). Recent advancements to study flowering time in almond and other *Prunus* species. *Frontiers in Plant Science, 5*, 1–7.

Sánchez-Pérez, R., Dicenta, F., & Martínez-Gómez, P. (2012). Inheritance of chilling and heat requirements for flowering in almond and QTL analysis. *Tree Genetics & Genomes, 8*, 379–389.

Santamaría, M. E., Hasbún, R., Valera, M. J., Meijón, M., Valledor, L., Rodríguez, J. L., et al. (2009). Acetylated H4 histone and genomic DNA methylation patterns during bud set and bud burst in *Castanea sativa*. *Journal of Plant Physiology, 166*, 1360–1369.

Santamaría, M. E., Rodríguez, R., Cañal, M. J., & Toorop, P. E. (2011). Transcriptome analysis of chestnut *(Castanea sativa)* tree buds suggests a putative role for epigenetic control of bud dormancy. *Annals of Botany, 108*, 485–498.

Sasaki, R., Yamane, H., Ooka, T., Jotatsu, H., Kitamura, Y., Akagi, T., et al. (2011). Functional and expressional analyses of PmDAM genes associated with endodormancy in Japanese apricot (*Prunus mume*). *Plant Physiology*. http://dx.doi.org/10.1104/pp.111.181982.

Schrader, J., Moyle, R., Bhalerao, R., Hertzberg, M., Lundeberg, J., Nilsson, P., et al. (2004). Cambial meristem dormancy in trees involves extensive remodelling of the transcriptome. *The Plant Journal, 40*, 173–187.

Shaver, S., Casa-Mollano, J. A., Cerny, R. L., & Ceruti, H. (2010). Origin of the Polycomb repressive complex 2 and gene silencing by an E(z) homolog in the unicellular alga *Chlamidomonas*. *Epigenetics, 5*, 301–312.

Shoot, C., & Rinne, P. (2011). Dormancy cycling at the shoot apical meristem: transitioning between self-organization and self-arrest. *Plant Science, 180*, 120–131.

Skrøppa, T., & Johnsen, Ø. (2000). Patterns of adaptive genetic variation in forest tree species; the reproductive environment as an evolutionary force in *Picea abies*. In C. Matyas (Ed.), *Forest genetics and Sustainability* (pp. 49–58). Dordrecht, The Netherlands: Kluwer Academic Publications.

Skrøppa, T., Kohmann, K., Johnsen, Ø., Steffenrem, A., & Edvardsen, Ø. M. (2007). Field performance and early test results of offspring from two Norway spruce seed orchards containing clones transferred to warmer climates. *Canadian Journal of Forest Research, 37*, 515–522.

Slavov, G. T., DiFazio, S. P., Martin, J., Schackwitz, W., Muchero, W., Rodgers-Melnick, E., et al. (2012). Genome resequencing reveals multiscale geographic structure and extensive linkage disequilibrium in the forest tree *Populus trichocarpa*. *New Phytologist, 196*, 713–725.

Socias i Company, R., Felipe, A. J., & Gomez Aparisi, J. (1996). Genetics of late blooming in almond. In *EUCARPIA Symp. Fruit breeding and genetics, Oxford.*

Srinivasan, C., Dardick, C., Callahan, A., & Scorza, R. (2012). Plum (*Prunus domestica*) trees transformed with poplar *FT1* result in altered architecture, dormancy requirement, and continuous flowering. *PLoS One*. http://dx.doi.org/10.1371/journal.pone.0040715.

Sung, S., & Amasino, R. M. (2004). Vernalization in *Arabidopsis thaliana* is mediated by the PHD finger protein VIN3. *Nature, 427*, 159–164.

Sunkar, R., Chinnusamy, V., Zhu, J., & Zhu, J. K. (2007). Small RNAs as big players in plant abiotic stress responses and nutrient deprivation. *Trends in Plant Science, 12*, 301–309.

Sunkar, R., & Zhu, J. K. (2004). Novel and stress-regulated microRNAs and other small RNAs from *Arabidopsis*. *Plant Cell, 16*, 2001–2019.

Tanino, K. K., Kalcsits, L., Silim, S., Kendall, E., & Gray, G. R. (2010). Temperature-driven plasticity in growth cessation and dormancy development in deciduous woody plants: a working hypothesis suggesting how molecular and cellular function is affected by temperature during dormancy induction. *Plant Molecular Biology, 73*, 49–65. http://dx.doi.org/10.1007/s11103-010-9610-y.

Teixeira, F. K., & Colot, V. (2010). Repeat elements and the *Arabidopsis* DNA methylation landscape. *Heredity, 105*, 14–23.

Topp, B. L., Sherman, W. B., & Rasera, M. C. B. (2008). Low-chill cultivars development. In D. R. Layne, & D. Bassi (Eds.), *The peach: botany, production and uses* (pp. 106–138). CAB International.

Ubi, B. E., Sakamoto, D., Ban, Y., Shimada, T., Ito, A., Nakajima, I., et al. (2010). Molecular cloning of dormancy-associated MADS-box gene homologs and their characterization during seasonal endodormancy transitional phases of Japanese pear. *Journal of the American Society for Horticultural Science, 135*, 174–182.

Ueno, S., Klopp, C., Leplé, J. C., Derory, J., Noirot, C., Léger, V., et al. (2013). Transcriptional profiling of bud dormancy induction and release in oak by next-generation sequencing. *BMC Genomics, 14*, 236.

Valledor, L., Hasbún, R., Meijon, M., Rodriguez, J. L., Santamaria, E., Viejo, M., et al. (2007). Involvement of DNA methylation in tree development and micropropagation. *Plant Cell, Tissue and Organ Culture, 91*, 75–86.

Vazquez, F., Legrand, S., & Windels, D. (2010). The biosynthetic pathways and biological scopes of plant small RNAs. *Trends in Plant Science, 15*, 337–345.

Verde, I., Abbott, A. G., Scalabrin, S., Jung, S., Shu, S., Marroni, F., et al. (2013). The high-quality draft genome of peach (*Prunus persica*) identifies unique patterns of genetic diversity, domestication and genome evolution. *Nature Genetics*. http://dx.doi.org/10.1038/ng.2586. Published on-line.

Vining, K. J., Pomraning, K. R., Wilhelm, L. J., Priest, H. D., Pellegrini, M., Mockler, T. C., et al. (2012). Dynamic DNA cytosine methylation in the *Populus trichocarpa* genome: tissue-level variation and relationship to gene expression. *BMC Genomics, 13*, 27.

Wang, Y., Georgi, L., Reighard, G. L., Scorza, R., & Abbott, A. G. (2002). Genetic mapping of the *evergrowing* gene in peach [*Prunus persica* (L) Batsch]. *Journal of Heredity, 93*(5), 352–358.

Wellinga, A., & Palva, E. T. (2006). Molecular control of cold acclimation in trees. *Physiologia Plantarum, 127*, 167–181.

Wu, R.-M., Walton, E. F., Richardson, A. C., Wood, M., Hellens, R. P., & Varkonyi-Gasic, E. (2012). Conservation and divergence of four kiwifruit SVP-like MADS-box genes suggest distinct roles in kiwifruit bud dormancy and flowering. *Journal of Experimental Botany, 63*, 797–807.

Yakovlev, I. A., Asante, D. K. A., Fossdal, C. G., Junttila, O., & Johnsen, Ø. (2011). Differential gene expression related to an epigenetic memory affecting climatic adaptation in Norway spruce. *Plant Science, 180*, 132–139.

Yakovlev, I. A., Fossdal, C. G., & Johnsen, Ø. (2010). MicroRNAs, the epigenetic memory and climatic adaptation in Norway spruce. *New Phytologist, 187*, 1154–1169.

Yamagishi, N., Sasaki, S., Yamagata, K., Komori, S., Nagase, M., Wada, M., et al. (2011). Promotion of flowering and reduction of a generation time in apple seedlings by ectopical expression of the *Arabidopsis thaliana* FT gene using the Apple latent spherical virus vector. *Plant Molecular Biology, 75*, 193–204.

Yamane, H., Ooka, T., Jotatsu, H., Hosaka, Y., Sasaki, R., & Tao, R. (2011). Expressional regulation of PpDAM5 and PpDAM6, peach (*Prunus persica*) dormancy-associated MADS-box genes, by low temperature and dormancy-breaking reagent treatment. *Journal of Experimental Botany, 62*, 3481–3488.

Yan, H., Kikuchi, S., Neumann, P., Zhang, W., Wu, Y., Chen, F., et al. (2010). Genome-wide mapping of cytosine methylation revealed dynamic DNA methylation patterns associated with genes and centromeres in rice. *Plant Journal, 63*, 353–365.

Yooyongwech, S., Horigane, A. K., Yoshida, M., Yamaguchi, M., Sekozawa, Y., Sugaya, S., et al. (2008). Changes in aquaporin expression and magnetic resonance imaging of water status in peach tree flower buds during dormancy. *Physiologia Plantarum, 134*, 522–533.

Yordanov, Y. S., Ma, C., Strauss, S. H., & Busov, V. B. (2014). *EARLY BUD-BREAK 1 (EBB1)* is a regulator of release from seasonal dormancy in poplar trees. *Proceedings of National Academy of Sciences of the United States of America, 111*(27), 10001–10006.

Zhang, J., Xu, Y., Huan, Q., & Chong, K. (2009). Deep sequencing of brachypodium small RNAs at the global genome level identifies microRNAs involved in cold stress response. *BMC Genomics, 10*, 449.

Zhebentyayeva, T. N., Fan, S., Chandra, A., Bielenberg, D. G., Reighard, G. L., Okie, W. R., et al. (2014). Dissection of chilling requirement and bloom date QTLs in peach using a whole genome sequencing of sibling trees from an F2 mapping population. *Tree Genetics & Genomes, 10*, 35–51.

Zhou, X., Wang, G., Sutoh, K., Zhu, J., & Zhang, W. (2008). Identification of cold-inducible microRNAs in plants by transcriptome analysis. *BBA-Gene Regulatory Mechanisms, 1779*, 780–788.

CHAPTER SEVEN

Tree Responses to Environmental Cues

Ingo Ensminger[*,§,¶,1], **Christine Yao-Yun Chang**[*,§], **Katharina Bräutigam**[§,||]

[*]Department of Biology, University of Toronto Mississauga, Mississauga, ON, Canada
[§]Department of Cell and Systems Biology, University of Toronto, Toronto, ON, Canada
[¶]Department of Ecology and Evolutionary Biology, Toronto, ON, Canada
[||]Department of Biological Sciences, University of Toronto Scarborough, Toronto, ON, Canada
[1]Corresponding author: E-mail: ingo.ensminger@utoronto.ca

Contents

Abstract

Trees use environmental cues such as drought, temperature and light to synchronize growth and development with seasonal changes, and to compete for resources with other plants. The ability of trees to effectively sense and respond to these signals requires molecular programs to enable them to track change rapidly and adapt accordingly. Advances in genomics, transcriptomics and metabolomics have improved our

Advances in Botanical Research, Volume 74
ISSN 0065-2296
http://dx.doi.org/10.1016/bs.abr.2015.05.003

understanding, but there are still considerable gaps and for trees in particular. This chapter reviews recent advances in understanding tree responses to episodes of drought, elevated temperature, low temperature and photoperiod, because these cues are amongst the most important environmental factors affecting tree performance over the course of the season and in a changing climate. For example, trees are expected to suffer more from extreme low temperature events during winter or increased frequency of heat waves during summer and, in many regions, from longer periods of drought. In addition, elevated temperature is expected to extend the length of the growing season, while responses to photoperiod might constrain positive responses to warmer spring or autumn. Finally, the interaction of multiple cues is reviewed and emerging areas of research are discussed such as the role of natural variation, adaptation, plasticity and epigenetics.

1. INTRODUCTION

Similar to all other higher plants, trees use environmental cues such as temperature, drought and light to synchronize growth and development with seasonal changes, and to compete for resources with other plants. The ability of trees to effectively sense and respond to these signals affects their distribution, development and productivity. Trees have evolved adaptations to monitor fluctuations of environmental cues and to acclimate to them individually or more commonly, in combination. Plants therefore require molecular programs to enable them to track change rapidly and adapt accordingly (Ahuja, de Vos, Bones, & Hall, 2010). Understanding the programs that sense environmental cues and coordinate plant responses has been a subject of research for decades. Advances in genomics, transcriptomics and metabolomics have provided an improved understanding mostly in herbaceous model species (Ahuja et al., 2010), but there are still considerable gaps and for trees in particular (but see e.g. Bizet et al., 2015; Tylewicz, Tsuji, & Miskolczi, 2015; Wisniewski, Norelli, & Artlip, 2015). Due to current and projected changes in climatic conditions it is important to predict how trees will cope with these changes and to identify trees with enhanced resistance or tolerance to stressful conditions. Conventional breeding has been successful for improving traits in trees, but because of their long generation time, large genome sizes and lack of genomic tools, improvement of forest trees is slow (Harfouche, Meilan, & Altman, 2014; Neale & Kremer, 2011). Recent advancements have provided important resources for genomic-based approaches in tree species that help better understand tree responses to environmental cues and support tree breeding and selection of trees for improving traits and survival in future climates. These include the poplar genome (Tuskan, Difazio, Jansson, & Bohlmann, 2006), the genome of *Eucalyptus grandis* (Myburg et al., 2014) and the release of the genomes of three conifer species, white spruce (Birol et al., 2013), Norway spruce (Nystedt et al., 2013), and loblolly pine (Neale et al., 2014). In addition, there is a growing number of transcriptome and expression studies revealing genes, gene networks involved in stress tolerance, adaptation in trees and epigenetic responses to environmental cues (Lorenz et al., 2011; Müller, Ensminger, & Schmid, 2012; Raj et al., 2011; Ríos, Leida, Conejero, & Badenes, 2014; Song, Chen, Ci, & Zhang, 2013). Furthermore, proteomic and metabolomic approaches are used to assess how gene expression affects proteins and metabolic pathways (Harfouche et al., 2012), and we should expect tremendous improvements in understanding signalling pathways and genes that encode and regulate proteins that are involved in response to environmental cues in trees.

In the following sections, this chapter will focus on recent advances in tree responses to episodes of drought, elevated temperature, low temperature and photoperiod, because these cues are amongst the most important environmental factors affecting tree performance over the course of the season and in a changing climate. For example, trees are expected to suffer more from extreme low temperature events during winter, an increased frequency of heat waves during summer and in many regions from longer periods of drought. In addition, elevated temperature is expected to extend the length of the growing season, while response to photoperiod might constrain positive responses to warmer spring or autumn temperature (Busch, Hüner, & Ensminger, 2007; Chang, Unda, Zubilewich, Mansfield, & Ensminger, 2015; Way & Montgomery, 2014). Each of the following sections will begin with a description of tree physiological responses to environmental cues. This will be followed by a review of how this change is sensed and which downstream signalling pathways are involved in mediating a molecular adaptive response. The interaction of multiple cues will then be reviewed in Section 6 based on some examples. Finally, Section 7 will explore emerging areas of research and examine the role of natural variation, adaptation, plasticity and epigenetics. This section will conclude with an overview on tree responses to climate change and how future research on tree abiotic stress tolerance can contribute to improve forest health, conservation, tree breeding and selection in a future climate.

2. DROUGHT

2.1 Physiological Responses to Drought

Trees have a variety of strategies at their disposal to respond to and contend with reduced water availability, comprising morphological, physiological and biochemical adjustments. While some of the physiological responses described below have been well characterized, much remains to be learned about mechanistic details, plasticity of responses and underlying regulatory networks in tree species (Ryan, 2011).

Drought avoidance is a strategy used by many tree species in order to prevent detrimental effects of drought by mechanisms providing sufficient water uptake and minimizing water loss. For example, an increase in the capacity for water uptake from drying soil can help sustain photosynthesis and growth under drought to a certain extent. Increased capacity for water uptake can be achieved through changes in biomass partitioning and increased investment of resources into larger and deeper root systems, as has been observed under water stress and in populations from habitats with limited water supply e.g. in maritime pine (*Pinus pinaster*) (Aranda, Alia, Ortega, Dantas, & Majada, 2010; Nguyen & Lamat, 1989). Similarly, reduction of water loss through transpiration can be achieved by a decrease in leaf area. When the cellular water content decreases, the turgor pressure against the cell wall is reduced due to a smaller cell volume. Turgor-dependent reduction in cell division and expansion reduces leaf expansion, which is an early response to water deficit in a variety of plants (Marron et al., 2003). Under severe stress conditions, leaf area adjustment can also be achieved by leaf senescence and shedding of the leaves.

Leaf stomata play a key role in controlling transpiration and CO_2 uptake. Physiological responses to water deficit mediated by stomatal regulation range from strategies maintaining a high leaf water potential (isohydric plants or 'water savers', with early stomatal closure) to those with reduced leaf water potential (anisohydric or 'water spenders', with longer maintenance of a high stomatal conductance) (Tardieu & Simonneau, 1998).

Drought tolerance entails mechanisms that include osmotic adjustment, but also molecular mechanisms involved in detoxification of reactive oxygen species (ROS). When soil water potential becomes very low, plant cells are at risk of losing cell turgor. An increase in solutes, however, alleviates water stress and helps sustain higher photosynthetic rates under drought. Compatible osmolytes in the cytoplasm do not interfere with enzyme function and cellular metabolism, and include, for example, amino acids (e.g. proline), quaternary ammonium compounds (e.g. glycine betaine) and soluble sugars or sugar alcohols (such as mannitol, galactinol or pinitol) (Ashraf & Foolad, 2007; Chaves & Oliveira, 2004). In addition to maintaining cell turgor, these compounds have been implicated in membrane protection, stabilization of proteins and protection against oxidation i.e. as osmoprotectants (Ashraf & Foolad, 2007). In the past, osmotic adjustment has received much attention with respect to its potential for crop improvement (Callister et al., 2008; Chaves & Oliveira, 2004). While significant improvements have been achieved in herbaceous species, a study in eucalyptus suggests that strategies other than increased osmotic adjustment might be more promising for tree improvement to drought such as direct selection for growth under water-limited conditions (Callister et al., 2008; Chaves & Oliveira, 2004).

Photosynthesis is a main target of water stress. An early physiological response to water limitation is typically stomatal closure, which prevents catastrophic hydraulic failure and maintains leaf hydration status. At the same time, it restricts CO_2 diffusion to the site of carboxylation and represents the main limiting factor for photosynthesis during early phases of drought stress. Based on the general relationship between stomatal conductance (g_s) and net CO_2 assimilation (A_N), a common pattern of progressive water stress has been observed in C_3 plants (Flexas & Medrano, 2002; Liu et al., 2010; Medrano, Escalona, Bota, Gulias, & Flexas, 2002). It is characterized by sequential downregulation of different photosynthetic processes as drought progresses and can be divided into three phases. Under mild water stress (1), reduced stomatal conductance determines the reduction in net carbon assimilation. Linear electron transport (ETR) and activities of photosynthetic enzymes are unaffected, while the rate of photorespiration increases due to reduced internal CO_2 concentrations. Under moderate stress (2), a further reduction in stomatal conductance is paralleled by a decrease in the rate of ETR and a slight increase in antioxidants and thermal energy dissipation, indicating preparation for severe stress. Severe stress (3) leads to drastically reduced photochemistry and carboxylation efficiency via decreased ribulose 1,5-bisphosphate carboxylase/oxygenase (Rubisco) activity and RuBP content, as well as permanent damage to the photosystems (Bota, Medrano, & Flexas, 2004; Flexas & Medrano, 2002; Galmes, Ribas-Carbo, Medrano, & Flexas, 2011; Medrano et al., 2002).

2.2 Signalling and Molecular Mechanisms of Drought Responses

The complex molecular mechanisms underlying drought responses have been well-characterized in a variety of species and extensive regulatory networks at the transcriptional level have been identified (Shinozaki & Yamaguchi-Shinozaki, 2007; Shinozaki, Yamaguchi-Shinozaki, & Seki, 2003; Wilkins, Bräutigam, & Campbell, 2010). Products of drought-regulated genes play key roles in drought tolerance mechanisms (e.g. key enzymes in osmolyte biosynthesis, detoxification enzymes, molecules with protective function such as late embryogenesis abundant (LEA) proteins and chaperones) or regulatory processes (e.g. as transcription factors, protein kinases, abscisic acid (ABA) biosynthesis enzymes) (Shinozaki & Yamaguchi-Shinozaki, 2007).

Water limitations cause a decrease in turgor pressure in plant cells and mediate the production of the phytohormone ABA. ABA in turn promotes stomatal closure (Morgan, 1984), but also affects gene expression. Moreover, both ABA-dependent and ABA-independent gene pathways have been described in response to drought (Figure 1). Conserved *cis*-elements were identified in the promoters of stress-responsive genes, namely the ABA-responsive element (ABRE) and the dehydration-responsive element (DRE)/C-repeat element, which are involved in ABA-dependent and ABA-independent pathways, respectively (Shinozaki & Yamaguchi-Shinozaki, 2007).

Enhanced ABA levels activate a number of protein kinase pathways, including pathways controlled by SNF-1-related protein kinases (SnRK) (Raghavendra, Gonugunta, Christmann, & Grill, 2010; Yoshida et al., 2010). SnRKs in turn regulate the activity of AREBs/ABFs (ABA-responsive element binding factor/protein), which are basic leucine zipper proteins that act as master transcription factors in ABA signalling under drought stress conditions (Yoshida et al., 2010; Figure 1). ABA also activates myelocytomatosis (MYC)/myeloblastosis (MYB) transcription factors, which regulate genes responsive to drought stress and salinity stress such as rd22 (Abe et al., 2003; Figure 1).

DREB2 transcription factors (drought response element binding factors) respond strongly to drought in an ABA-independent manner and bind to the *cis*-acting DRE/C-repeat elements found in the promoter regions of stress response genes such as rd29A. Intriguingly, several stress gene promoters (e.g. rd29A) contain both DRE/C-repeat and ABRE elements,

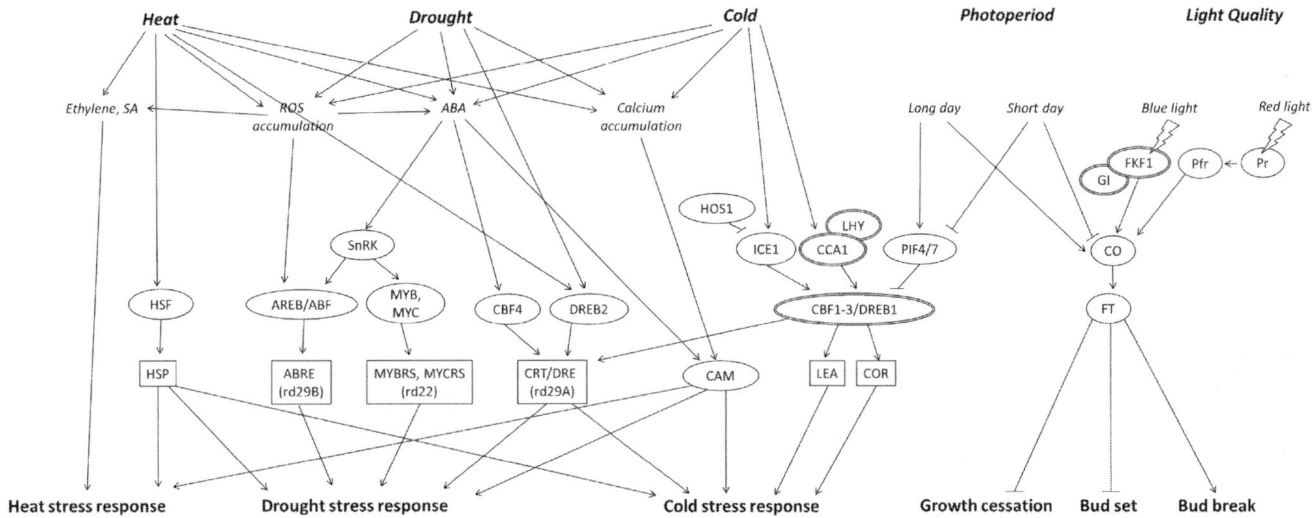

Figure 1 Signalling pathways involved in responses to environmental cues. Heat, drought, cold and light signals induce a network of stress response pathways. Italics indicate pathway activators; ovals indicate key signaling mechanisms; double-lined ovals indicate pathway components associated with the circadian clock mechanisms; rectangles indicate target *cis*-acting elements. ABA, abscisic acid; ABRE, ABA-responsive element; AREB/ABF, ABA-responsive binding protein; CAM, calmodulin; CBF, C-repeat binding factor; CCA1, circadian clock-associated 1; CO, CONSTANS transcription factor; COR, cold responsive; CRT/DRE, C-repeat/drought-responsive element; DREB, DRE binding protein; FKF1, flavin-binding kelch repeat F-box 1; FT, flowering locus T; GI, GIGANTEA protein; HOS1, high expression of osmotically responsive gene 1; HSF, heat shock factor; HSP, heat shock protein; ICE1, inducer of CBF expression 1; LHY, late elongated hypocotyl; LEA, late embryogenesis abundant; MYB/MYC, plant homologues of vertebrate MYB (myeloblastosis) and MYC (myelocytomatosis) oncogenes; MYBRS/MYCRS, MYB- and MYC-recognition sites; Pfr, far-red light-responsive phytochrome isoform; Pr, red light-responsive phytochrome isoform; PIF, phytochrome-interacting factor; ROS, reactive oxygen species; SA, salicylic acid; SnRK, SNF1-related kinase. *Figure represents a composite pathway based on: Gong et al. (1998), Imaizumi et al. (2003), Shinozaki et al. (2003), Larkindale et al. (2005), Chinnusamy et al. (2007), Jiao et al. (2007), Kotak et al. (2007), Sawa et al. (2007), Catalá, Medina, and Salinas (2011), Andrés and Coupland (2012), Petterle et al. (2013).*

thus integrating ABA-dependent and ABA-independent regulatory signals (Shinozaki & Yamaguchi-Shinozaki, 2007). Extensive studies in the model species *Arabidopsis* have revealed considerable transcriptional crosstalk in abiotic stress responses, such as the activation of rd29A by the DREB1/CBF transcription factors under cold and drought (Figure 1; Shinozaki & Yamaguchi-Shinozaki, 2007).

Recently, intracellular Ca^{2+} levels have also been implicated in the drought, cold and salt stress response of *Populus euphratica* with the discovery of a calcium-dependent protein kinase (PeCPK10) which confers drought and freezing tolerance as well as increased sensitivity to ABA (Chen, Xue, Xia, & Yin, 2013).

The molecular mechanisms underlying tree responses to different levels of water deprivation described above have been investigated in a classical study in loblolly pine (*Pinus taeda*) by Watkinson et al. (2003). Intriguingly, transcriptome patterns revealed little overlap between genes responding to mild and severe stress, highlighting that different levels of stress characterized by different physiological responses initiate different response pathways at the molecular level. Genes induced under mild drought stress included genes coding for LEA group 2 proteins and enzymes, which are involved in core carbon metabolism. In contrast, severe drought stress-regulated genes included genes coding for polyamine biosynthesis enzymes and senescence-related genes, which are directly or indirectly involved in the regulation of senescence and dormancy. In addition, the number of drought stress-responsive genes increased over repeated cycles of mild drought stress, while the rate of CO_2 assimilation recovered in subsequent drought cycles. This provides one of several compelling examples for photosynthetic acclimation to drought (Watkinson et al., 2003).

3. HEAT

3.1 Physiological Responses to Heat

Exposure to high temperature poses an important constraint to tree growth and development and heat stress has been implicated in climate-driven tree mortality (Anderegg, Kane, & Anderegg, 2013). In plants, a number of strategies have evolved to avoid and reduce or even tolerate negative effects of heat stress. In habitats with high light intensity and high temperatures, for example, plants can grow reflective trichomes and waxes, or change their leaf orientation (Shepherd & Griffiths, 2006; Wahid,

Gelani, Ashraf, & Foolad, 2007). Moreover, plants in full sunlight rely on evaporative cooling to avoid excessive heating of the leaves, especially in C3 plants (Wahid et al., 2007).

Photosynthesis is one of the most sensitive processes affected by increased temperature. Past its temperature optimum, photosynthesis decreases with increasing temperature. Higher temperatures promote photorespiration and favour the competing oxygenation reaction of Rubisco over the carboxylation reaction required for carbon fixation (Ogren, 1984), thus generating hydrogen peroxide and increasing intracellular ROS (Sharkey, 2005). Heat stress also suppresses linear electron transport through photosystem II (PSII) and instead stimulates cyclic electron transport around PSI (Sharkey, 2005). At sufficiently high temperatures, the structure and function of the thylakoid membrane and photosynthetic proteins can also become damaged (Sharkey, 2005). During heat stress, cellular respiration is affected much later than photosynthesis (Samuelson & Teskey, 1991; Saxe, Cannell, Johnsen, Ryan, & Vourlitis, 2001). Thus, when the temperature rises above the temperature/carbon compensation point, where CO_2 assimilation through photosynthesis equals CO_2 release from cellular respiration, carbohydrate reserves can be depleted with potentially detrimental effects for the plant.

The membrane system is another early target of temperature stress. Elevated temperature usually increases membrane fluidity, which can result in ion leakage. Membrane fluidity can be reduced through increasing the protein content of the membrane or by altering the relative amount of saturated lipids, thus preventing ion leakage under high temperatures (Saidi et al., 2010; Somerville & Browse, 1991).

Cell injury, characterized by impaired cellular function and compromised cell and tissue integrity, is induced by excessive exposure to elevated temperature. While influenced by a variety of factors, heat stress in plants is generally considered to be induced by a transient increase in temperature of about 10–15 °C above ambient (Wahid et al., 2007). At the cellular level, detrimental effects of high temperature include inactivation of enzymes, protein denaturation and aggregation, membrane dysfunction and de-organization of microtubules, which can eventually lead to breakdown of cell integrity (Smertenko, Draber, Viklicky, & Opatrny, 1997; Wahid et al., 2007).

Thermotolerance allows plants to mitigate the effects of heat stress. For most plant tissues and growing cells, a sudden and prolonged exposure to temperatures above a threshold temperature (often ca. 45 °C) is lethal and destroys basic cellular function and organization. A brief exposure to a nonlethal heat stress, however, can prepare plants to subsequently contend with otherwise lethal high temperatures (Vierling, 1991). This acquired thermotolerance is a vital but cost-intensive strategy, which involves synthesis of heat shock proteins (HSP), as well as ABA, ROS and salicylic acid (SA) pathways (Wahid et al., 2007).

3.2 Signalling and Molecular Mechanisms of Heat Responses

A hallmark of the 'heat shock response' (HSR) is the accumulation of HSP, many of which acting as molecular chaperones with protective functions (Figure 1). The classical view of HSR regulation involves key heat shock transcription factors (HSFs), which are activated under stress conditions through feedback mechanisms triggered by unfolded proteins (Voellmy & Boellmann, 2007). Recently, this view has been expanded, and additional mechanisms have been implicated in heat sensing and signalling.

Membranes respond very early and sensitively to elevated temperature, and calcium channels in the plasma membrane as well as changes in membrane lipid composition were found to mediate heat responses in plants (Saidi et al., 2010). Intracellular calcium levels spike transiently following exposure to heat stress (Larkindale & Knight, 2002). Inhibiting calcium signalling increased mortality and electrolyte leakage in *Zea mays* (Gong, Li, & Chen, 1998); inhibiting calmodulin function increased oxidative damage and mortality in heat-stressed *Arabidopsis* (Larkindale & Knight, 2002). Together, these findings suggest that calcium signalling is involved in the heat stress response (Figure 1). In addition, ROS and nucleosomes containing the alternative histone H2A.Z were found to act as temperature-responsive regulators in *Arabidopsis* (Kumar & Wigge, 2010; Volkov, Panchuk, Mullineaux, & Schoffl, 2006).

Despite the overall complexity of heat stress responses, a common feature among organisms is the accumulation of HSPs (Vierling, 1991). While many HSPs are induced in response to abiotic stresses, including heat, osmotic, cold and oxidative stress (Wang, Vinocur, Shoseyev, & Altman, 2004), some HSPs are also expressed in the absence of stress and are required for growth and development (Vierling, 1991). During heat stress, HSPs improve protein stability by binding to and supporting existing protein structures, preventing denaturation and improving protein repair, thereby increasing thermotolerance (Saidi, Finka, & Goloubinoff, 2011). HSPs are classified into five major families based on approximate molecular weight: Hsp100, which functions to resolubilize protein aggregates; Hsp90, which interacts with steroid hormone receptors and is involved in disease resistance and genetic buffering (Sangster & Queitsch, 2005); Hsp70, which is involved in protein refolding, molecular chaperoning and signal transduction; Hsp60, which is involved in protein folding and small HSPs which act as protein stabilizers and prevent aggregation (Vierling, 1991; Wang et al., 2004). The HSPs are controlled upstream by HSFs (Figure 1).

Aside from inducing the HSP response, heat stress induces the accumulation of phytohormones including SA, ethylene and ABA (Kotak et al., 2007; Larkindale & Knight, 2002). Mutants that were unable to synthesize ABA or that accumulated SA exhibited impaired thermotolerance, although they accumulated HSPs at levels similar to wild-type heat stress response (Larkindale, Hall, Knight, & Vierling, 2005). This indicates that these hormones mediate HSP-independent thermotolerance pathways. Ethylene and ABA have been implicated in antioxidant biosynthesis (Gong et al., 1998; Kotak et al., 2007; Larkindale & Knight, 2002), as ROS are typically accumulated under abiotic stress. Heat stress also activates the ABA-independent drought response pathway controlled by DREB2A (Figure 1), which induces drought and salt tolerance genes, as well as HSFs, thereby activating HSP-mediated thermotolerance (Sakuma et al., 2006).

A recent study in a heat stress-tolerant tree species (*P. euphratica*) highlights the relevance of the different molecular mechanisms and the gene network contributing to heat stress responses in long-lived trees (Chen, Yin, & Xia, 2014). After a pre-acclimation to 37 °C, plants were exposed to a heat stress at 45 °C. The transcriptome response revealed a pronounced induction of genes coding for HSP, predominantly of the HSP70 family, while genes coding for polyubiquitins (heat shock, protein degradation) and hormone-mediated signalling were also enriched (Chen et al., 2014).

4. COLD

4.1 Physiological Responses to Cold

Cold stress and cold hardiness in trees has been studied extensively with regards to horticulture in order to discern the effects of frost episodes on fruit trees and cold storage of fruits for export. In an ecological context, cold stress is often investigated in order to identify constraints on the geographic distribution of species based on their tolerance to low temperature, or the physiological processes affected by low temperature (Ensminger, Berninger, & Streb, 2012).

Chilling stress and injury is experienced by chilling-sensitive plants when temperatures decrease below the photosynthetic optimum, but are still above freezing. In plants which are not cold tolerant, injury and responses to chilling are typically induced at the range of 10–15 °C. Sensitivity to chilling is largely exhibited by tropical and subtropical species, e.g. banana, orange and avocado (Lyons, 1973). Cold-hardy species from temperate climates, such as apple, are considered chilling-resistant species, and can survive exposure to chilling temperatures as low as 0–4 °C (Ensminger et al., 2012; Graham & Patterson, 1982). Exposure to chilling temperatures inhibits growth, photosynthesis and metabolism (Ensminger et al., 2012). On the cellular level, chilling induces protein aggregation, inhibits enzymatic activity and increases membrane rigidity (Crosatti, Rizza, Badeck, Mazzucotelli, & Cattivelli, 2013; Graham & Patterson, 1982). The inhibition of enzymatic activity disrupts metabolic processes and also impedes the synthesis of proteins that normally undergo rapid turnover. Therefore, even if low temperature does not directly induce protein degradation, chilling can induce loss of proteins over time (Graham & Patterson, 1982; Levitt, 2012). Mechanical damage to membranes associated with membrane rigidification induces plasmolysis and results in solute and electrolyte leakage (Levitt, 2012). Consequently, determination of the percentage of electrolyte leakage has become a widely used method of assessing chilling and freezing tolerance in plants (Burr et al., 2001). Membrane-bound organelles, and particularly the chloroplast thylakoids, are especially sensitive to changes in temperature (Crosatti et al., 2013; Levitt, 2012); thus, the chloroplast redox state has been posited as a sensor of cold stress (Ensminger, Busch, & Hüner, 2006; Huner et al., 1996; Kurepin et al., 2013). Visible symptoms of chilling injury include browning or development of spots on fruit and leaves and the induction of leaf senescence (Lyons, 1973). Depending on the susceptibility of the species, chilling injury can be incurred within a few hours of chilling exposure, or more slowly following exposure for several days (Levitt, 2012). While tropical and subtropical species usually do not experience low temperatures in their native environment and therefore have not evolved protective mechanisms against low temperature, deciduous tree species from temperate and boreal environments undergo leaf senescence during autumn and develop new leaves in the spring, thereby minimizing exposure of sensitive leaf tissues to low temperature conditions.

Freezing stress, injury and cryoprotective mechanisms are induced once temperatures fall below 0 °C. Plants exposed to freezing temperatures are challenged by a combination of stresses: (1) cold stress, via increased membrane rigidification and inhibition of metabolic activity, (2) mechanical stress, due to the formation of intracellular or extracellular ice crystals and increased brittleness of tissues and (3) osmotic stress, since atmospheric and groundwater sources are trapped in the form of ice (Guy, 1990; Thomashow, 1999). Species adapted to survive freezing in higher latitudes, including trees such as *Picea*, *Pinus*, *Larix*, *Betula*, *Pseudotsuga* and *Populus*, have evolved extensive freezing defences. Proper development of cold hardiness allows these species to survive temperatures as low as −40 °C. *Pinus strobus* seedlings, grown in a field experiment in southern Ontario, were shown to easily survive exposure to −55 °C during late autumn (Chang et al., 2015); in *Pinus sylvestris*, freezing resistance to −196 °C was observed (Sakai & Weiser, 1973). Cryoprotective mechanisms include protein stabilization, increase of solute concentrations to lower the freezing point and adjustment of the composition of membrane lipids to improve membrane fluidity (Crosatti et al., 2013; Guy, 1990). Under low temperature conditions, plants also enhance sucrose biosynthesis (Chang, Fréchette, Unda, Mansfield, & Ensminger, in preparation; Chang et al., 2015; Schrader & Sauter, 2002), which contributes to improved freezing resistance (Gilmour, Sebolt, Salazar, Everard, & Thomashow, 2000). Other important soluble sugars that accumulate in response to low temperature include fructose, stachyose, raffinose and other oligosaccharides (Dauwe, Holliday, Aitken, & Mansfield, 2012). Studies on isolated thylakoid membranes and chloroplasts demonstrated that glucose, sucrose and raffinose enhance freezing tolerance (Uemura & Steponkus, 2003). Recently, raffinose was suggested to act as a stabilizer for PSII (Knaupp, Mishra, Nedbal, & Heyer, 2011).

Photoprotection in overwintering evergreen conifers develops in addition to freezing stress. Low temperatures exacerbate an imbalance between the source of energy and the metabolic sink, thus requiring adjustments of photosynthesis to maintain the balance of energy flow (Ensminger et al., 2006). Conifers adjust to this imbalance by downregulation of photosynthesis and increasing photoprotective mechanisms (Demmig-Adams, Cohu, Muller, & Adams, 2012; Ensminger et al., 2006; Öquist & Huner, 2003). This downregulation involves the inactivation of PSII reaction centres via degradation of the D1 protein, and a decrease in the size of the PSII light-harvesting complexes (Ottander, Campbell, & Öquist, 1995). Cyclic electron transport at PSI is subsequently enhanced to provide ATP in place of linear electron transport (Ensminger et al., 2006). In addition, photoprotective mechanisms are induced to quench excess light energy, such as increased production of accessory carotenoid leaf pigments and nonphotochemical quenching via increased thermal dissipation mediated by xanthophyll pigments (Demmig-Adams et al., 2012; Ensminger et al., 2006).

4.2 Signalling and Molecular Mechanisms of Cold Responses

Though several candidates exist, the primary sensor for low temperature is not yet elucidated. Membrane fluidity has long been considered a potential environmental sensor; a number of cold-responsive genes in plants have been identified as fatty acid desaturases, which function to increase membrane fluidity and function under low temperatures (Los & Murata, 2004). Örvar, Sangwan, Omann, and Dhindsa (2000) demonstrated that artificially induced membrane rigidification at 25 °C induced freezing tolerance, and conversely that membrane fluidity at 4 °C prevented development of freezing tolerance in *Medicago sativa*. Vaultier et al. (2006) showed that the activation of diacylglycerol kinase, which functions in cold signal transduction, was correlated with membrane fluidity. Calcium may also act as an initial low temperature signal in plants (Figure 1). Ca^{2+} is observed to spike during initial cold exposure, and is required for the development of freezing tolerance in *Arabidopsis* (Knight, Trewavas, & Knight, 1996). Monroy and Dhindsa (1995) demonstrated that an influx of calcium ions was correlated with the expression of cold acclimation-specific genes in *M. sativa* protoplasts.

The primary regulators of cold response genes, first described in *Arabidopsis*, are the CBF/DREB1 family of transcription factors (Shinozaki et al., 2003), which are rapidly activated following 15 min of exposure to 4 °C (Thomashow, 2010). CBF1–3 are ABA-independent and function upstream of a series of cold-responsive (COR) and LEA genes, including those that promote the synthesis of cryoprotective compounds such as dehydrins (Kjellsen, Yakovlev, Fossdal, & Strimbeck, 2013), raffinose (Cook, Fowler, Fiehn, & Thomashow, 2004) and proline (Dörffling, Dörffling, & Luck, 2008). In *Arabidopsis*, overexpression of DREB1a (also known as CBF3) has been shown to enhance overall freezing tolerance as well as proline and soluble sugar levels (Gilmour et al., 2000). The CBF family is regulated upstream by ICE1 (inducer of CBF expression 1), which induces CBF3 expression upon cold exposure, and also inhibits the MYB15 transcription factor which functions as a repressor of CBF (Chinnusamy, Zhu, & Zhu, 2007; Thomashow, 2010). Similar CBF regulons have been described in *Populus* (Benedict et al., 2006) and *Betula* (Welling & Palva, 2008).

ABA biosynthesis is upregulated during exposure to low temperatures in part as a result of glucose signalling, as glucose is accumulated during low temperature stress (Dörffling et al., 2008; Li, Junttila, Ernstsen, Heino, & Palva, 2003). Interestingly, the recently discovered *Arabidopsis* gene CBF4 is a homologue of DREB1 but is both drought activated and ABA dependent. Like the other CBFs, CBF4 binds to the promoter of rd29A; RD29A then acts upstream of drought and cold stress-responsive genes (Figure 1; Haake et al., 2002). Furthermore, ABA-deficient *Arabidopsis* mutants demonstrated increased susceptibility to freezing (Gilmour & Thomashow, 1991). Thus, although the ABA-independent DREB1 family is recognized as the major cold response regulator, ABA still plays a significant role in cold response.

Dehydrins are involved in the development of freezing tolerance. Dehydrins are a subset of the LEA genes and consist of a diverse family of osmo- and cryoprotective proteins (Rorat, 2006). In *Arabidopsis*, the Cor15a dehydrin protein reduces freezing damage and forms oligomers that bind to enzymes in the chloroplast stroma and protect them against freeze-induced inactivation (Nakayama et al., 2007). Dehydrins that prevent the coagulation of macromolecules such as membrane proteins (Close, 1997) were shown to be involved in antifreeze activity in *Prunus persica* (Wisniewski et al., 1999). In *Citrus unshiu*, the dehydrin CuCOR19 was shown to enhance cold tolerance by scavenging oxygen radicals produced during cold or drought stress (Hara, Fujinaga, & Kuboi, 2004). Dehydrins accumulated during the development of freezing tolerance in *Picea obovata* (Kjellsen et al., 2013) and in *P. strobus* (Chang et al., in preparation).

5. LIGHT

5.1 Physiological Responses to Photoperiod

Photoperiod controls spring and autumn phenology of trees, and also regulates developmental processes such as vegetative development and flowering. As will be discussed in the subsequent section, plants measure day length from the amount of blue and red light absorbed during the day (Jiao, Lau, & Deng, 2007); thus, plants integrate photoperiod and light quality in order to determine the day length (Figure 1). Though both spring and autumn involve perception of changes in temperature and photoperiod, processes involved in spring phenology tend to be primarily temperature responsive while autumn processes are under stronger control of photoperiod.

Cessation of growth is induced by short photoperiod in a vast number of woody species, including e.g. *Pinus, Acer, Catalpa, Liriodendron, Paulownia*

(Downs & Borthwick, 1956), *Picea* (Heide, 1974), *Betula* (Downs & Borthwick, 1956; Li et al., 2003; Welling, Kaikuranta, & Rinne, 1997) and *Populus* (Böhlenius et al., 2006; Rohde, Bastien, & Boerjan, 2011). The molecular signal that initiates cessation of growth originates in the leaves, where photoreceptors are located. This signal is mediated by CO and FT (Figure 1 and section below), and is subsequently transmitted to apical meristems (Rohde & Bhalerao, 2007). During this time, buds undergo targeted desiccation, which coincides with the induction of dehydrins (Welling et al., 1997). Exposure to low temperature can accelerate cessation of growth in trees (Rohde et al., 2011; Tanino, Kalcsits, Silim, Kendall, & Gray, 2010). Furthermore, several studies suggest that exposure to short photoperiod increases bud sensitivity to ABA (Welling & Palva, 2006), since ABA accumulates in dormant buds (Li et al., 2003). However, in a separate study, ABA did not induce cessation of growth, and was instead correlated with freezing tolerance (Welling et al., 1997).

Bud dormancy transitions between three states during the year: (1) paradormancy, or apical dominance, in which bud development is suppressed by hormone signals from tissues outside of the meristems; (2) endodormancy, in which buds enter an endogenously enforced dormant state and (3) ecodormancy, in which dormancy is no longer genetically controlled, but is maintained by unfavourable environmental conditions (Basler & Körner, 2014; Horvath, Anderson, Chao, & Foley, 2003; Rohde & Bhalerao, 2007; Welling & Palva, 2006). These states are associated with different seasons: paradormancy typically occurs during summer, endodormancy is induced during autumn and ecodormancy occurs during winter (Horvath et al., 2003). The release from each of these stages depends upon the environmental cues photoperiod and temperature. In particular, the release of paradormancy to endodormancy, also known as bud set, occurs shortly after the cessation of growth and is induced in response to photoperiod and temperature cues (Horvath et al., 2003). Species from higher latitudes, including *Populus*, *Betula*, *Salix* and *Picea*, are strongly reliant upon photoperiod for the induction of bud dormancy, while fruit trees native to more temperate zones are primarily temperature dependent (Cooke, Eriksson, & Junttila, 2012). An extensive transcriptional and metabolite analysis performed in *Populus tremula* × *Populus alba* demonstrated the existence of an ordered sequence of activation for the different regulatory pathways involved in bud dormancy, beginning with induction of photoperiod response pathways, followed by activation of ethylene and ABA signalling pathways (not shown in Figure 1; Ruttink et al., 2007). This study also revealed the sequence of physiological events that occur in response to short photoperiod, beginning with a transient adjustment of soluble carbohydrate levels and the remobilisation of starch in leaves within the first few days, followed by bud formation, development of bud scales, enhanced cold and desiccation tolerance and finally cessation of growth and dormancy after 3–4 weeks (Ruttink et al., 2007).

Autumn leaf senescence in deciduous trees allows the remobilisation of nutrients and avoids winter stress at the leaf level. Like other phenological processes, autumn leaf senescence is controlled by photoperiod and temperature. Studies in European aspen (*P. tremula*) demonstrated that senescence is induced by a decrease in photoperiod. The critical day length required for induction of senescence varies with latitude of origin; exposure to low temperature changed the rate at which leaves senesced, but did not affect the critical day length for the induction of senescence (Fracheboud et al., 2009; Keskitalo, Bergquist, Gardeström, & Jansson, 2005). However, studies on apple (*Malus pumila* Mill.) and pear (*Pyrus communis* L.) cultivars suggested that leaf senescence in these species was driven by low temperature and not photoperiod (Heide & Prestrud, 2005). Once initiated, leaf senescence occurs in a highly regulated sequence of events, beginning with the degradation of chlorophylls and export of nutrients, particularly nitrogen and carbon. The breakdown and conversion of macromolecules into transportable elements is highly energy intensive; therefore, mitochondrial respiration is among the last processes to be disassembled in the senescing leaf (Keskitalo et al., 2005). As chlorophylls are degraded, there is a strong increase in photoprotective pigment synthesis, including carotenoids and anthocyanins, which results in the characteristic colours of autumn leaves (Junker & Ensminger, submitted for publication; Keskitalo et al., 2005). Starch and sucrose content decrease during the earlier stages of senescence (Keskitalo et al., 2005). Glucose and fructose accumulate during later stages, presumably to fuel respiration (Keskitalo et al., 2005); these sugars may also play a role in senescence-related sugar signalling (Pourtau, Jennings, Pelzer, Pallas, & Wingler, 2006). Up to 80% of nitrogen and phosphorus content is recovered prior to leaf abscission (Keskitalo et al., 2005).

5.2 Signalling and Molecular Mechanisms Controlling Responses to Photoperiod

Plants detect light using photoreceptors, which are predominantly sensitive to red, far-red and blue wavelengths. Phytochromes (Phy) function mainly as red and far-red photoreceptors and are involved in seed germination, photomorphogenesis, flowering induction and dormancy (Franklin, Larner, & Whitelam, 2005; Smith, 2000). Exposure of phytochromes to red or far-red wavelengths causes conversion to active (Pfr) or inactive (Pr) isoforms, respectively (Figure 1; Smith, 2000). Once activated, Pfr triggers kinase cascades in the cytosol or nucleus, depending on localization (Smith, 2000). Cryptochromes (Cry) and phototropins (Phot) are both blue/UV-A photoreceptors (Franklin et al., 2005). Cryptochromes are involved in growth, flowering and regulating the circadian clock (Franklin et al., 2005). Phototropins are involved in plant responses to light quantity, including phototropism, chloroplast movement and stomatal opening, and also participate in circadian clock regulation (Franklin et al., 2005).

Photoperiod is sensed via integration of photoreceptor-mediated light signals and the endogenous circadian clock (Figure 1; Franklin et al., 2005). The circadian oscillator exists as a network of negative feedback loops that function upstream of the highly conserved CONSTANS (CO)/FLOWERING LOCUS T (FT) module (Andrés & Coupland, 2012; Azeez, Miskolcz, Tylewicz, & Bhalerao, 2014; Cooke et al., 2012). CO is constantly synthesized, but is degraded during the night (Andrés & Coupland, 2012). Over the course of a long day, CO accumulates sufficiently to activate FT transcription in leaves, which is stabilized by the activation of PhyA and Cry2 photoreceptors (Andrés & Coupland, 2012). Once synthesized, the FT protein is transported to shoot apical meristems where it activates downstream transcriptional pathways that result in flowering and vegetative growth (Andrés & Coupland, 2012).

In photoperiod-sensitive species, such as *Populus*, *Betula* and *Salix*, interactions of phytochromes with the CO/FT module determine the critical day length required to induce cessation of growth and bud set (Böhlenius et al., 2006; Cooke et al., 2012; Petterle, Karlberg, & Bhalerao, 2013). *Populus* clones overexpressing PhyA were unable to downregulate FT and did not initiate bud set in response to short photoperiod, while clones in which FT was downregulated by RNA interference experienced increased sensitivity to shortening photoperiod and began to set buds within 2 weeks after transition from 16 to 15 h day length (Böhlenius et al., 2006).

Recently, the FLAVIN-BINDING, KELCH REPEAT, F-BOX 1 (FKF1) and GIGANTEA (GI) proteins were also shown to regulate the CO/FT module and were found to be required for photoperiod signalling in *Arabidopsis* (Imaizumi, Tran, Swartz, Briggs, & Kay, 2003; Sawa, Nusinow, Kay, & Imaizumi, 2007). The expression of both FKF1 and GI proteins are controlled by the circadian clock and upregulated by blue light

(Imaizumi et al., 2003; Sawa et al., 2007). FKF1 is considered a putative photoreceptor for blue wavelengths, as it exhibits similar absorption spectra to the *Arabidopsis* phototropin phot1 and is highly blue-light-responsive (Imaizumi et al., 2003). Accumulation of FKF1 during long-day conditions results in the formation of a FKF1−GI complex which regulates expression of CO and enables expression of FT (Figure 1; Sawa et al., 2007).

6. INTERACTIONS OF MULTIPLE CUES

6.1 Interaction of Drought and Heat Stress/Elevated Temperature

Reduced water availability and increased temperature are often concurrent stress factors in nature (Anderegg et al., 2013; Wahid et al., 2007). In fact, widespread tree mortality due to dry and hot climate has been well-documented (Allen et al., 2010; van Mantgem et al., 2009). Under conditions of climate change, an increase in forest dieback driven by a higher frequency of drought and heat stress events has been predicted (Anderegg et al., 2013). While the majority of studies investigating stress responses and tree mortality has focussed on individual stresses, only a few studies explore tree responses to the combined effect of elevated temperature and drought (Adams et al., 2009; Kolb & Robberecht, 1996; Yeaman et al., 2014; Zhao, Hartmann, Trumbore, Ziegler, & Zhang, 2013). In trees, most work on the combined effects of multiple stresses has focused on physiological responses. For example, the sensitivity of *Pinus edulis* to drought was strongly increased under warmer temperature (ca. +4 °C above ambient), and the time to drought-induced mortality was reduced by ca. 30% compared to drought-stressed trees under ambient temperature (Adams et al., 2009). Possible mechanisms to explain this observation include the following: (1) stomatal closure to maintain the hydraulic status of the tree at the risk of strongly reduced photosynthesis and eventually carbon starvation and/or (2) maintenance of stomatal conductance to maintain transpiration and photosynthesis at the risk of catastrophic hydraulic failure. Intriguingly, respiration cost and likely carbon starvation were identified as main causes for the observed temperature-dependent differences in tree mortality (Adams et al., 2009). Similar observations were made in *Thuja occidentalis*, when water was withheld under different temperatures (15, 25 and 35 °C). Higher temperatures induced a negative carbon balance (respiration higher than net photosynthesis) and ultimately tree death earlier and at higher soil moisture content (Zhao et al., 2013). This suggests that the risk of carbon starvation for trees increases with elevated temperature even under mild drought (Adams et al., 2009; Zhao et al., 2013). In addition, interactive effects of hydraulic failure and carbon starvation, biomass partitioning between leaves and roots or species-specific characteristics can play key roles in shaping tree responses to the combined influence of elevated temperature and water limitation (Saxe et al., 2001; Way & Oren, 2010; Zhao et al., 2013).

To date, only a limited number of reports have investigated the regulatory mechanisms underlying plant responses to a combination of stresses (Prasch & Sonnewald, 2013; Rasmussen et al., 2013; Rizhsky et al., 2004; Yeaman et al., 2014). A pioneering study by Rizhsky et al. (2004) investigated the effect of drought, heat stress and the combination of the two stresses on the transcriptome of *Arabidopsis*. Considerable overlap between transcripts regulated by either heat or drought stress, or the combination of both stresses was observed. However, many transcripts showed distinct regulatory patterns in response to the combined impacts of heat and drought that differed clearly from the individual responses to either drought or heat (Rizhsky et al., 2004). This finding was substantiated by an extended study on transcriptome responses to a number of individual stresses (salt, cold, heat, high light, flagellin) as well as their combinations in *Arabidopsis* (Rasmussen et al., 2013). A large proportion of transcriptome changes observed for a combination of stresses were not predictable from the responses to the individual stress treatments pointing towards the complexity of regulatory networks and the importance to investigate stresses in combination (Rasmussen et al., 2013). While pathways and components regulating individual stress responses are relatively well characterized, we are just beginning to understand how combined stresses induce shifts in the complex signalling network and how responses to combined stresses are regulated (Adams et al., 2009; Prasch & Sonnewald, 2013; Rasmussen et al., 2013). In particular, very little information is available for molecular responses to combined stresses in trees. One of the few studies that explore molecular responses to combinations of stresses in trees focussed on two common conifers in western North America, lodgepole pine (*Pinus contorta*) and interior spruce (*Picea glauca* × *Picea engelmannii*), and transcriptome response to different combinations of temperature, moisture and light regime were investigated (Yeaman et al., 2014). Among other patterns, groups of genes were identified that responded similarly during bud set and to heat stress (under both, well-watered and water-limited conditions) indicating potential overlap of response pathways. Intriguingly, these groups of genes were enriched in genes coding for HSP, which comprise key components of the HSR or are regulated during development (Yeaman et al., 2014). Moreover, despite the large evolutionary distance between lodgepole pine and interior spruce, a substantial proportion of genes responded similarly to the environmental treatments in both 'species' indicating a high degree of conservation in the regulation of gene expression (Yeaman et al., 2014).

6.2 Interaction of Low Temperature and Photoperiod

In higher latitudes, changes in photoperiod and temperature are synchronized during spring and autumn. Thus, many phenological responses reflect the interaction of these two environmental cues.

Bud burst — As outlined in Section 5, the release from ecodormancy and subsequent bud burst involves a chilling requirement during the temperature increase from freezing to chilling. When maintained at chilling temperatures for a genetically determined period of time, metabolism is reactivated and carbohydrates are remobilised to fuel bud development. Subsequently, the first stage of spring bud development involves swelling of buds and is accompanied by a loss of freezing tolerance, followed by temperature-induced bud burst (Basler & Körner, 2014). The control of photoperiod over bud burst is more common in late-successional trees, e.g. *Fagus*, *Quercus*, *Abies* and *Picea*, as opposed to early-successional trees such as *Populus*, *Betula*, *Acer*, *Larix* and *Sorbus* which are photoperiod-insensitive in the spring (Körner & Basler, 2010; Laube et al., 2014). However, in an experiment testing 36 woody species, the chilling requirement was unequivocally found to supersede the effect of photoperiod (Laube et al., 2014). Recently, microarray analysis of *P. glauca* buds at different stages of development identified a number of photoperiod-responsive genes whose expression correlated with developmental events (El Kayal et al., 2011). Notably, the study observed upregulation of several categories of genes, which were downregulated during later stages of bud formation (El Kayal et al., 2011). These genes included those involved in the photoperiod signalling pathway mediated by CONSTANS, photosynthesis and carbohydrate metabolism-related genes associated with photoperiod-induced adjustments of source-sink dynamics, and genes associated with cold and desiccation stress (El Kayal et al., 2011).

Cold hardiness — Changes in photoperiod are sufficient to induce increased cold tolerance in trees (Chang et al., in preparation). A comparison of freezing tolerance in *Arabidopsis* grown under 16 h vs 8 h day length revealed that short-day plants had significantly increased freezing tolerance compared to long-day plants (Lee & Thomashow, 2012). Rostad, Granhus, Fløistad, and Morgenlie (2006) demonstrated that *Picea abies* seedlings grown at 8 h day length also exhibited higher freezing tolerance than counterparts grown at 16–18 h day length. *Pinus strobus* seedlings grown in southern Ontario demonstrated that decreasing photoperiod caused freezing tolerance to increase from August to September, when air temperature was well above freezing (Chang et al., 2015). However, once night-time temperature decreased below 0 °C in October, *P. strobus* exhibited a major increase in freezing resistance, demonstrating the strong interactive effect of temperature and photoperiod on freezing tolerance (Chang et al., 2015).

7. PERSPECTIVES

As reviewed in the previous sections, research on responses of trees to environmental cues includes a wide range of areas and many important aspects are still largely unclear. Some of the most fascinating aspects emerge around questions regarding how responses to environmental cues vary due to natural variation, adaptation, plasticity and epigenetics. Closely linked to the previous questions is the response of trees to changes in climatic patterns and growing season length. We therefore conclude with an overview of how future research can contribute to improve forest health, conservation, tree breeding and selection in a future climate.

7.1 Adaptation and Natural Variation of Drought Responses – An Example

The natural range of forest tree species can cover wide geographic areas with considerable variation in local habitat characteristics, suggesting a high degree of adaptive genetic variability within and among populations (Aitken, Yeaman, Holliday, Wang, & Curtis-McLane, 2008). As an example, we focus here on responses to drought; however, similar patterns can be observed in variation and adaptive plasticity for other stress as well (Aspinwall et al., 2014). Modulation of species-specific responses to drought by local populations and genotypes has been intensively studied in a number of trees including e.g. *Betula* (Aspelmeier & Leuschner, 2004, 2006), *Quercus* (Kremer, 2010), *Populus* (Hamanishi et al., 2010) or *Pinus* (Aranda et al., 2010), indicating that functional responses to water deficit are at least in part under genetic control. The response to drought is a complex process, and combined genomic and phenotypic analyses can enhance our understanding of this fundamental response as exemplified by two studies in poplar. To investigate the genetic basis of adaptive drought responses, Street et al. (2006) made use of a transgressive F2 population that was highly variable in their drought response i.e. exhibiting phenotypic traits above and below parental values. This mapping population was generated by crossing two poplar species with contrasting physiological responses to drought (*Populus deltoides* and *Populus trichocarpa*). Quantitative trait loci for a late-onset indicator of drought, leaf abscission, were identified, that is, genetic loci that likely contribute to the observed leaf abscission phenotype could be determined. In addition, genotypes that differed strongly in leaf abscission behaviour under drought were found to induce the expression of different sets of genes, suggesting adaptation at the level of gene expression (Street et al., 2006). Intraspecific variation of drought responses in balsam poplar (*Populus balsamifera*) was investigated by comparing phenotypic traits with gene expression profiles (Hamanishi et al., 2010). Genotypes that were able to sustain growth under drought conditions also exhibited a higher magnitude of stress-induced changes in gene expression. In addition, genotypes with similar drought transcriptomes also had fewer genetic differences, indicating that genetic relatedness contributes to similar drought responses.

7.2 Phenotypic Plasticity and Epigenetic Responses

Over their long life spans ranging from decades to centuries, trees are likely to experience a variety of variable environmental conditions, seasonal and annual change and episodic exposure to severe stress events. A high degree of phenotypic plasticity is crucial for survival under such variable conditions, and there is a complex interplay between phenotypic plasticity and local adaptation. In addition to a genetic contribution, an epigenetic component to phenotypic plasticity has been proposed (Bossdorf, Richards, & Pigliucci, 2008; Bräutigam et al., 2013; Chinnusamy & Zhu, 2009; Raj et al., 2011). Epigenetic phenomena induce (via mitosis or meiosis) heritable changes in gene function without altering the DNA sequence itself, and they are mediated by chromatin modifications, RNA species (such as miRNAs or long ncRNAs) or changes in DNA methylation. Recently, the influence of environmental factors on epigenetic phenomena has attracted considerable interest. However, we are currently just beginning to understand their relevance in long-lived tree species (Bräutigam et al., 2013; Chinnusamy & Zhu, 2009).

Remarkable phenotypic plasticity has been described in the mangrove *Laguncularia racemosa* with tall treelike individuals in a riverside habitat and abnormal shrublike morphology in a nearby salt marsh. Despite strong morphological differences, little genetic variation but abundant differences in DNA methylation were detected, indicating that epigenetic variation can play important roles in long-term adaptation to different environments (Lira-Madeiros et al., 2010). In Norway spruce, temperature and photoperiod period during embryogenesis rather than the identity and origin of the parent ecotype were found to influence bud phenology in progeny; moreover, these influences were found to persist for as long as 20 years (Skrøppa, Tollefsrud, Sperisen, & Johnsen, 2010; Yakovlev, Fossdal, & Johnsen, 2010). At the molecular level, epigenetically different progeny showed differential expression of miRNA pathway and transposon-related genes as well as genes involved in the epigenetic machinery (Yakovlev et al., 2010, 2014). Modulation of drought responses by epigenetic processes was further investigated in poplar. A study of six hybrid genotypes detected a correlation between DNA methylation and plant productivity under well-watered conditions, and genotype-specific fine-tuning of plant responses has been suggested (Gourcilleau et al., 2010). Recently, the modulation of drought responses by lasting effects of previous environmental experience (clone history) was investigated in poplar (Raj et al., 2011). Individuals of the same genotype were obtained from geographically distinct locations, grown in a common environment, and subjected to drought stress. Intriguingly, differences in transcript abundance patterns in response to drought, based on the differences in geographic origin, were detected in clonally propagated trees. These patterns were paralleled by differences in total DNA methylation, indicating a potential role for epigenetic mechanisms in the adaptation of long-lived plants to the local environment (Raj et al., 2011). The potential for changes in DNA methylation to contribute to rapid adaptation to changes in the environment is further substantiated by a study in *Quercus ilex* (Rico, Ogaya, Barbeta, & Penuelas, 2014).

7.3 Are Tree Responses to Environmental Cues Altered by Climatic Change?

The exact response of forests to global climate change is of paramount ecological and economic importance. Major challenges for trees and forests are pest outbreaks (Cullingham et al., 2011; Geils, Hummer, & Hunt, 2010) and rapid changes in temperature and precipitation (Joyce & Rehfeldt, 2013), with implications on forest productivity (Grant et al., 2007), survival (Birdsey & Pan, 2011) and changes in the distribution of tree species (Joyce & Rehfeldt, 2013; McKenney, Pedlar, Lawrence, Campbell, & Hutchinson,

2007). Trees are particularly vulnerable to the rapid rate of climatic change, as it contrasts with the life cycle of most long-living tree species (Joyce & Rehfeldt, 2013; Loarie et al., 2009). As a result, evolutionary adaptation to changing climate is difficult, if not impossible (Körner & Basler, 2010). On a shorter timescale, non-evolutionary responses to climate change may be facilitated by acclimation. A major task is to assess the vulnerability of current forests to climate change and the availability of tools to predict which trees will grow optimally and where in a future climate. It will be important to assess how well tree species and provenances fare under future climatic conditions and if phenotypic plasticity allows for physiological acclimation to the expected future climate. In order to answer such questions, it is important to better understand adaptation and acclimation and to develop tree genomic tools for selection, breeding and conservation strategies adapted to future climates.

Forest management strategies have started to consider climate change issues. The development of tree genomic tools for selection, breeding and conservation strategies adapted to future climate is underway to predict tolerance or sensitivity of long-lived tree species to anticipated environmental conditions. Traditionally, provenance trials and common garden experiments were the methods of choice to generate information on complex adaptive traits in forest trees (e.g. Morgenstern, D'Eon, & Penner, 2006). However, provenance trials are hampered by significant constraints since they focus on a subset of tree species and require the use of a very large number of populations. Furthermore, since forest trees are long-lived, it takes a long time to collect the information needed for decision-making. Sork et al. (2013) argue that the existing wealth of provenance trials can be used to develop new tools to the analysis of these experiments. They suggest a combination of traditional forest genetic approaches including provenance trials with landscape and functional genomic studies in order to assess mechanisms responsible for adaptive traits such as sensitivity to drought and increased temperature. In another study, provenance trials were used to identify candidate genes and SNP resources for Douglas-fir (Hess, Wildhagen, & Ensminger, 2013; Müller et al., 2012). In an elegant follow-up study, Hess, Wildhagen, Junker, and Ensminger (in preparation) assessed gene expression in 50-year-old Douglas-fir trees and isolated the transcriptome responses to photoperiod, water availability and temperature in two different genotypes. Using microarrays, Holliday, Ralph, White, Bohlmann, and Aitken (2008) assessed transcriptome dynamics at the population level in Sitka spruce in response to photoperiod and temperature. These types of studies extend beyond the traditional investigation of responses of trees to individual cues, and provide insight to more realistic scenarios involving interactions of multiple cues. These approaches will bring together analysis of genotype and phenotype characteristics in relation with bioclimatic variables, which can then help assess adaptive traits that can be used for gene-assisted selection or for refining models used to predict species distribution and productivity under climate change.

Recent advances in sequencing technologies and analysis tools will facilitate future studies of tree response to environmental cues. As outlined in the introduction, major progress has been made in sequencing and annotating a number of tree genomes including *E. grandis* (Myburg et al., 2014), and the large conifer genomes from white spruce (*P. glauca)* (Birol et al., 2013), Norway spruce (*P. abies)* (Nystedt et al., 2013) and loblolly pine (*P. taeda)* (Neale et al., 2014), providing resources for genome-wide analysis for high resolution analyses of multifactorial traits and responses in trees.

REFERENCES

Abe, H., Urao, T., Ito, T., Seki, M., Shinozaki, K., & Yamaguchi-Shinozaki, K. (2003). *Arabidopsis* AtMYC2 (bHLH) and AtMYB2 (MYB) function as transcriptional activators in abscisic acid signaling. *Plant Cell, 15*(1), 63–78.

Adams, H. D., Guardiola-Claramonte, M., Barron-Gafford, G. A., Villegas, J. C., Breshears, D. D., Zou, C. B., et al. (2009). Temperature sensitivity of drought-induced tree mortality portends increased regional die-off under global-change-type drought. *Proceedings of the National Academy of Sciences of the United States of America, 106*(17), 7063–7066.

Ahuja, I., de Vos, R. C., Bones, A. M., & Hall, R. D. (2010). Plant molecular stress responses face climate change. *Trends in Plant Science, 15*(12), 664–674.

Aitken, S. N., Yeaman, S., Holliday, J. A., Wang, T., & Curtis-McLane, S. (2008). Adaptation, migration or extirpation: climate change outcomes for tree populations. *Evolutionary Applications, 1*(1), 95–111.

Allen, C. D., Macalady, A., Chenchouni, H., Bachelet, D., McDowell, N., Vennetier, M., et al. (2010). A global overview of drought and heat-induced tree mortality reveals emerging climate change risks for forests. *Forest Ecology and Management, 259*(4), 660–684.

Anderegg, W. R., Kane, J. M., & Anderegg, L. D. (2013). Consequences of widespread tree mortality triggered by drought and temperature stress. *Nature Climate Change, 3*, 30–36.

Andrés, F., & Coupland, G. (2012). The genetic basis of flowering responses to seasonal cues. *Nature Reviews Genetics, 13*(9), 627–639.

Aranda, I., Alia, R., Ortega, U., Dantas, A. K., & Majada, J. (2010). Intra-specific variability in biomass partitioning and carbon isotopic discrimination under moderate drought stress in seedlings from four *Pinus pinaster* populations. *Tree Genetics and Genomes, 6*(2), 167–178.

Ashraf, M., & Foolad, M. R. (2007). Roles of glycine betaine and proline in improving plant abiotic stress resistance. *Environmental and Experimental Botany, 59*(2), 206–216.

Aspelmeier, S., & Leuschner, C. (2004). Genotypic variation in drought response of silver birch (*Betula pendula*): leaf water status and carbon gain. *Tree Physiology, 24*(5), 517–528.

Aspelmeier, S., & Leuschner, C. (2006). Genotypic variation in drought response of silver birch (*Betula pendula* Roth): leaf and root morphology and carbon partitioning. *Trees, 20*(1), 42–52.

Aspinwall, M. J., Loik, M. E., Resco, D. E., Dios, V., Tjoelker, M. G., Payton, P. R., et al. (2014). Utilizing intraspecific variation in phenotypic plasticity to bolster agricultural and forest productivity under climate change. *Plant, Cell and Environment*. http://dx.doi.org/10.1111/pce.12424.

Azeez, A., Miskolcz, P., Tylewicz, S., & Bhalerao, R. (2014). A tree ortholog of *APETALA1* mediates photoperiodic control of seasonal growth. *Current Biology, 24*(7), 717–724.

Basler, D., & Körner, C. (2014). Photoperiod and temperature responses of bud swelling and bud burst in four temperate forest tree species. *Tree Physiology, 34*(4), 377–388.

Benedict, C., Skinner, J. S., Meng, R., Chang, Y., Bhalerao, R., Huner, N. P. A., et al. (2006). The CBF1-dependent low temperature signalling pathway, regulon and increase in freeze tolerance are conserved in *Populus* spp. *Plant, Cell and Environment, 29*(7), 1259–1272.

Birdsey, R., & Pan, Y. (2011). Ecology: drought and dead trees. *Nature Climate Change, 1*, 444–445.

Birol, I., Raymond, A., Jackman, S. D., Pleasance, S., Coope, R., Taylor, G. A., et al. (2013). Assembling the 20 Gb white spruce (*Picea glauca*) genome from whole-genome shotgun sequencing data. *Bioinformatics, 29*(12), 1492–1497.

Bizet, F., Triboulot, M., Montpied, P., Christophe, A., Ningre, N., Cohen, D., et al. (2015). Phenotypic plasticity toward water regime: response of leaf growth and underlying candidate genes in *Populus*. *Physiologia Plantarum, 154*(1), 39–53.

Böhlenius, H., Huang, T., Charbonnel-Campaa, L., Brunner, A. M., Jansson, S., Straus, S. H., et al. (2006). CO/FT regulatory module controls timing of flowering and seasonal growth cessation in trees. *Science, 312*(5776), 1040–1043.

Bossdorf, O., Richards, C. L., & Pigliucci, M. (2008). Epigenetics for ecologists. *Ecology Letters, 11*(2), 106–115.

Bota, J., Medrano, H., & Flexas, J. (2004). Is photosynthesis limited by decreased Rubisco activity and RuBP content under progressive water stress? *New Phytologist, 162*(3), 671–681.

Bräutigam, K., Vining, K. J., Lafon-Placette, C., Fossdal, C. G., Mirouze, M., Marcos, J. G., et al. (2013). Epigenetic regulation of adaptive responses of forest tree species to the environment. *Ecology and Evolution, 3*(2), 399–415.

Burr, K. E., Hawkins, C. D. B., L'Hirondelle, S. J., Binder, W. D., George, M. F., & Repo, T. (2001). Methods for measuring cold hardiness of conifers. In F. J. Bigras, & S. J. Colombo (Eds.), *Conifer cold hardiness* (pp. 369–401). Netherlands: Springer.

Busch, F., Hüner, N. P. A., & Ensminger, I. (2007). Increased air temperature during simulated autumn conditions does not increase photosynthetic carbon gain but affects the dissipation of excess energy in seedlings of the evergreen conifer Jack pine. *Plant Physiology, 143*(3), 1242–1251.

Callister, A. N., Arndt, S. K., Ades, P. K., Merchant, A., Rowell, D., & Adams, M. A. (2008). Leaf osmotic potential of *Eucalyptus* hybrids responds differently to freezing and drought, with little clonal variation. *Tree Physiology, 28*(8), 1297–1304.

Catalá, R., Medina, J., & Salinas, J. (2011). Integration of low temperature and light signaling during cold acclimation response in *Arabidopsis*. *Proceedings of the National Academy of Sciences of the United States of America, 108*(39), 16475–16480.

Chang, C. Y., Fréchette, E., Unda, F., Mansfield, S. D., Ensminger, I. *Autumn simulations reveal sensitivity of Eastern white pine (Pinus strobus) seedlings to photoperiod, temperature, and elevated* CO_2. (in preparation).

Chang, C. Y., Unda, F., Zubilewich, A., Mansfield, S. D., & Ensminger, I. (2015). Sensitivity of cold acclimation to elevated autumn temperature in field-grown *Pinus strobus* seedlings. *Frontiers in Plant Science, 6*, 165.

Chaves, M. M., & Oliveira, M. M. (2004). Mechanisms underlying plant resilience to water deficits: prospects for water-saving agriculture. *Journal of Experimental Botany, 55*(407), 2365–2384.

Chen, J., Xue, B., Xia, X., & Yin, W. (2013). A novel calcium-dependent protein kinase gene from *Populus euphratica*, confers both drought and cold stress tolerance. *Biochemical and Biophysical Research Communications, 441*(3), 630–636.

Chen, J., Yin, W., & Xia, X. (2014). Transcriptome profiles of *Populus euphratica* upon heat shock stress. *Current Genomics, 15*(5), 326–340.

Chinnusamy, V., Zhu, J., & Zhu, J.-K. (2007). Cold stress regulation of gene expression in plants. *Trends in Plant Science, 12*(10), 444–451.

Chinnusamy, V., & Zhu, J. K. (2009). Epigenetic regulation of stress responses in plants. *Current Opinion in Plant Biology, 12*(2), 133–139.

Close, T. J. (1997). Dehydrins: a commonalty in the response of plants to dehydration and low temperature. *Physiologia Plantarum, 100*(2), 291–296.

Cooke, J. E. K., Eriksson, M. E., & Junttila, O. (2012). The dynamic nature of bud dormancy in trees: environmental control and molecular mechanisms. *Plant, Cell and Environment, 35*(10), 1707–1728.

Cook, D., Fowler, S., Fiehn, O., & Thomashow, M. F. (2004). A prominent role for the CBF cold response pathway in configuring the low-temperature metabolome of *Arabidopsis*. *Proceedings of the National Academy of Sciences of the United States of America, 101*(42), 15243–15248.

Crosatti, C., Rizza, F., Badeck, F. W., Mazzucotelli, E., & Cattivelli, L. (2013). Harden the chloroplast to protect the plant. *Physiologia Plantarum, 147*, 55–63.

Cullingham, C. I., Cooke, J. E. K., Dang, S., Davis, C. S., Cooke, B. J., & Coltman, D. W. (2011). Mountain pine beetle host-range expansion threatens the boreal forest. *Molecular Ecology, 20*(10), 2157–2171.

Dauwe, R., Holliday, J. A., Aitken, S. N., & Mansfield, S. D. (2012). Metabolic dynamics during autumn cold acclimation within and among populations of Sitka spruce (*Picea sitchensis*). *New Phytologist, 194*(1), 192–205.

Demmig-Adams, B., Cohu, C. M., Muller, O., & Adams, W. W., III (2012). Modulation of photosynthetic energy conversion efficiency in nature: from seconds to seasons. *Photosynthesis Research, 113*(1–3), 75–88.

Dörffling, K., Dörffling, H., & Luck, E. (2008). Improved frost tolerance and winter hardiness in proline overaccumulating winter wheat mutants obtained by in vitro-selection is associated with increased carbohydrate, soluble protein and abscisic acid (ABA) levels. *Euphytica, 165*(3), 545–556.

Downs, R. J., & Borthwick, H. A. (1956). Effects of photoperiod on growth of trees. *Botanical Gazette, 117*(4), 310–326.

El Kayal, W., Allen, C. C. G., Ju, C. J.-T., Adams, E., King-Jones, S., Zaharia, L. I., et al. (2011). Molecular events of apical bud formation in white spruce, *Picea glauca*. *Plant, Cell and Environment, 34*(3), 480–500.

Ensminger, I., Berninger, F., & Streb, P. (2012). Response of photosynthesis to low temperature. In J. Flexas, et al. (Eds.), *Terrestrial photosynthesis in a changing environment* (pp. 272–289). Cambridge: Cambridge University Press.

Ensminger, I., Busch, F., & Hüner, N. P. A. (2006). Photostasis and cold acclimation: sensing low temperature through photosynthesis. *Physiologia Plantarum, 126*, 28–44.

Flexas, J., & Medrano, H. (2002). Drought-inhibition of photosynthesis in C3 plants: stomatal and non-stomatal limitations revisited. *Annals of Botany, 89*(2), 183–189.

Fracheboud, Y., Luquez, V., Björkén, L., Sjödin, A., Tuominen, H., & Jansson, S. (2009). The control of autumn senescence in European aspen. *Plant Physiology, 149*(4), 1982–1991.

Franklin, K. A., Larner, V. S., & Whitelam, G. C. (2005). The signal transducing photoreceptors of plants. *International Journal of Developmental Biology, 49*(5–6), 653–664.

Galmes, J., Ribas-Carbo, M., Medrano, H., & Flexas, J. (2011). Rubisco activity in Mediterranean species is regulated by the chloroplastic CO_2 concentration under water stress. *Journal of Experimental Botany, 62*(2), 653–665.

Geils, B., Hummer, K., & Hunt, R. (2010). White pines, Ribes, and blister rust: a review and synthesis. *Forest Pathology, 40*(3–4), 145–417.

Gilmour, S. J., Sebolt, A. M., Salazar, M. P., Everard, J. D., & Thomashow, M. F. (2000). Overexpression of the *Arabidopsis* CBF3 transcriptional activator mimics multiple biochemical changes associated with cold acclimation. *Plant Physiology, 124*(4), 1854–1865.

Gilmour, S. J., & Thomashow, M. F. (1991). Cold acclimation and cold-regulated gene expression in ABA mutants of *Arabidopsis thaliana*. *Plant Molecular Biology, 17*(6), 1233–1240.

Gong, M., Li, Y.-J., & Chen, S.-Z. (1998). Abscisic acid-induced thermotolerance in maize seedlings is mediated by calcium and associated with antioxidant systems. *Journal of Plant Physiology, 153*(3–4), 488–496.

Gourcilleau, D., Bogeat-Triboulot, W.-B., Le Thiec, D., Lafon-Placette, C., Delaunay, A., El-Soud, W. A., et al. (2010). DNA methylation and histone acetylation: genotypic variations in hybrid poplars, impact of water deficit and relationships with productivity. *Annals of Forest Science, 67*(2), 208.

Graham, D., & Patterson, B. D. (1982). Responses of plants to low, nonfreezing temperatures: proteins, metabolism, and acclimation. *Annu Review of Plant Physiology, 33*, 347–372.

Grant, R. F., Barr, A. G., Black, T. A., Gaumont-Guay, D., Iwashita, H., Kidson, J., et al. (2007). Net ecosystem productivity of boreal jack pine stands regenerating from clearcutting under current and future climates. *Global Change Biology, 13*(7), 1423–1440.

Guy, C. L. (1990). Cold acclimation and freezing stress tolerance: role of protein metabolism. *Annual Review of Plant Physiology and Plant Molecular Biology, 41*, 187–223.

Haake, V., Cook, D., Riechmann, L., Pineda, O., Thomashow, M. F., & Zhang, J. Z. (2002). Transcription factor CBF4 is a regulator of drought adaptation in *Arabidopsis*. *Plant Physiology, 130*(2), 639–648.

Hamanishi, E. T., Raj, S., Wilkins, O., Thomas, B. R., Mansfield, S. D., Plant, A. L., et al. (2010). Intraspecific variation in the *Populus balsamifera* drought transcriptome. *Plant, Cell and Environment, 33*(10), 1742–1755.

Hara, M., Fujinaga, M., & Kuboi, T. (2004). Radical scavenging activity and oxidative modification of citrus dehydrin. *Plant Physiology and Biochemistry, 42*(7–8), 657–662.

Harfouche, A., Meilan, R., & Altman, A. (2014). Molecular and physiological responses to abiotic stress in forest trees and their relevance to tree improvement. *Tree Physiology, 34*(11), 1181–1198.

Harfouche, A., Meilan, R., Kirst, M., Morgante, M., Boerjan, W., Sabatti, M., et al. (2012). Accelerating the domestication of forest trees in a changing world. *Trends in Plant Science, 17*(2), 6472.

Heide, O. M. (1974). Growth and dormancy in Norway spruce ecotypes (*Picea abies*) I. Interaction of photoperiod and temperature. *Physiologia Plantarum, 30*(1), 1–12.

Heide, O. M., & Prestrud, A. K. (2005). Low temperature, but not photoperiod, controls growth cessation and dormancy induction and release in apple and pear. *Tree Physiology, 25*(1), 109–114.

Hess, M., Wildhagen, H., & Ensminger, I. (2013). Suitability of Illumina deep mRNA-sequencing for reliable gene expression profiling in a non-model conifer species (*Pseudotsuga menziesii*). *Tree Genetics and Genomes, 9*(6), 1513–1527.

Hess, M., Wildhagen, H., Junker, L. V., & Ensminger, I. *Transcriptome responses to a complex natural environment in field-grown divergent Douglas-fir*. (in preparation).

Holliday, J. A., Ralph, S. G., White, R., Bohlmann, J., & Aitken, S. N. (2008). Global monitoring of autumn gene expression within and among phenotypically divergent populations of Sitka spruce (*Picea sitchensis*). *New Phytologist, 178*(1), 103–122.

Horvath, D. P., Anderson, J. V., Chao, W. S., & Foley, M. E. (2003). Knowing when to grow: signals regulating bud dormancy. *Trends in Plant Science, 8*(11), 534–540.

Huner, N. P. A., Maxwell, D. P., Gray, G. R., Savitch, L. V., Krol, M., Ivanov, A. G., et al. (1996). Sensing environmental temperature change through imbalances between energy supply and energy consumption: redox state of photosystem II. *Physiologia Plantarum, 98*(2), 358–364.

Imaizumi, T., Tran, H. G., Swartz, T. E., Briggs, W. R., & Kay, S. A. (2003). FKF1 is essential for photoperiod-specific light signalling in *Arabidopsis*. *Nature, 426*(6964), 302–306.

Jiao, Y., Lau, O. S., & Deng, X. W. (2007). Light-regulated transcriptional networks in higher plants. *Nature Reviews Genetics, 8*(3), 217–230.

Joyce, D. G., & Rehfeldt, G. E. (2013). Climatic niche, ecological genetics, and impact of climate change on eastern white pine (*Pinus strobus* L.): guidelines for land managers. *Forest Ecology and Management, 295*, 173–192.

Junker, L. V., & Ensminger, I. *Autumn phenology in Acer saccharum leaves revealed by chlorophyll fluorescence, pigment changes and leaf optical properties*. (submitted for publication to *Tree Physiology*).

Keskitalo, J., Bergquist, G., Gardeström, P., & Jansson, S. (2005). A cellular timetable of autumn senescence. *Plant Physiology, 139*(4), 1635–1648.

Kjellsen, T. D., Yakovlev, I. A., Fossdal, C. G., & Strimbeck, G. R. (2013). Dehydrin accumulation and extreme low-temperature tolerance in Siberian spruce (*Picea obovata*). *Tree Physiology, 33*(12), 1354–1366.

Knaupp, M., Mishra, K. B., Nedbal, L., & Heyer, A. G. (2011). Evidence for a role of raffinose in stabilizing photosystem II during freeze-thaw cycles. *Planta, 234*(3), 477–486.

Knight, H., Trewavas, A. J., & Knight, M. R. (1996). Cold calcium signaling in *Arabidopsis* involves two cellular pools and a change in calcium signature after acclimation. *Plant Cell, 8*(3), 489–503.

Kolb, P. F., & Robberecht, R. (1996). High temperature and drought stress effects on survival of *Pinus ponderosa* seedlings. *Tree Physiology, 16*(8), 665–672.

Körner, C., & Basler, D. (2010). Phenology under global warming. *Science, 327*(5972), 1461–1462.

Kotak, S., Larkindale, J., Lee, U., von Koskull-Döring, P., Vierling, E., & Scharf, K.-D. (2007). Complexity of the heat stress response in plants. *Current Opinion in Plant Biology, 10*(3), 310–316.

Kremer, A. (2010). Evolutionary responses of European oaks to climate change. *Irish Forestry, 67*, 53–66.

Kumar, S. V., & Wigge, P. A. (2010). H2A.Z-containing nucleosomes mediate the thermosensory response in *Arabidopsis*. *Cell, 140*(1), 136–147.

Kurepin, L. V., Dahal, K. P., Savitch, L. V., Singh, J., Bode, R., Ivanov, A. G., et al. (2013). Role of CBFs as integrators of chloroplast redox, phytochrome and plant hormone signaling during cold acclimation. *International Journal of Molecular Sciences, 14*(6), 12729–12763.

Larkindale, J., Hall, J. D., Knight, M. R., & Vierling, E. (2005). Heat stress phenotypes of *Arabidopsis* mutants implicate multiple signaling pathways in the acquisition of thermotolerance. *Plant Physiology, 138*(2), 882–897.

Larkindale, J., & Knight, M. R. (2002). Protection against heat stress-induced oxidative damage in *Arabidopsis* involves calcium, abscisic acid, ethylene, and salicylic acid. *Plant Physiology, 128*(2), 682–695.

Laube, J., Sparks, T. H., Estrella, N., Höfler, J., Ankerst, D. P., & Menzel, A. (2014). Chilling outweighs photoperiod in preventing precocious spring development. *Global Change Biology, 20*(1), 170–182.

Lee, C.-M., & Thomashow, M. F. (2012). Photoperiodic regulation of the C-repeat binding factor (CBF) cold acclimation pathway and freezing tolerance in *Arabidopsis thaliana*. *Proceedings of the National Academy of Sciences of the United States of America, 109*(37), 15054–15059.

Levitt, J. (2012). *Responses of plants to environmental stresses volume I: Chilling, freezing and high temperature stresses* (2nd ed.). New York: Academic Press, Inc.

Li, C., Junttila, O., Ernstsen, A., Heino, P., & Palva, E. T. (2003). Photoperiodic control of growth, cold acclimation and dormancy development in silver birch (*Betula pendula*) ecotypes. *Physiologia Plantarum, 117*(2), 206–212.

Lira-Medeiros, C. F., Parisod, C., Fernandes, R. A., Mata, C. S., Cardoso, M. A., & Ferreira, P. C. (2010). Epigenetic variation in mangrove plants occurring in contrasting natural environment. *PLoS One, 5*(4), e10326.

Liu, C. C., Liu, Y. G., Guo, K., Zheng, Y. R., Li, G. Q., Yu, L. F., et al. (2010). Influence of drought intensity on the response of six woody karst species subjected to successive cycles of drought and rewatering. *Physiologia Plantarum, 139*(1), 39–54.

Loarie, S. R., Duffy, P. B., Hamilton, H., Asner, G. P., Field, C. B., & Ackerly, D. D. (2009). The velocity of climate change. *Nature, 462*, 1052–1055.

Lorenz, W. W., Alba, R., Yu, Y.-S. S., Bordeaux, J. M., Simões, M., & Dean, J. F. (2011). Microarray analysis and scale-free gene networks identify candidate regulators in drought-stressed roots of loblolly pine (*P. taeda* L.). *BMC Genomics, 12*, 264.

Los, D. A., & Murata, N. (2004). Membrane fluidity and its roles in the perception of environmental signals. *Biochimica et Biophysica Acta, 1666*, 142–157.

Lyons, J. M. (1973). Chilling injury in plants. *Annual Review of Plant Physiology, 24*, 445–466.

van Mantgem, P. J., Stephenson, N. L., Byrne, J. C., Daniels, L. D., Franklin, J. F., Fule, P. Z., et al. (2009). Widespread increase of tree mortality rates in the western United States. *Science, 323*(5913), 521–524.

Marron, N., Dreyer, E., Boudouresque, E., Delay, D., Petit, J. M., Delmotte, F. M., et al. (2003). Impact of successive drought and re-watering cycles on growth and specific leaf area of two *Populus* × *canadensis* (Moench) clones, 'Dorskamp' and 'Luisa_Avanzo'. *Tree Physiology, 23*(18), 1225–1235.

McKenney, D., Pedlar, J. H., Lawrence, K., Campbell, K., & Hutchinson, M. F. (2007). Potential impacts of climate change on the distribution of North American trees. *BioScience, 57*(11), 939–948.

Medrano, H., Escalona, J. M., Bota, J., Gulias, J., & Flexas, J. (2002). Regulation of photosynthesis of C3 plants in response to progressive drought: stomatal conductance as a reference parameter. *Annals of Botany, 89*, 895–905.

Monroy, A. F., & Dhindsa, R. S. (1995). Low-temperature signal transduction: Induction of cold acclimation-specific genes of alfalfa by calcium at 25 °C. *Plant Cell*, 7(3), 321–331.

Morgan, J. M. (1984). Osmoregulation and water stress in higher plants. *Annual Review of Plant Physiology, 35*, 299–319.

Morgenstern, E. K., D'Eon, S. P., & Penner, M. (2006). White spruce growth to age 44 in a provenance test at the Petawawa Research Forest. *Forestry Chronicle, 82*, 572–578.

Müller, T., Ensminger, I., & Schmid, K. (2012). A catalogue of putative unique transcripts from Douglas-fir (*Pseudotsuga menziesii*) based on 454 transcriptome sequencing of genetically diverse, drought stressed seedlings. *BMC Genomics, 13*, 673.

Myburg, A. A., Grattapaglia, D., Tuskan, G. A., Hellsten, U., Hayes, R. D., Grimwood, J., et al. (2014). The genome of *Eucalyptus grandis*. *Nature, 510*(7505), 356–362.

Nakayama, K., Okawa, K., Kakizaki, T., Honma, T., Itoh, H., & Inaba, T. (2007). *Arabidopsis* Cor15am is a chloroplast stromal protein that has cryoprotective activity and forms oligomers. *Plant Physiology, 144*(1), 513–523.

Neale, D. B., & Kremer, A. (2011). Forest tree genomics: growing resources and applications. *Nature Reviews Genetics, 12*(2), 111–122.

Neale, D. B., Wegrzyn, J. L., Stevens, K. A., Zimin, A. V., Puiu, D., Crepeau, M. W., et al. (2014). Decoding the massive genome of loblolly pine using haploid DNA and novel assembly strategies. *Genome Biology, 15*(3), R59.

Nguyen, A., & Lamat, A. (1989). Variation in growth and osmotic regulation of roots of water-stressed maritime pine (*Pinus pineaster* Ait.) provenances. *Tree Physiology, 5*(1), 123–133.

Nystedt, B., Street, N. R., Wetterbom, A., Zuccolo, A., Lin, Y. C., Scofield, D. G., et al. (2013). The Norway spruce genome sequence and conifer genome evolution. *Nature, 497*(7451), 579–584.

Ogren, W. L. (1984). Photorespiration: pathways, regulation, and modification. *Annual Review of Plant Physiology, 35*, 415–442.

Öquist, G., & Huner, N. P. A. (2003). Photosynthesis of overwintering evergreen plants. *Annual Review of Plant Biology, 54*, 329–355.

Örvar, B. L., Sangwan, V., Omann, F., & Dhindsa, R. S. (2000). Early steps in cold sensing by plant cells: the role of actin cytoskeleton and membrane fluidity. *Plant Journal, 23*(6), 785–794.

Ottander, C., Campbell, D., & Öquist, G. (1995). Seasonal changes in photosystem II organisation and pigment composition in *Pinus sylvestris*. *Planta, 197*(1), 176–183.

Petterle, A., Karlberg, A., & Bhalerao, R. P. (2013). Daylength mediated control of seasonal growth patterns in perennial trees. *Current Opinion in Biotechnology, 16*(3), 301–306.

Pourtau, N., Jennings, R., Pelzer, E., Pallas, J., & Wingler, A. (2006). Effect of sugar-induced senescence on gene expression and implications for the regulation of senescence in *Arabidopsis*. *Planta, 224*(3), 556–568.

Prasch, C. M., & Sonnewald, U. (2013). Simultaneous application of heat, drought, and virus to *Arabidopsis* plants reveals significant shifts in signaling networks. *Plant Physiology, 162*(4), 1849–1866.

Raghavendra, A. S., Gonugunta, V. K., Christmann, A., & Grill, E. (2010). ABA perception and signalling. *Trends in Plant Science, 15*(7), 395–401.

Raj, S., Bräutigam, K., Hamanishi, E. T., Wilkins, O., Thomas, B. R., Schroeder, W., et al. (2011). Clone history shapes *Populus* drought responses. *Proceedings of the National Academy of Sciences of the United States of America, 108*(30), 12521–12526.

Rasmussen, S., Barah, P., Suarez-Rodriguez, M. C., Bressendorff, S., Friis, P., Costantino, P., et al. (2013). Transcriptome responses to combinations of stresses in *Arabidopsis*. *Plant Physiology, 161*(4), 1783–1794.

Rico, L., Ogaya, R., Barbeta, A., & Penuelas, J. (2014). Changes in DNA methylation fingerprint of *Quercus ilex* trees in response to experimental field drought simulating projected climate change. *Plant Biology (Stuttgart), 16*(2), 419–427.

Ríos, G., Leida, C., Conejero, A., & Badenes, M. (2014). Epigenetic regulation of bud dormancy events in perennial plants. *Frontiers in Plant Science, 5*, 247.

Rizhsky, L., Liang, H., Shuman, J., Shulaev, V., Davletova, S., & Mittler, R. (2004). When defense pathways collide. The response of *Arabidopsis* to a combination of drought and heat stress. *Plant Physiology, 134*(4), 1683–1696.

Rohde, A., Bastien, C., & Boerjan, W. (2011). Temperature signals contribute to the timing of photoperiodic growth cessation and bud set in poplar. *Tree Physiology, 31*(5), 472–482.

Rohde, A., & Bhalerao, R. P. (2007). Plant dormancy in the perennial context. *Trends in Plant Science, 12*(5), 217–223.

Rorat, T. (2006). Plant dehydrins — tissue location, structure and function. *Cellular and Molecular Biology Letters, 11*(4), 536–556.

Rostad, H., Granhus, A., Fløistad, I. S., & Morgenlie, S. (2006). Early summer frost hardiness in *Picea abies* seedlings in response to photoperiod treatment. *Canadian Journal of Forest Research, 36*(11), 2966–2973.

Ruttink, T., Arend, M., Morreel, K., Storme, V., Rombauts, S., Fromm, J., et al. (2007). A molecular timetable for apical bud formation and dormancy induction in poplar. *Plant Cell, 19*(8), 2370–2390.

Ryan, M. G. (2011). Tree responses to drought. *Tree Physiology, 31*(3), 237–239.

Saidi, Y., Finka, A., & Goloubinoff, P. (2011). Heat perception and signalling in plants: a tortuous path to thermotolerance. *New Phytologist, 190*(3), 556–565.

Saidi, Y., Peter, M., Finka, A., Cicekli, C., Vigh, L., & Goloubinoff, P. (2010). Membrane lipid composition affects plant heat sensing and modulates Ca(2+)-dependent heat shock response. *Plant Signaling and Behavior, 5*(12), 1530–1533.

Sakai, A., & Weiser, C. J. (1973). Freezing resistance of trees in North America with reference to tree regions. *Ecology, 54*(1), 118–126.

Sakuma, Y., Maruyama, K., Qin, F., Osakabe, Y., Shinozaki, K., & Yamaguchi-Shinozaki, K. (2006). Dual function of an *Arabidopsis* transcription factor DREB2A in water-stress-responsive and heat-stress-responsive gene expression. *Proceedings of the National Academy of Sciences of the United States of America, 103*(49), 18822–18827.

Samuelson, L. J., & Teskey, R. O. (1991). Net photosynthesis and leaf conductance of loblolly pine seedlings in 2 and 21% oxygen as influenced by irradiance, temperature and provenance. *Tree Physiology, 8*(2), 205–211.

Sangster, T. A., & Queitsch, C. (2005). The HSP90 chaperone complex, an emerging force in plant development and phenotypic plasticity. *Current Opinion in Plant Biology, 8*(1), 86–92.

Sawa, M., Nusinow, D. A., Kay, S. A., & Imaizumi, T. (2007). FKF1 and GIGANTEA complex formation is required for day-length measurement in *Arabidopsis. Science, 318*(5848), 261–265.

Saxe, H., Cannell, M. G. R., Johnsen, O., Ryan, M. G., & Vourlitis, G. (2001). Tree and forest functioning in response to global warming. *New Phytologist, 149*(3), 369–400.

Schrader, S., & Sauter, J. (2002). Seasonal changes of sucrose-phosphate synthase and sucrose synthase activities in poplar wood (*Populus × canadensis Moench <robusta>*) and their possible role in carbohydrate metabolism. *Journal of Plant Physiology, 159*(8), 833–843.

Sharkey, T. D. (2005). Effects of moderate heat stress on photosynthesis: importance of thylakoid reactions, rubisco deactivation, reactive oxygen species, and thermotolerance provided by isoprene. *Plant, Cell and Environment, 28*(3), 269–277.

Shepherd, T., & Griffiths, D. W. (2006). The effects of stress on plant cuticular waxes. *New Phytologist, 171*(3), 469–499.

Shinozaki, K., & Yamaguchi-Shinozaki, K. (2007). Gene networks involved in drought stress response and tolerance. *Journal of Experimental Botany, 58*(2), 221–227.

Shinozaki, K., Yamaguchi-Shinozaki, K., & Seki, M. (2003). Regulatory network of gene expression in the drought and cold stress responses. *Current Opinion in Plant Biology, 6*(5), 410–417.

Skrøppa, A. G., Tollefsrud, M. M., Sperisen, C., & Johnsen, Ø. (2010). Rapid change in adaptive performance from one generation to the next in *Picea abies*—Central European trees in a Nordic environment. *Tree Genetics and Genomes, 6*(1), 93–99.

Smertenko, A., Draber, P., Viklicky, V., & Opatrny, Z. (1997). Heat stress affects the organization of microtubules and cell division in *Nicotiana tabacum* cells. *Plant, Cell and Environment, 20*(12), 1534–1542.

Smith, H. (2000). Phytochromes and light signal perception by plants — an emerging synthesis. *Nature, 407*, 585–591.

Somerville, C., & Browse, J. (1991). Plant lipids: metabolism, mutants and membranes. *Science, 252*(5002), 80–87.

Song, Y., Chen, Q., Ci, D., & Zhang, D. (2013). Transcriptome profiling reveals differential transcript abundance in response to chilling stress in *Populus simonii*. *Plant Cell Reports, 32*(9), 1407–1425.

Sork, V. L., Aitken, S. N., Dyer, R. J., Eckert, A. J., Legendre, P., & Neale, D. B. (2013). Putting the landscape into the genomics of trees: approaches for understanding local adaptation and population responses to changing climate. *Tree Genetics and Genomes, 9*(4), 901–911.

Street, N. R., Skogström, O., Sjödin, A., Tucker, J., Rodriguez-Acosta, M., Nilsson, P., et al. (2006). The genetics and genomics of the drought response in *Populus*. *Plant Journal, 48*(3), 321–341.

Tanino, K. K., Kalcsits, L., Silim, S., Kendall, E., & Gray, G. R. (2010). Temperature-driven plasticity in growth cessation and dormancy development in deciduous woody plants: a working hypothesis suggesting how molecular and cellular function is affected by temperature during dormancy induction. *Plant Molecular Biology, 73*(1–2), 49–65.

Tardieu, F., & Simonneau, T. (1998). Variability among species of stomatal control under fluctuating soil water status and evaporative demand: modelling isohydric and anisohydric behaviours. *Journal of Experimental Botany, 49*, 419–432.

Thomashow, M. F. (1999). Plant cold acclimation: freezing tolerance genes and regulatory mechanisms. *Annual Review of Plant Physiology, 50*, 571–599.

Thomashow, M. F. (2010). Molecular basis of plant cold acclimation: insights gained from studying the CBF cold response pathway. *Plant Physiology, 154*(2), 571–577.

Tuskan, G. A., Difazio, S., Jansson, S., & Bohlmann, J. (2006). The genome of black cottonwood, *Populus trichocarpa* (Torr. & Gray). *Science, 312*(5793), 1596–1604.

Tylewicz, S., Tsuji, H., & Miskolczi, P. (2015). Dual role of tree florigen activation complex component FD in photoperiodic growth control and adaptive response pathways. *Proceedings of the National Academy of Sciences of the United States of America, 112*(10), 3140–3145.

Uemura, M., & Steponkus, P. L. (2003). Modification of the intracellular sugar content alters the incidence of freeze-induced membrane lesions of protoplasts isolated from *Arabidopsis thaliana* leaves. *Plant, Cell and Environment, 26*(7), 1083–1096.

Vaultier, M.-N., Cantrel, C., Vergnolle, C., Justin, A.-M., Demandre, C., Benhassaine-Kesri, G., et al. (2006). Desaturase mutants reveal that membrane rigidification acts as a cold perception mechanism upstream of the diacylglycerol kinase pathway in *Arabidopsis* cells. *FEBS Letters, 580*(17), 4218–4223.

Vierling, E. (1991). The roles of heat shock proteins in plants. *Annual Review of Plant Physiology and Molecular Biology, 42*, 579–620.

Voellmy, R., & Boellmann, F. (2007). Chaperone regulation of the heat shock protein response. *Advances in Experimental Medicine and Biology, 594*, 89–99.

Volkov, R. A., Panchuk, , II, Mullineaux, P. M., & Schoffl, F. (2006). Heat stress-induced H(2)O(2) is required for effective expression of heat shock genes in *Arabidopsis*. *Plant Molecular Biology, 61*(4–5), 733–746.

Wahid, A., Gelani, S., Ashraf, M., & Foolad, M. R. (2007). Heat tolerance in plants: an overview. *Environmental and Experimental Botany, 61*(3), 199–223.

Wang, W., Vinocur, B., Shoseyev, O., & Altman, A. (2004). Role of plant heat-shock proteins and molecular chaperones in the abiotic stress response. *Trends in Plant Science, 9*(5), 244–252.

Watkinson, J. I., Sioson, A. A., Vasquez-Robinet, C., Shukla, M., Kumar, D., Ellis, M., et al. (2003). Photosynthetic acclimation is reflected in specific patterns of gene expression in drought-stressed loblolly pine. *Plant Physiology, 133*(4), 1702–1716.

Way, D., & Montgomery, R. (2014). Photoperiod constraints on tree phenology, performance and migration in a warming world. *Plant, Cell and Environment*. http://dx.doi.org/10.1111/pce.12431.

Way, D. A., & Oren, R. (2010). Differential responses to changes in growth temperature between trees from different functional groups and biomes: a review and synthesis of data. *Tree Physiology, 30*(6), 669–688.

Welling, A., Kaikuranta, P., & Rinne, P. (1997). Photoperiodic induction of dormancy and freezing tolerance in *Betula pubescens*. Involvement of ABA and dehydrins. *Physiologia Plantarum, 100*(1), 119–125.

Welling, A., & Palva, E. T. (2006). Molecular control of cold acclimation in trees. *Physiologia Plantarum, 127*(2), 167–181.

Welling, A., & Palva, E. T. (2008). Involvement of CBF transcription factors in winter hardiness in birch. *Plant Physiology, 147*(3), 1199–1211.

Wilkins, O., Bräutigam, K., & Campbell, M. M. (2010). Time of day shapes *Arabidopsis* drought transcriptomes. *Plant Journal, 63*(5), 715–727.

Wisniewski, M., Norelli, J., & Artlip, T. (2015). Overexpression of a peach CBF gene in apple: a model for understanding the integration of growth, dormancy, and cold hardiness in woody plants. *Frontiers in Plant Science, 6*, 85.

Wisniewski, M., Webb, R., Balsamo, R., Close, T. J., Yu, X.-M., & Griffith, M. (1999). Purification, immunolocalization, cryoprotective, and antifreeze activity of PCA60: a dehydrin from peach (*Prunus persica*). *Physiologia Plantarum, 105*(4), 600–608.

Yakovlev, I. A., Fossdal, C. G., & Johnsen, O. (2010). MicroRNAs, the epigenetic memory and climatic adaptation in Norway spruce. *New Phytologist, 187*(4), 1154–1169.

Yakovlev, I. A., Lee, Y. K., Rotter, B., Olsen, J. E., Skrøppa, A. G., Johnsen, Ø., et al. (2014). Temperature-dependent differential transcriptomes during formation of an epigenetic memory in Norway spruce embryogenesis. *Tree Genetics and Genomes, 10*(2), 355–366.

Yeaman, S., Hodgins, K. A., Suren, H., Nurkowski, K. A., Rieseberg, L. H., Holliday, J. A., et al. (2014). Conservation and divergence of gene expression plasticity following c. 140 million years of evolution in lodgepole pine (*Pinus contorta*) and interior spruce (*Picea glauca* × *Picea engelmannii*). *New Phytologist, 203*(2), 578–591.

Yoshida, T., Fujita, Y., Sayama, H., Kidokoro, S., Maruyama, K., Mizoi, J., et al. (2010). AREB1, AREB2, and ABF3 are master transcription factors that cooperatively regulate ABRE-dependent ABA signaling involved in drought stress tolerance and require ABA for full activation. *Plant Journal, 61*(4), 672–685.

Zhao, J., Hartmann, H., Trumbore, S., Ziegler, W., & Zhang, Y. (2013). High temperature causes negative whole-plant carbon balance under mild drought. *New Phytologist, 200*(2), 330–339.

CHAPTER EIGHT

Molecular Signatures of Adaptation and Selection in Forest Trees

Juan P. Jaramillo-Correa[*,1], **Julien Prunier**[§], **Alejandra Vázquez-Lobo**[*,†], **Stephen R. Keller**[¶], **Alejandra Moreno-Letelier**[*,||]

[*]Department of Evolutionary Ecology, Institute of Ecology, Universidad Nacional Autónoma de México, México City, Mexico
[§]Department of Forestry, Centre for Forest Research and Institute for Systems and Integrative Biology, Université Laval, Québec, QC, Canada
[†]Research Centre for Biodiversity and Conservation, Universidad Autónoma del Estado de Morelos, Cuernavaca, Mexico
[¶]Department of Plant Biology, University of Vermont, Burlington, VT, USA
[||]Botanical Garden, Institute of Biology, Universidad Nacional Autónoma de México, México City, Mexico
[1]Corresponding author: E-mail: jaramillo@ecologia.unam.mx

Contents

Abstract

Uncovering the genes and molecular basis of phenotypic variation and adaptation is a major goal in conservation and evolutionary genetics; it also sets the basis for future operational breeding in commercial species, like forest trees. These taxa are characterized by their large size, growth habit and longevity, which hampers the use of reverse-genetic approaches (i.e. from gene function to phenotype) to pinpoint adaptive molecular variants. In this chapter, we summarize the basis of the forward-genetic approaches (i.e. from phenotype to gene function) currently used in forest trees. For each strategy, we provide a brief overview of the statistical approaches employed to identify candidate genes, and then highlight the main findings of landmark studies that provide

Advances in Botanical Research, Volume 74
ISSN 0065-2296
http://dx.doi.org/10.1016/bs.abr.2015.04.003

evidence for adaptation in forest trees. Adaptive and commercial traits are generally well inherited in trees, although they are mostly affected by the variation of multiple genes, each one accounting for a small part of the phenotypic variance of each character. However, some individual and important genes involved in growth, phenology, drought resistance and cold hardiness have been identified; many of them showing evidence of selection across multiple taxa (sometimes including angiosperms and gymnosperms). Future challenges for detecting the signatures and understanding the molecular basis of adaptation in trees include more adequate and precise phenotype assessment in natural populations, and the inclusion of gene interactions and epigenetic variations in current models. The implications of these findings in conservation and breeding of forest trees are finally discussed.

1. INTRODUCTION

Trees are woody vascular plants, angiosperms or gymnosperms that share attributes like large size, growth habit, longevity and high reproductive outcome (reviewed by Petit and Hampe (2006)). Although they do not belong to a unique phylogenetic group, they do conform to a particular life form that has independently arisen at different geological times from non-tree ancestors by convergent evolution across unrelated clades (reviewed by Niklas (1997), Groover (2005)). Individuals sharing this life form can achieve dominance in a variety of terrestrial environments, forming the base of entire ecosystems, which are often characterized by a tremendous biomass accumulation and productivity (Petit & Hampe, 2006; Whiteman, 2014). For such reason, trees have become the base of a whole industry relying on wood extraction, fruit production or the exploitation of other nonwood products such as latex for rubber manufacturing, metabolites with medical purposes (e.g. taxol) or oils for biofuel, among many others (reviewed by Whiteman (2014)). While often seen as undomesticated species, forest trees have been submitted to genetic improvement through selective breeding for over 50 years (Neale & Savolainen, 2004). More recently, they have been the subject of genomic-based studies aiming to understand the architecture of traits with commercial and adaptive interest, such as wood quality, cold hardiness or pest resistance (Neale & Kremer, 2011). Some of these studies have focused on the capacity of forest trees for rapid physiological and molecular adaptation at different temporal and spatial scales, and on their ability to undergo rapid range shifts in response to environmental changes (Savolainen, Pyhäjärvi, & Knurr, 2007).

While angiosperm and coniferous trees share many attributes that affect their evolutionary trajectories (Petit & Hampe, 2006), they also differ in several fundamentally important ways that are relevant to the search for the genetic basis of adaptively important traits. Several trait differences separate angiosperm from conifer trees, including their reproductive biology (e.g. flowering and fruiting vs cone production), pollination syndromes, their perennial or deciduous habits, as well as wood architectural traits such as the composition of lignin and the ratio of cellulose/hemicellulose, as well as the use of tracheids and vessels for water transport (Niklas, 1997). Some of these traits have begun to be addressed by association genetics studies in both gymnosperm and angiosperm trees by using different strategies (e.g. whole genome vs candidate gene studies), while others still await comparative genomic data. From a strictly genomic point of view, the most striking difference between gymnosperm and angiosperm trees is the size and content of their genomes, which has great relevance for performing studies aimed at discovering the genomic signatures of adaptation. For instance, conifers have huge and highly complex genomes (>20 Gb) characterized by the proliferation of repetitive sequences from transposable elements (e.g. Birol et al., 2013; Neale et al., 2014; Nystedt et al., 2013), while angiosperm trees generally have more compact and less redundant genomes (e.g. Myburg et al., 2014; Tuskan et al., 2006). Indeed, while conifer genomics has recently received a boost from published draft genome assemblies of Norway spruce (*Picea abies*; Nystedt et al., 2013), white spruce (*Picea glauca*; Birol et al., 2013), and loblolly pine (*Pinus taeda*; Neale et al., 2014), angiosperm tree genomics has benefited for nearly a decade from the first forest tree to have its whole genome sequenced and annotated – the black cottonwood, *Populus trichocarpa* (Salicaceae) (Tuskan et al., 2006).

Progress in forest tree genomics appears nevertheless limited when compared to that in model species such as *Arabidopsis* or maize. Advances in forest tree genomics are hampered by the long generation times of species and the lack of well-defined mutants for specific genes, which constrains the use of reverse-genetic approaches. Such strategies rely on the *prior* knowledge of gene functions and the study of their effects on the phenotype of mutants for those particular genes (but see Séguin, Lachance, Désjardins, Leplé, and Pilate (2014)). Instead, dissecting the genomic architecture of complex phenotypic traits in forest trees has principally relied on forward-genetic approaches (Figure 1), and it is following that logic that this chapter is organized. Forward genetics is based on the observation, selection and intensive study of individuals exhibiting contrasting phenotypes (e.g. height, wood quality, drought or pest resistance, etc.) to uncover the genes

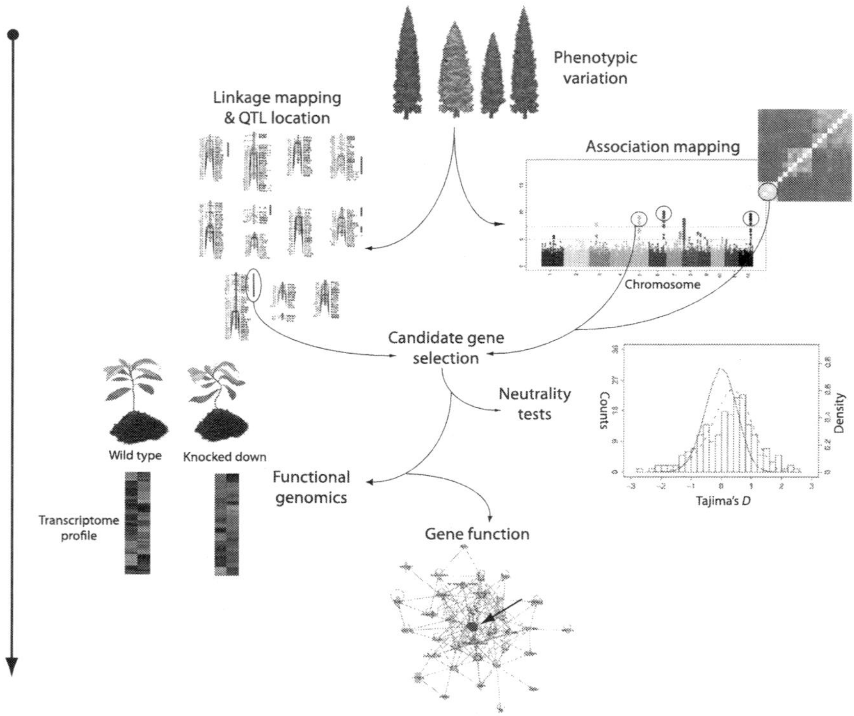

Figure 1 Schematic representation and illustrations of some the forward-genetics approaches used to detect candidate genes involved in adaptation in forest trees that are discussed throughout this chapter. (See colour plate)

underlying such phenotype variation. Forward-genetic approaches can vary from quantitative trait loci (QTL) detection and association mapping to comparative transcriptome profiling (Figure 1). In this chapter, we summarize the rationale behind each one of these methods, together with an overview of the statistical approaches that are currently used to identify candidate genes/alleles underlying trait variation. Then, we highlight the main findings of landmark studies that provide evidence for adaptation in forest trees.

2. QUANTITATIVE TRAITS VARIATION

Adaptation mostly arises from selection on phenotypes and their underlying genetic variants (i.e. alleles) that increase fitness in a given set of environmental conditions (Kremer & Le Corre, 2012; Le Corre & Kremer, 2003).

Traditionally, genetic variation in adaptively important phenotypes in forest trees has been studied by using reciprocal transplant experiments or common garden tests (also called provenance trials in the forestry literature) using a quantitative genetics framework (e.g. Falconer & MacKay, 1996) that allows estimation of trait heritability within populations and trait genetic differentiation among populations (Q_{ST}). These approaches allow explicit quantification of genetic versus environmental effects on the phenotypes, and can indirectly test if trait evolution has been driven by local adaptation (Latta, 1998).

Most quantitative genetics studies in forest trees have long focused in traits of economic interest (González-Martinez, Ersoz, Brown, Wheeler, & Neale, 2006), although height, diameter, pest resistance, cold hardiness, wood quality and bud phenology (mostly bud flush and bud set timings) are the most commonly surveyed characters (e.g. Holliday, Ralph, White, Bohlmann, & Aitken, 2008; Jaramillo-Correa, Beaulieu, & Bousquet, 2001; Rehfeldt, Ying, Spittlehouse, & Hamilton, 1999). Height and diameter can be affected by many factors, such as local climate or soils, and reflect the overall performance of a tree in a test site. Because of this, heritability estimates for these traits are usually low to moderate (i.e. $h^2 < 0.2$) and vary across sites, populations and tree age (e.g. Benowicz & El-Kassaby, 1999; Li, Beaulieu, Corriveau, & Bousquet, 1993; Li, Beaulieu, Daoust, & Plourde, 1997). On the other hand, characters such as pest resistance, cold hardiness, wood density and bud phenology usually exhibit moderate to high heritabilities (i.e. h^2 between 0.3 and 0.8; Howe et al., 2003; Porth, Klápště, Skyba, Hannemann et al., 2013), which suggests that abundant genetic variation exists for these traits, and therefore can be shaped by selection.

Quantitative genetic theory shows that trait differentiation among populations (Q_{ST}) is expected to be equivalent to allele frequency differentiation at molecular markers (F_{ST} or G_{ST}; Nei, 1977; Wright, 1951) if selection is not acting; thus, the comparison of these two estimates is thought to test for the occurrence of local adaptation in phenotypic traits (Leinonen, McCairns, O'Hara, & Merilä, 2013; Leinonen, O'Hara, Cano, & Merilä, 2008; McKay & Latta, 2002). As with heritability, most traits exhibiting high Q_{ST} estimates (i.e. above 0.3) are those related to phenology and defence, while traits associated with the general form of an individual (i.e. diameter, branch number) mostly have lower Q_{ST} values (below 0.2; reviewed by Leinonen et al. (2008), Savolainen, Lascoux, and Merilä (2013)). It must be noted, however, that several authors argue that Q_{ST} versus F_{ST} comparisons may not be valid under all situations (McKay & Latta, 2002). The discrepancy between these two measures may reflect differences in their genetic architecture. That is,

Q_{ST} for most quantitative traits results from the combined action (and interaction) of many genetic variants, each of which may contribute a small amount to the total variance of a trait. These many small effect variants may experience strong patterns of covariance or statistical linkage disequilibrium (LD) across populations that shift trait means more than would be predicted from the combined frequency changes at individual loci (Kremer & Le Corre, 2012; McKay & Latta, 2002). On the other hand, F_{ST} is typically estimated from physically unlinked molecular markers and does not incorporate covariances or statistical LD among loci.

Comparisons of Q_{ST} estimates across studies must be interpreted with caution, since they will necessarily depend on the breadth of population sampling and the age and ecological conditions at which quantitative traits were evaluated (Hamilton, Lexer, & Aitken, 2013). For instance, after sampling *Picea mariana* populations in a latitudinal transect in northeastern Canada, a Q_{ST} of 0.91 was estimated for bud set (Morgenstern, 1969), while a much lower estimate ($Q_{ST} = 0.22$) was obtained for this trait when surveying more southern populations (Prunier, Laroche, Beaulieu, & Bousquet, 2011). Similarly, low Q_{ST} values were determined for cold hardiness (0.02−0.32) in *Picea sitchensis* populations from an island (Campbell, Pawuk, & Harris, 1989), while a much higher estimate (0.89) was obtained when the whole species distribution was surveyed (Mimura & Aitken, 2007). Variation in Q_{ST} estimates has also been observed among test sites when evaluating the same provenances. For instance, phenotypic differentiation for susceptibility to *Mycosphaerella* leaf disease varied between 0.05 and 0.25 in five different trials of *Eucalyptus globulus* in Tasmania; these values were partially correlated to the severity of the infection at each test site (Hamilton et al., 2013). Nevertheless, in spite of these caveats, Q_{ST} is still regarded as a useful mean to distinguish whether population differentiation in complex polygenic traits has been driven by selective or stochastic factors, particularly in genome-wide contexts that allow establishing a more accurate link between molecular and phenotypic variation (see Leinonen et al. (2013) for a review).

3. LINKAGE MAPPING AND QTL DETECTION

While a quantitative genetics framework allows studying genetic variation and inferring selection on phenotypes, a more mechanistic understanding of adaptive evolution requires knowing the genomic architecture of adaptive traits and linking molecular and phenotypic variation. From this point of view, identifying QTL in a linkage map is often the first step. This relies on identifying genetic markers that co-segregate with morphological variation in a linkage genetic map built from the progeny of individuals exhibiting contrasting phenotypes (usually an F2 or a backcross; Falconer & MacKay, 1996). However, this method can be troublesome for long-lived species like forest trees, for which the establishment of experimental mapping populations can be expensive and attaining the number of generations can take from years to decades.

The first linkage maps for forest trees (and other taxa) were built based on dominant markers, like RAPDs or AFLPs (e.g. Bradshaw & Stettler, 1995; Scalfi, Troggio, Piovani, et al., 2004). Later, they incorporated codominant and gene-derived markers such as SSRs, ESTPs and SNPs, which facilitated the co-location of QTLs and the identification of the putative coding genes involved; they also allowed comparisons across closely related taxa in cases where markers were transferable among them (e.g. Brondani, Williams, Brondani, & Grattapaglia, 2006; Paolucci et al., 2010; Pelgas et al., 2006). These studies have mainly aimed to decipher the genetic architecture and location of genome regions underlying quantitative traits with economical interest, including wood yield and quality or pest resistance (e.g. Bradshaw & Stettler, 1995; Scalfi et al., 2004). However, QTLs for adaptation to abiotic conditions like temperature or water availability have been also investigated. For instance, in *Picea glauca*, a genetic map that has been steadily developed since 2006 (Pelgas et al., 2006; Pelgas, Bousquet, Meirmans, Ritland, & Isabel, 2011) has allowed the identification of several regions involved in cold adaptation; these regions contain from 8 to 38 loci, and are spread across the genome (Pelgas et al., 2011). Similarly, between 11 and 15 QTLs involved in autumn and spring cold hardiness were located in a *Pseudotsuga menziensii* linkage map (Jermstad et al., 2001). Many other examples of adaptive QTL studies exist for conifers (e.g. Prunier et al., 2013; reviewed by Ritland et al. (2011)); however, most of them have the unfortunate drawback that each individual QTL identified usually accounts only for a very small part of the total adaptive variation, and altogether they rarely explain more than 10% of the phenotypic variance (PVE; see Figure 2; see also Marguerit et al. (2014) for a remarkable exception of a QTL for water-use efficiency accounting for 60% of the phenotypic variation). This last point should be particularly highlighted given that the rather low population size used in these studies (<300 trees) have certainly led to overestimated QTL effects (Xu, 2003).

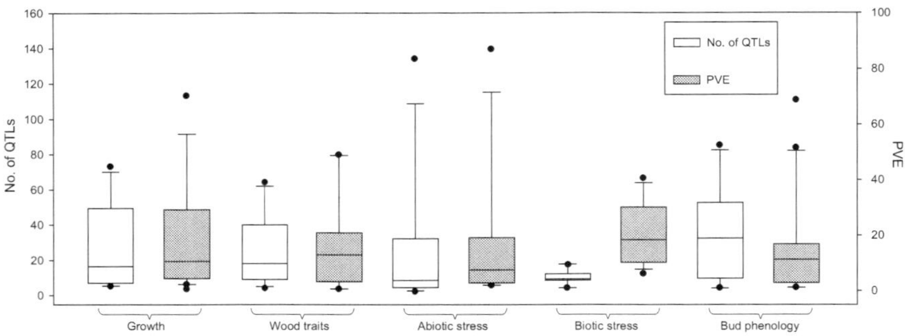

Figure 2 Number of quantitative trait loci (QTL), and percentage of the variance explained (PVE) by each individual QTL for various phenotypic and adaptive traits in a collection of 85 studies in forest trees reported herein and in Muranty et al. (2014; Table 2). Boxes denote interquartile ranges, whiskers 95% confidence intervals, and horizontal lines within boxes represent the mean.

In angiosperm trees, several studies have obtained similar results to those in conifers. For instance, 32 QTLs related to bud burst timing were placed on a *Quercus robur* linkage map, each one explaining 3–11% of the phenotypic variance (Derory et al., 2010), while 31 loci involved in bud set were identified in *Castanea sativa*, with weak to medium PVE (4–17%) (Casasoli et al., 2004). Other studies have found fewer QTLs associated with adaptive traits (Savolainen et al., 2007). For example, in *Q. robur*, 2–10 QTLs related to water-logging tolerance were identified (Parelle et al., 2007), while six regions associated with cold adaptation were mapped in a *Populus* hybrid cross, each one having a low to medium phenotypic effect (PVE = 5–18%; Frewen et al., 2000). Equally low QTL numbers but with higher PVE were observed for other traits, such as resistance to leaf blight in the tropical rubber tree *Hevea* spp. (5 QTLs with PVE = 4–44%; Lespinasse et al., 2000), or to pathogen moulds in *Theobroma cacao* (6 QTLs with PVE = 12–28%; Risterucci, Paulin, Ducamp, N'goran, & Lanaud, 2003). Thus, the number of QTLs involved, and their effect size, seems more related to variability among different traits than to differences in angiosperm/conifer evolution.

Altogether, QTL mapping studies have shown that the genes related to adaptation in forest trees are often spread over the entire genome, they are seldom co-located across closely related species and each one of them has a low to medium phenotypic effect. However, given the usually large genomes of many tree taxa, and the few generations of recombination possible in most linkage mapping studies in trees (due to long generation time), this approach is best suited to identifying broad linkage blocks associated with an adaptive trait, and rarely allows pinpointing the specific polymorphisms, sequences or even genes involved in adaptation. They are thus mostly useful for delineating candidate genomic regions encompassing adaptive variation that can be targeted with other approaches, such as genotype–phenotype association mapping.

4. ASSOCIATION MAPPING

Also known as LD mapping, this method has proven to be a good alternative to QTL mapping in forest trees, as it can be readily used in natural populations without prior controlled crosses or experimental designs. Association mapping methods take advantage of the genetic diversity existing in natural stands to test for genotype–phenotype correlations in individuals, while controlling for background effects of population structure and kinship. The presence of historical recombination in natural populations allows for much greater precision at identifying genotype–phenotype associations, since markers must be in very tight LD with causal variants (Soto-Cerda & Cloutier, 2010). Such covariation is usually detected by surveying loci scattered all over the genome (Neale & Savolainen, 2004), although association studies performed with a few specific (and carefully selected) candidate genes or genomic regions have also provided good results for traits for which the potential causal genes are known (e.g. González-Martínez et al., 2006; Ingvarsson, García, Hall, Luquez, & Jansson, 2006).

When used in combination with dense high-throughput genotyping (e.g. SNPs identified from next-generation sequencing), genome-wide association studies (GWAS) can detect loci closely associated with causal variants, or even the quantitative trait nucleotides themselves, especially for lineages with high recombination rates like forest trees (Soto-Cerda & Cloutier, 2010). However, caution is needed, given that GWAS inferences can be biased by population structure and admixture, genetic drift and/or environmental variation through time (e.g. De Mita et al., 2013), which must all be taken into account through a series statistical corrections to avoid the detection of false positives. For example, population structure can produce a high rate of false positives in association studies when there is a correspondence between trait variation and the axes of ancestry across populations (e.g. Ingvarsson & Street, 2011). Consequently, the use of generalized linear models (GLM), unified mixed models (UMM; Yu et al., 2006) or Bayesian approaches (Günther & Coop, 2013), that allow controlling for population and family structure, has become regular practice in association studies (e.g. Cappa et al., 2013; Kang et al., 2008; Uchiyama et al., 2013). For instance, Cappa et al. (2013) greatly decreased the number of spurious SNP-associations with growth and wood properties in *E. globulus* after performing such corrections with an UMM.

Another way to increase statistical power for identifying genotype—phenotype associations is to consider haplotypes rather than single-SNP variation (Akey, Jin, & Xion, 2001). For example, only six marker—trait pairs explaining between 0.1 and 15% of the phenotypic variation were detected in *Populus nigra* when analyzing associations between chemical wood properties and individual SNPs, while the number of significant associations and PVE largely increased when these variants were concatenated in haplotypes (Guerra et al., 2013). Similar results were obtained in *P. trichocarpa* when using the same strategy (Wegrzyn et al., 2010).

In forest trees, GWAS employing thousands to millions of markers spread across the genome have allowed the identification of multiple loci related to traits with ecological and economic importance. Therefore, the discovery of associations is not dictated by the choice of candidate genes, but instead by the density of the markers employed. The most powerful GWAS study performed to date in a tree was based on whole genome re-sequencing (c. 18 million SNPs) in *P. trichocarpa* (Evans et al., 2014). In this study, the strongest candidate selected regions associations were the *FT2* paralog of *FLOWERING LOCUS T*, which correlated with continued height growth and bud set, and the *PHYTOCHROME AND FLOWERING TIME 1* gene (*PFT1*), which was associated with bud flush timing. This study further showed that the genetic architecture of most trait associations involved SNPs in intronic or intergenic regions, suggesting that regulatory regions make up a large fraction of the standing genetic variance influencing adaptive phenotypes in natural populations.

Further surveys in *Populus* include extensive phenotyping efforts followed by GWAS, which made it possible to infer the genetic architecture underlying complex traits in provenance trials and common gardens (McKown, Guy, et al., 2014; McKown, Klápště, et al., 2014; Porth, Klápště, Skyba, Hannemann et al., 2013; Porth, Klápště, Skyba, Lay et al., 2013). For example, a series of association tests for phenological traits revealed a high degree of genetic complexity in *P. trichocarpa* involving over 70 genes, each one explaining 1–13% of the phenotypic variance of each trait (McKown, Klápště, et al., 2014). Other studies identified genes underlying fungal resistance, leaf drop and cellulose content (*REVOLUTA*; Porth, Klápště, Skyba, Lay et al., 2013), or stomatal variation (*PHABULOSA*, *BRASSINOSTEROID-INSENSITIVE 2*) and wound/disease response (*GLUTAMATE–CYSTEINE LIGASE*; McKown, Guy, et al., 2014), while a GWAS involving 29,333 SNPs across 3543 candidate genes reported significant associations with cell wall sugars, lignin and ultrastructure traits for 141 SNPs distributed across all 19 *Populus* chromosomes (Porth, Klápště, Skyba, Hannemann et al., 2013). This large number of associations, each explaining a small portion of the phenotypic variance (4–6%), is consistent with the highly polygenic architecture for wood characteristics.

The strong genetic basis of the traits above suggests that prospects are excellent for ongoing GWAS studies on whole genome re-sequencing in species that are candidates for woody biomass production and that have newly developed reference genome sequences, such as willow (*Salix purpurea* v1.0, DOE-JGI) or *Eucalyptus* (Myburg et al., 2014). However, GWAS can also be used to disentangle the complex genetic architecture underlying phenotypic traits when a reference genome is not available. For example, Parchman et al. (2012) used 95,000 SNPs identified by genotyping by sequencing and Bayesian GLM to identify 11 loci associated with serotiny in lodgepole pine (*Pinus contorta*). Serotiny is considered a key adaptation in conifers inhabiting environments with frequent fire regimes, and its variability and inheritance seems to be far more complex than previously thought (Parchman et al., 2012).

Association tests using fewer markers (i.e. a few 100–~1000 SNPs) have also retrieved interesting associations with phenotypic traits. For example, six SNPs were related to wood quality and pollen cone production in *Cryptomeria japonica* (Uchiyama et al., 2013), while 24 markers exhibited associations with the phenotypic variation of 20 secondary metabolites in *P. taeda* (Eckert et al., 2012). When considered jointly in a multi-locus approach, these 24 SNPs explained more than 50% of the phenotypic variation of some compounds' concentration, thus indicating that the interaction of genes may be more important to explain metabolome phenotypes than individual genes. In *Eucalyptus*, Cappa et al. (2103) identified 18 loci related to growth and wood properties, two of which physically mapped closely to a gene coding for the enzyme *F5H* (coniferaldehyde 5-hydroxylase) that has a key role in the lignin synthesis pathway.

Studies using a more modest quantity of markers (i.e. tens to hundreds of SNPs) can also provide promising results when these are located in carefully selected candidate genes or in regions encompassing QTL. For example, SNPs located in three candidates from the photoperiod pathway accounted for up to 15% of the phenotypic variation in growth cessation in *Populus tremula* (Ma, Hall, Onge, Jansson, & Ingvarsson, 2010), while SNPs in the *CAD* and *SAM*-2 genes accounted for 7–20% of the total genetic variance for earlywood specific gravity in *P. taeda* (González-Martínez, Wheeler, Ersoz, Nelson, & Neale, 2007). Similarly, haplotype variation in a single gene (*CESA*4) explained 2–5% of the variance of growth and some wood properties in *Populus tomentosa* (Du et al., 2013), while another gene (*CCR*) explained over 3% of the lignin composition in *Eucalyptus urophylla* (Mandrou et al., 2012). This last gene also exhibited strong associations with microfibril angle in two other *Eucalyptus* taxa (Thumma, Nolan, Evans, & Moran, 2005). The candidate gene approach can also deliver interesting results for more complex characters when a larger number of genes are considered. For instance, in *P. taeda*, associations of 1–2% per SNP were observed when the variation at 41 stress-inducible genes was correlated with carbon isotope discrimination (González-Martínez, Huber, Ersoz, Davis, & Neale, 2008). However, for such traits involving countless metabolic pathways, a whole-genome re-sequencing in a much larger sample of individuals might be required to account for larger portions of the phenotypic variance.

5. GENOME SCANS

Association methods can also be used at the range-wide level to identify clinal variation between allele frequencies and environmental variables. This approach is often complemented by methods that test markers for extreme population genetic structure (i.e. outlier methods). Both kinds of strategies rely on the premise that differential selective pressures in populations subjected to contrasting environmental conditions result in a significant differentiation of allele frequencies at the loci targeted by selection (and their linked loci). The outlier methods address this by estimating genetic differentiation (e.g. F_{ST}) across loci and looking for those exhibiting significantly higher or lower values than those expected under various stochastic neutral models; these outlier loci are then assumed to be potentially affected by directional or balancing selection, respectively (Luikart, England, Tallmon, Jordan, & Taberlet, 2003). Genotype–environment association (GEA) methods, on the other hand, allow the inference of local adaptation along environmental gradients, perhaps even when selection is not strong enough to produce F_{ST} outliers (Joost et al., 2007). Both, the outlier and GEA methods have the drawback of being prone to significant bias when populations are not at demographic equilibrium, for example when they are hierarchically structured, show isolation by distance and/or when mutation rates are highly heterogeneous across loci (e.g. De Mita et al., 2013). Some approaches are available to correct some of this potential bias (e.g. Günther & Coop, 2013), but currently, most authors prefer to combine different strategies when scanning the genome for putatively adaptive genes and give greater confidence to those loci that are identified by multiple methods (e.g. Alberto et al., 2013; Prunier, Gerardi, Beaulieu, & Bousquet, 2012; Tsumura et al., 2012).

Another caveat that both the outlier and GEA approaches have in common is that they are most sensitive to detecting locally adapted alleles that are evolving through antagonistic pleiotropy; that is, alleles that are beneficial in one environment but deleterious in another one (Tiffin & Ross-Ibarra, 2014). However, studies in model species have shown that locally favoured alleles actually tend to be neutral elsewhere in the species range (e.g. Fournier-Level et al., 2011; Lowry, Hall, Salt, & Willis, 2009). In such conditional neutrality circumstances, the power for detecting local adaptation will be greatly diminished for both kinds of methods (see Tiffin and Ross-Ibarra (2014)). A potential way to circumvent this problem could be to test for environmental association and deviations from neutral expectations on individual genes, populations or gene clusters (e.g. Grivet, Sebastiani, González-Martínez, & Vendramin, 2009; Jaramillo-Correa et al., 2015; Prunier et al., 2012).

The first genome scan that employed an outlier approach in forest trees was performed to identify regions potentially involved in the divergence of two closely related oak species in Europe (Scotti-Saintagne et al., 2004). In this study, the authors pinpointed over 47 loci (a combination of AFLPs, isozymes and SSRs) exhibiting significantly high F_{ST} values between taxa, and which were distributed in most of the linkage groups of the species pair. Other early surveys in forest trees include one in *P. abies*, where 37 outliers (AFLPs, ESTPs and SSRs) distributed across its genome were detected (Acheré, Favre, Besnard, & Jeandroz, 2005), another one in *C. japonica*, that found five CAPs (cleaved amplified polymorphisms developed from ESTs) exhibiting significantly high F_{ST} values (Tsumura et al., 2007), and a survey in populations of *P. glauca* from eastern Canada, that reported a list of 49 outlier SNPs from coding genes (Namroud, Beaulieu, Juge, Laroche, & Bousquet, 2008), among others. However, as these early studies did not take population structure or differences in the mutation rate across loci into account, the candidates they report should be taken with caution.

More recent studies (summarized in Table 1) that do correct for such potential bias tend to combine outlier and environmental association approaches and report lists of candidates that overlap across methods. In conifers, these surveys have been performed for different species in genus *Pinus* and *Picea*, and for *C. japonica* (Table 1); although an interesting comparative study in coniferous taxa from the Alps also included *Abies alba* and *Larix decidua* (Mosca et al., 2012). In angiosperms other than oaks, genome scans have been carried out in *Fagus*, *Eucalyptus*, *Populus*, *Alnus* and in the tropical tree *Eperua falcata*, among others (Table 1). The identity of candidates generally differs across studies and species, even when they are co-distributed in the same region. For instance, in the above-mentioned survey by Mosca et al. (2012) no candidate outlier SNPs were shared by the four Alpine taxa analyzed, while there were only seven loci in common between species pairs, only one of which (located in a heat-shock protein 101) was associated with the same environmental variable in two pines (Mosca et al., 2012). Similarly, two spruce species co-distributed in eastern Canada only shared three of the 25 and 49 outlier genes respectively detected for each taxa (Prunier et al., 2011), while two environmental gradients surveyed for associations in *Quercus petrae* only had a few SNPs in common (Alberto et al., 2013).

When candidate comparisons across taxa and studies are expanded to the gene family or function levels, several interesting patterns begin to arise. For instance, transcription factors, especially those of the *MYB*, *C3HL*, *C2H2* zinc finger and *HD-ZIP* gene families recurrently appear as outliers in species as varied as *P. mariana*, *Q. petrae* or *C. japonica*, among others, where they further show significant correlations with temperature and precipitation (Alberto et al., 2013; Prunier et al., 2011; Tsumura et al., 2012). Other gene families such as the dehydrins, the 60*S* ribosomal and heat-shock proteins or *4CL* also have shown associations with environment across taxa (e.g. González-Martínez et al., 2007; Grivet et al., 2009), while other genes coding for proteins involved in membrane transport, abiotic stress response, lignin biosynthesis or plant development are also recurrent outliers in many studies (Table 2).

Table 1 Number of Candidate Gene-Makers Detected with Outlier (i.e. F_{ST} Based) and Correlation (i.e. Genotype–Environment Associations; GEA) Methods in Various Studies in Forest Trees. The F_{ST} Range and the Identity of the Environmental Factors Associated with the Markers Are also Included

Species	No. Outliers (F_{ST})[a]/ No. Markers	F_{ST} Range	No. Candidates (GEA)[b]	Environmental Factor	References
Angiosperms					
Alnus glutinosa	57/1990	0.05–0.12	20	Temperature	De Kort et al. (2014)
			2	Precipitation	
Castanopsis fargesii	7/32	0.15–0.20	5	Precipitation	Li, Sun, Huang, and Cannon (2014)
Eperua falcata	8/74	0.04–0.15	4	Habitat type	Audigeos, Brousseau, Traissac, Scotti-Saintagne, and Scotti (2013)
Eucalyptus camaldulensis	5/76	0.21–0.42	4	Climate	Dillon et al. (2014)
Fagus sylvatica	5/546	0.27–0.30	—	Elevation	Csilléry et al. (2014)
Populus balsamifera	26/747	0.22/0.94	45	Climate	Keller, Levsen, Olson, and Tiffin (2012)
Populus trichocarpa	1130/$\approx 18 \times 10^6$	0.12–0.37	194	Temperature	Evans et al. (2014)
			157	Precipitation	
Quercus petraea	28/384	0.01–0.11	5	Elevation	Alberto et al. (2013)
			3	Latitude	
Quercus robur–Q. petraea	60/262	0.22–0.93	—	—	Guichoux et al. (2013)
Conifers					
Abies alba	N.A./249	N.A.	4	Climate	Mosca et al. (2012)
Cryptomeria japonica	20/1026	0.11–0.22	9	Temperature	Tsumura et al. (2012)
			2	Precipitation	
			4	Sun radiation	
Larix decidua	N.A./267	N.A.	6	Climate	Mosca et al. (2012)

(*Continued*)

Table 1 Number of Candidate Gene-Makers Detected with Outlier (i.e. F_{ST} Based) and Correlation (i.e. Genotype–Environment Associations; GEA) Methods in Various Studies in Forest Trees. The F_{ST} Range and the Identity of the Environmental Factors Associated with the Markers Are also Included—cont'd

Species	No. Outliers (F_{ST})[a]/ No. Markers	F_{ST} Range	No. Candidates (GEA)[b]	Environmental Factor	References
Picea abies	29/445	0.10–0.17	18	Latitude	Chen et al. (2012)
Picea glauca	20/768	0.04–0.13	–	–	Namroud et al. (2008)
P. glauca–Picea engelmannii	7/311	0.08–0.18	17	Climate	De la Torre, Roberts, and Aitken (2014)
Picea mariana	10/99	0.04–0.16	14	Precipitation	Prunier et al. (2012)
			10	Temperature	
Picea obovata	13/356	0.03–0.05	22	Latitude	Chen et al. (2014)
Picea rubens	4/33	0.03–0.12	7	Climate	Bashalkhanov, Eckert, and Rajora (2013)
			3	Pollution	
Pinus banksiana	1/472	N.A.	3	Temperature	Cullingham et al. (2014)
Pinus cembra	N.A./459	N.A.	11	Climate	Mosca et al. (2012)
Pinus contorta	8/472	N.A.	14	Temperature	Cullingham et al. (2014)
Pinus mugo	N.A./693	N.A.	18	Climate	Mosca et al. (2012)
Pinus taeda	24/3059	0.14–0.27	5	Aridity	Eckert et al. (2010)

[a]Only markers detected with Bayesian (i.e. Bayescan) methods at $P < 0.01$ (i.e. BF = 2) or higher are reported.
[b]Only markers detected with regression methods that correct for population structure are included.

Table 2 Short List of Candidate Genes or Gene Families Associated with Adaptive Traits in Various (i.e. At least Two) Forest Tree Species through Some of the Methods Discussed in the Text

Gene/Gene Family	Biological Function	Species	Associated Trait	Evidence[a]	References
4CL	Plant defence/lignin biosynthesis	*Picea glauca*	Growth, pest resistance	Gene expression/ QTL mapping	Porth et al. (2011)
		Picea rubens	Air pollution	GWAS	Bashalkhanov et al. (2013)
		Picea sitchensis	Cold hardiness	Gene expression	Holliday et al. (2008)
		Pinus pinaster	Drought resistance	GEA	Jaramillo-Correa et al. (2015)
Aquaporin	Membrane transport	*Quercus petraea*	Bud phenology	GWAS	Alberto et al. (2013)
		Eucalyptus camaldulensis	Drought resistance	GEA	Dillon et al. (2014)
AUX-IAA	Growth regulation	*Picea mariana*	Bud set	GWAS	Prunier et al. (2013)
		Picea sitchensis	Bud set	GWAS	Holliday, Ritland, and Aitken (2010)
		Picea sitchensis	Cold hardiness	GWAS	Holliday et al. (2010)
α- and β-*TUBULIN*	Membrane transport	*Picea glauca*	Wood quality	GWAS, gene expression	Beaulieu et al. (2011)
		Pinus pinaster	—	GEA	Jaramillo-Correa et al. (2015)
		Pinus taeda	Wood quality	GWAS, NT	González-Martínez et al. (2007)
		Populus nigra	Wood quality	GWAS	Guerra et al. (2013)
CAD	Plant defence/lignin biosynthesis	*Picea glauca*	Growth, pest resistance	Gene expression/ QTL mapping	Porth et al. (2011)
		Pins taeda	Wood quality	GLM, NT	González-Martínez et al. (2007)

(Continued)

Table 2 Short List of Candidate Genes or Gene Families Associated with Adaptive Traits in Various (i.e. At least Two) Forest Tree Species through Some of the Methods Discussed in the Text—cont'd

Gene/Gene Family	Biological Function	Species	Associated Trait	Evidence[a]	References
CCoAOMT	Plant defence	*Picea glauca*	Growth, pest resistance	Gene expression/ QTL mapping	Porth et al. (2011)
		Pinus taeda	Drought resistance	GWAS	González-Martínez et al. (2008)
		Populus nigra	Wood quality	GWAS	Guerra et al. (2013)
		Popuus trichocarpa	Bud set, growth	GWAS	McKown, Guy, et al. 2014
CCR	Plant defence/lignin biosynthesis	*Eucalyptus grandis*	Wood quality	LD-mapping, GWAS	Thumma et al. (2005)
		Eucalyptus nitens	Wood quality	LD-mapping, GWAS	Thumma et al. (2005)
		Eucalyptus urophylla	Wood quality	GWAS	Mandrou et al. (2012)
CESA	Cellulose synthesis/ plant defence	*Pinus pinaster*	—	NT	Pot et al. (2005)
		Pinus radiata	—	NT	Pot et al. (2005)
		Populus nigra	Wood quality	GWAS	Guerra et al. (2013)
		Populus tomentosa	Growth, wood quality	GWAS	Du et al. (2013)
		Popuus trichocarpa	Bud set, growth, biomass	GWAS	McKown, Klápště, et al. (2014)
DHN	Macromolecule protection	*Eucalyptus gunnii*	Cold hardiness	Gene expression	Keller et al. (2009)
		Picea sitchensis	Cold hardiness	Gene expression	Holliday et al. (2008)
		Pinus pinaster	Drought resistance	GWAS	Eveno et al. (2008)
		Pinus taeda	Photosynthesis efficiency	GWAS	González-Martínez et al. (2008)
		Populus tremula × *Populus tremuloides*	Cold hardiness	Gene expression	Ruttink et al. (2007)
		Popuus trichocarpa	Leaf drop	GWAS	McKown, Klápště, et al. (2014)
		Prunus persica	Cold hardiness	Gene expression	Artlip et al. (1997)

		Pseudotsuga menziesii	—	NT	Eckert et al. (2009)
α- and β-*EXPANSIN*	Growth regulation	*Picea glauca*	Wood quality	GWAS, gene expression	Beaulieu et al. (2011)
		Picea sitchensis	Cold hardiness	GWAS	Holliday et al. (2010)
		Pinus radiata	Wood quality	GWAS	Dillon et al. (2009)
FT2/FTL2/PTFT	Photoperiod response, flowering time	*Picea abies* *Picea obovata*	Bud set	GEA, GWAS	Chen et al. (2012, 2014)
		Picea abies	Bud set	Gene expression	Gyllenstrand et al. (2007)
		Picea glauca	Bud set	Gene expression	El Kayal et al. (2011)
		Pinus sylvestris	Bud set	Gene expression/GWAS	Avia, Kärkkäinen, Lagercrantz, and Savolainen (2014)
		Populus tremula × *P. tremuloides*	Growth, bud phenology, flowering	Transgenic plants, gene expression	Böhlenius et al. (2006)
		Populus trichocarpa	Bud set, growth	GEA/GWAS/Gene expression	Evans et al. (2014)
		Prunus domestica	Growth, bud phenology, flowering	Transgenic plants	Srinivasan et al. (2012)
GIGANTEA	Circadian clock	*Picea abies* *Picea obovata*	Bud set	GEA, GWAS	Chen et al. (2012, 2014)
		Picea sitchensis	Bud set	GWAS	Holliday et al. (2010)
		Populus balsamifera		GWAS	

(*Continued*)

Table 2 Short List of Candidate Genes or Gene Families Associated with Adaptive Traits in Various (i.e. At least Two) Forest Tree Species through Some of the Methods Discussed in the Text—cont'd

Gene/Gene Family	Biological Function	Species	Associated Trait	Evidence[a]	References
			Bud set, cold hardiness		Keller et al. (2012); Olson et al. (2013)
Heat-shock protein	Drought resistance	*Picea rubens*	Air pollution	GWAS	Bashalkhanov et al. (2013)
		Picea mariana	Growth	GWAS	Prunier et al. (2013)
		Picea sitchensis	Cold hardiness	Gene expression	Holliday et al. (2008)
		Pinus pinaster	Drought resistance	GEA	Jaramillo-C et al. (2015)
MYB	Gene regulation	*Cryptomeria japonica*	Growth	GEA	Tsumura et al. (2012)
		Eucalyptus gunnii	Cold hardiness	Gene expression	Keller et al. (2009)
		Picea glauca	Growth, pest resistance	Gene expression/ QTL mapping	Porth et al. (2011)
		Picea glauca	Growth, plant defense	Gene expression/ transgenic plants	Bedon et al. (2010)
		Picea mariana	Bud set, growth	GWAS	Prunier et al. (2013)
		Picea sitchensis	Cold hardiness	Gene expression	Holliday et al. (2008)
		Pinus pinaster	Drought resistance, photosynthesis efficiency	Gene expression/ QTL mapping	De Miguel et al. (2014)
		Populus tremula × *P. tremuloides*	Cold hardiness	Gene expression	Ruttink et al. (2007)
		Popuus trichocarpa	Bud set, growth	GWAS	McKown, Guy, et al. (2014)
PAL	Plant defence/lignin biosynthesis	*Picea sitchensis*	Bud set	GWAS	Holliday et al. (2010)
		Picea glauca	Growth/Pest resistance	Gene expression/ QTL mapping	Porth et al. (2011)

		Pinus radiata	Wood density	GWAS	Dillon et al., (2009)
PHYA/PHYB	Circadian clock	*Picea sitchensis*	Bud set	GWAS	Holliday et al. (2010)
		Populus tremula	Bud set	GWAS, NT	Ingvarsson et al. (2006, 2008)
		Populus tremula × *P. tremuloides*	Gorwth, cold hardiness	Transgenic plants	Olsen et al. (1997)
PRR/PaRR	Circadian clock	*Picea abies/Picea obovata*	Bud set	GEA, GWAS	Chen et al. (2012, 2014)
		Picea abies	—	NT	Källman et al. (2014)
		Populus trichocarpa	Temperature response	GEA	Geraldes et al. (2014)
		Popuus trichocarpa	Bud set, growth, biomass	GWAS	McKown, Guy, et al. (2014)
Xiloglucan glucosyl transferase	Plant defence	*Picea sitchensis*	Bud set	GWAS	Holliday et al. (2010)
		Picea sitchensis	Cold hardiness	GWAS	Holliday et al. (2010)
		Populus tremula × *P. tremuloides*	Wood formation	Gene expression/ transgenic plants	Derba-Maceluch et al. (2015)
		Popuus trichocarpa	Bud set, growth, biomass	GWAS	McKown, Guy, et al. (2014)
Ribosomal protein	Gene regulation	*Eucalyptus gunnii*	Cold hardiness	Gene expression	Keller et al. (2009)
		Quercus petraea	Bud phenology	GWAS	Alberto et al. (2013)
		Pinus pinaster	Drought resistance	GEA	Jaramillo-Correa et al. (2015)

[a]GEA = Genotype-environment associations (e.g. Bayenv, MatSAM, etc.); GWAS = Genome-wide association studies (e.g. general linear models, linear-mixed models, unfied-mixed models, etc.); NT = Neutrality tests; QTL = Quantitative trait loci.

As briefly mentioned above, genome scans are powerful tools for studying the introgression and divergence patterns between closely related species pairs. For instance, other than the early survey by Scotti-Saintagne et al. (2004), introgression between the early successional *Q. robur* and the late-successional *Q. petrae* was analysed with outlier and non-outlier loci, with the former showing a pattern that is coincident with the predicted introgression from the late-successional species into the early successional one, and the later showing no particular pattern (Guichoux et al., 2013). An analogous study in the hybrid zone between the coastal *Picea sitchensis* and the continental *P. glauca* in western Canada revealed a far more complex story, with separate sets of outliers introgressing more than expected under a neutral pattern from the coast into the continent and vice versa, and other loci exhibiting less introgression than predicted (Hamilton, Lexer, et al., 2013). Similarly complex patterns were observed on the other side of the Canadian Rockies between the mountain species *Pinus contorta* var. *latifolia* and the interior taxa *P. baksiana* (Cullingham, Cooke, & Coltman, 2014), while in Europe, an in-depth genome scan in the sympatric but ecologically divergent *Populus alba* and *P. tremula* revealed long shared genomic blocks, particularly within a region containing an incipient sex chromosome, which suggests that introgression is being favoured by balancing selection in this species pair (Stölting et al., 2013).

6. SURVEYING SEQUENCE VARIATION AT CANDIDATE GENES

Much of the studies discussed so far focused on a genome-wide strategy, in which a large amount (i.e. from several hundreds to some thousands) of markers are surveyed to identify either genomic regions (i.e. QTL mapping) or specific polymorphisms covarying with phenotypic traits or environmental conditions (GWAS, GEA). However, most pinpointed regions/polymorphisms are usually assumed to be linked to instead of being the causal variant targeted by selection, which often remains unknown (Neale & Savolainen, 2004). The candidate gene approach is a powerful way to fill this gap, as it focuses on these specific regions, particularly those containing genes of known functions. In addition, this strategy can be directly applied in natural populations of undomesticated species with no other prior knowledge than the location (i.e. flanking DNA sequence) of the region/gene itself. The candidate gene approach has been favoured in forest trees because of their mostly outcrossing nature and high genome-wide recombination rates, which makes LD decay rapidly, mostly within gene limits (Neale & Kremer, 2011). Therefore, if one assumes that genomes evolve in haplotype blocks, the blocks to explore should be smaller in forest trees than in other species with different biological traits, like selfers or restricted endemics (Neale & Savolainen, 2004).

Early candidate gene studies focused on model organisms and the knowledge of biochemical and regulatory basis of specific traits to target the genes involved (Zhu & Zhao, 2007). However, in non-model organisms these methods are not easily applied, and candidate gene discovery has taken several complementary approaches for these taxa, like QTL cloning, or comparative and functionally based genomics. QTL cloning focusses on the extensive sequencing of regions previously associated with quantitative trait variation in linkage mapping surveys. This approach has been useful in species with small and completely sequenced genomes, but for taxa with larger genomes, this method can be applied only when LD is strong (Zhu & Zhao, 2007) and QTL intervals are narrow enough to contain only a few genes. Unfortunately, the size of such intervals in trees make positional cloning impractical, even for species with completely sequenced genomes like *Eucalyptus*, *Populus*, *Picea* or *Pinus*. So far, only the annotation of genes contained in QTLs can provide short lists of candidates to survey by direct sequencing (e.g. Ingvarsson et al., 2006; Thumma et al., 2005).

Another strategy is to survey the genomic resources already available for forest trees (e.g. annotated genes and transcriptomes) and identify potential candidate genes by comparison with model species. For instance, if one assumes that the orthologous gene coding for an enzyme involved in a specific biochemical pathway in a model species, like the biosynthesis of lignin or other secondary metabolites, could be affecting the variation of wood quality or pest resistance in a forest tree, then the genomic resources of the model taxa can be used to identify and design primers to test this ortholog for selection signals in the forest tree species (e.g. González-Martínez et al., 2007; Ingvarsson et al., 2006; Külheim, Yeoh, Maintz, Foley, & Moran, 2009). This comparative genomics approach is, however, limited by the availability of genomic resources in the target species, the gene transferability (i.e. ortholog identification) across unrelated taxa and the prior knowledge of the biochemical pathways in the model species. Consequently, if a gene that plays a major adaptive role in a target species belongs to a gene family or a biochemical pathway that does not exist in a model species, it cannot be surveyed until it has been identified de novo through functional surveys.

For instance, the diversity and adaptive value of taxol-related genes in *Taxus* could only be evaluated once the genes involved in its biosynthetic pathways were completely outlined (e.g. Burgarella et al., 2012; Jennewein, Wildung, Chau, Walker, & Croteau, 2004).

Once a potential candidate gene has been identified, there is a number of methods or neutrality tests (NT) to infer if its variation deviates from the expectations of the neutral theory of molecular evolution (see Wright and Gaut (2005), González-Martínez et al. (2011) for reviews). Some involve an examination of its site-frequency spectrum (SFS) and compare different estimators of diversity (θ) with those produced by coalescent simulations that the take the underlying demographic history of the species/population of interest into account (e.g. Grivet et al., 2009), while others rely on the ratio of non-synonymous to synonymous substitution rates ($\omega = d_N/d_S$), and allow detecting selection at different evolutionary timescales and acting within or among species (e.g. Palmé, Pyhäjärvi, Wachowiak, & Savolainen, 2009).

The earliest candidate gene studies in forest trees date from about a decade ago and, as with previous approaches, they mostly concentrated in economically important characters, like growth, wood formation, stress tolerance and phenology, and focused on commercial taxa such as pines, spruces, *Eucalyptus* and poplars. For instance, Brown, Gill, Kuntz, Langley, and Neale (2004) tested for selection on the SFS (using Tajima's *D*) in 19 candidate genes in *P. taeda* and found no deviations from neutral expectations. Later on, some of these genes and other candidates for drought tolerance (selected after verifying homology with drought-response genes in model taxa) were surveyed in this same species through additional NT, and signals of selective sweeps and balancing selection were inferred for the *ERD*3 and *CCOAMT*-1 genes, respectively (González-Martínez et al., 2006). A similar study performed on wood formation candidates in two other pines (*Pinus pinaster* and *Pinus radiata*) also revealed the putative action of balancing selection in a cellulose-synthase gene (*CESA*3) and of positive selection in a member of the *KORRIGAN* gene family (Pot et al., 2005), while an extensive study based on compound tests and ABC simulations on photoperiod-related genes in *P. abies* showed that the circadian clock gene *PaPRR3* is a strong candidate for divergent selection and local adaptation (Källman et al., 2014). This same approach was used to infer positive selection in six cold hardiness candidates in *Pseudotsuga menziesii*, most of them encoding structural components or proteins associated with the cellular membrane (Eckert et al., 2009).

In *P. tremula*, the candidate gene approach uncovered strong associations between bud set and the *PHYB2* and the *LATE-ELONGATED HYPOCOTYL* genes involved in the circadian clock (paralogs *LHY1* and *LHY2*) (Ingvarsson et al., 2006, Ingvarsson, Garcia, Luquez, Hall, & Jansson, 2008; Ma et al., 2010). The *PHYB2* gene, which was selected from previous QTL surveys, additionally exhibited a strong latitudinal allele frequency cline that mirrored the phenotypic clinal variation in bud set measured in a common garden (Ingvarsson et al., 2006). Further comparisons across species, either on phylogenetic or coalescent frameworks, also allowed inferring the action of positive selection on particular candidate genes (see Table 2), such as *CCOAMT*-1 (caffeoylCoA O-methyltransferase) in *P. trichocarpa* and *P. nigra*, *CESA*3 (cellulose synthase A3) in *P. tomentosa* and several members of the dehydrin and *KNOX* families in pines and spruces, respectively (e.g. Du et al., 2013; Grivet et al., 2011; Guerra et al., 2013; Namroud, Guillet-Claude, Mackay, Isabel, Bousquet, 2010; Wegrzyn et al., 2010). Similarly, the comparison of 23 genes from four biosynthetic pathways of secondary metabolites across four *Eucalyptus* taxa revealed that two of them (both from the 1-deoxyxylulose-5-phosphate synthase family) were likely affected by purifying selection in all species (Külheim et al., 2009). Altogether, these results highlighted how subfunctionalization can foster the action of selection in recently duplicated genes by allowing the quick evolution of adaptation across taxa.

There is evidence suggesting that convergent evolution often involves separate genes and various evolutionary paths. For instance, closely related *Populus* species showed different genetic architecture for bud set, despite exhibiting similar phenotypic clines with latitude. Unlike the abovementioned example in *P. tremula*, *P. balsamifera* showed little segregating covariation with bud set in phytochrome genes, including *PHYB2* (Keller, Levsen, Ingvarsson, Olson, & Tiffin, 2011); instead, significant associations were most prominent for SNPs in the *GIGANTEA* (*GI*), *EARLY FLOWERING 3* (*ELF3*) and *LEAFY* (*LFY*) genes (Olson et al., 2013). This curious case for parallel evolution became even more interesting when *P. trichocarpa* was surveyed (Evans et al., 2014; McKown, Guy, et al., 2014; McKown, Klápště, et al., 2014). This species has recently diverged from *P. balsamifera* (c. 75,000 years), with whom it still shares high amounts of ancestral polymorphism (Levsen, Tiffin, & Olson, 2012). However, in *P. trichocarpa* there was a notable absence of associations between bud set and the previously implicated genes in either *P. balsamifera* or the more distantly related *P. tremula* (*PHYB2*, *LHY1*, *LHY2*, *GI*, *ELF3*, *LFY*). A similar case was observed for associations with climate in co-distributed boreal spruces, which only shared a few number of genes showing homologous adaptive responses in Canada and Eurasia (Chen et al., 2014; Prunier et al., 2011).

7. FUNCTIONAL GENETICS OF TREE ADAPTATION

Once a gene has been identified as candidate and has shown evidence to be affected by selective forces, it must be validated through different strategies. A powerful approach to validation is provided by functional genetics, which focuses on the expression levels or the phenotypic and physiological effects of specific genes to decipher the genetic basis of adaptation. For instance, functional genetics can be used to infer if a candidate gene changes its expression levels when an organism is submitted to particular conditions and/or to measure the extent of phenotypic or physiological variations in an individual that has the expression of this gene modified through transgenic manipulations (e.g. Flachowsky, Hnake, Peil, Strauss, & Fladung, 2009). The rationale behind functional genomics studies is that the genes involved in an adaptive response to a particular factor should be over-expressed during the exposition to this factor, while other genes should keep their expression levels constant or even see them diminished. On the other hand, individuals that have been modified to over-express these adaptive genes should perform better when exposed to such a factor than those with the same genes suppressed or expressing at normal levels (e.g. Flachowsky et al., 2009; Peña & Séguin, 2001; Séguin et al., 2014).

This last approach is usually employed with cloned plants, where the physiological responses can be evaluated using a constant genomic background except for the modified gene. Such a procedure always relies on previous knowledge regarding the candidate gene to evaluate (see above), which is usually gathered from model species like *Arabidopsis*. For instance, knowing that bud set timing is an important adaptive trait in boreal and temperate trees, Olsen et al. (1997) produced transgenic clones of *P. tremula* × *P. tremuloides* that over-expressed a gene (*PHYA*) known for its role in the photoperiod control of flowering in *Arabidopsis thaliana*. Similarly, the functional roles of genes involved in dormancy have been evaluated in *Prunus domestica* (Srinivasan, Dardick, Callahan, & Scorza, 2012), and those implicated in growth cessation and bacterial resistance have been surveyed in *Populus* spp. (e.g. Böhlenius et al., 2006; Peña & Séguin, 2001), while those conferring resistance to insects have been studied in *Pinus radiata* (Grace, Charity, Gresham, Kay, & Walter, 2005), among many other examples (reviewed by Séguin et al. (2014)).

Changes in transcript accumulation through time is usually investigated among phenotypically different individuals submitted to specific controlled conditions, and by surveying various tissues to verify if the response occurs at the organ or organismal levels. Transcript accumulation of candidate genes has been studied through classic methods, like northern blot or in situ hybridization, or by using more recent ones, like reverse-transcription quantitative PCR (RT-qPCR). Examples (summarized in Table 2) include the expression of a dehydrin gene that had differential transcript accumulations in evergreen and deciduous genotypes of *Prunus persica* (Artlip, Callahan, Bassett, & Wisniewski, 1997), some *FT*-like genes that are involved in bud phenology and temperature adaptation in conifers, which changed their transcript profiles under different photoperiod and thermal conditions in *Picea abies* (Gyllenstrand, Clapham, Källman, & Lagercrantz, 2007), or members of the subgroup 4 of the R2R3-MYBs that were over-expressed after wounding, jasmonic acid or cold treatments in *P. glauca* and *Pinus taeda* (Bedon et al., 2010), among many others.

Functional genomic approaches can also be used to discover new adaptive genes or those specific to non-model taxa, through microarray or whole transcriptome sequencing (EST or RNA-seq) studies. Experiments based on EST measurements can be implemented by characterizing differentially expressed EST libraries (e.g. Porth, Koch, Berenyi, Burg, & Burg, 2005), although the most currently used strategy is to first construct a gene catalog from EST libraries of a variety of tissues and development stages that cover as much genes as possible (e.g. Keller et al., 2009; Lesur et al., 2015; Pavy et al., 2005). Then, microarrays are developed by embedding probes that target all or most of these genes, and where RNA samples from individuals submitted to various conditions can be tested (Bouck & Vision, 2007). Such kind of resources are now available and have been used for a great variety of tree species of the genus *Pinus*, *Picea*, *Populus*, *Eucalyptus*, *Quercus* and *Castanea* (Neale & Kremer, 2011).

Among many interesting results, the microarray approach has allowed the clustering of genes in various expression profiles during the development of freezing and cold tolerance of *Populus tremula* × *P. alba* and *Eucalyptus gunnii*, the development of bud set in *Picea glauca* or the physiological responses that occurred between late summer and early winter in *Picea sitchensis* (El Kayal et al., 2011; Holliday et al., 2008; Keller et al., 2009; Ruttink et al.,

2007), among many others. These studies have pinpointed hundreds to thousands of genes involved in cold response, including dehydrins, metabolic proteins, signal transduction related genes, photoreceptors, transcription factors and other metabolism regulators, which help corroborate the results of previous GWAS or candidate gene-based studies (see Table 2). These surveys have also identified additional de novo genes with strong adaptive or commercial value. This is the case in a recent study in *P. glauca* that discovered a β-glucosidase gene, *Pgβglu-1*, that confers resistance against spruce budworm (*Choristoneura fumiferana*; Mageroy et al., 2015).

It must be noted though, that the study of transcriptome profiles quickly brings us to the concept of internal phenotype, which involves all internal states of an organism throughout its life, from the molecular and cellular levels to its tissue and physiological properties (see Houle, Govindaraju, and Omholt (2010)). As such, the rates of transcript variation can be seen as an internal phenotype, with its own heritability, and that is submitted to the same genetic and environmental controls as any quantitative trait. Nevertheless, microarray or RNA-seq studies rarely track the fraction of transcript expression that is transmitted across generations, which is where natural selection acts, thus leaving unexplored the genetic basis of individual differences in gene expression. However, as with any phenotype, the loci that co-segregate with expression variation can be pinpointed in a linkage map by surveying the expression levels of particular genes in a mapping population. In an adaptive context, these regulating hotspots can be seen as keystone genes having pleiotropic effects on a more complex external phenotype, for example, bud phenology. QTL mapping of these loci (eQTLs) is currently being conducted in trees. To date, a few hotspots related to defence response against white pine weevil (*Pissodes strobi*) have been found by eQTL mapping in *Picea* (Porth, Hamberger, White, & Ritland, 2011), but examples are still scarce in the forest trees literature. Hopefully, eQTL mapping will become more common in the near future since next-gen sequencing technologies applied to cDNA, such as RNA-seq, simultaneously yield transcript abundance and sequence polymorphisms of the genes expressed in a sample.

8. APPLICATIONS IN BREEDING AND TREE SELECTION

Breeding is one of the first steps to maximize the income potential of any renewable resource. In forestry and agriculture (among others), it is an iterative process that starts with the identification of individuals with outstanding phenotypes for economically important or adaptive characters (Muranty et al., 2014), and then uses selective crossing to increase the frequency of desirable traits. However, as previously discussed, the long generation times of forest trees have made selection and breeding tedious and slow, particularly for complex characters with medium to low heritability such as height, diameter or wood quality (Neale & Savolainen, 2004). Traditionally, the selection of individuals is made based on their phenotypes; however, with the development of high-throughput genotyping techniques and the strategies for detecting candidate genes discussed so far, it has become possible to predict the breeding value of individuals at an earlier age based on genetic marker data (Isik, 2014). This is particularly appealing for forest tree breeders, given that marker-assisted selection (MAS) can shorten the rotation age by more than 30%, diminish the required area to establish field trials, limit phenotyping costs and open the possibility for controlling both the inbreeding and gene diversity within populations (Muranty et al., 2014).

There are several strategies for MAS (reviewed by Hospital (2009)), which range from simply detecting the individuals to be kept in a stand based on their genotypes (i.e. population screening), to specific frameworks of directed and recurrent crossings (or back crossings) of trees with particular genotypes aiming to maximize the number of alleles of interest in their progeny (e.g. gene pyramiding). These strategies can further involve the inclusion of marker and environmental information into predictive models for breeding values and expected phenotypes (Hospital, 2009), among others.

The use of MAS has been particularly successful in fruit trees, for which there are various examples of improved lines developed from candidate gene variation, particularly for agronomic characters in the Rosaceae (e.g. Dirlewanger et al., 2004). However, the empirical examples of MAS in forest trees are still scarce compared to their fruit crop counterparts. The issues underlying this lack of MAS programs in forest trees are thoroughly discussed elsewhere (Muranty et al., 2014), but causes include their early stage of domestication and the absence of sound cost/benefit analyses supporting the advantage of MAS over other silvicultural techniques. In addition, there are only a few examples of studies that report subsets of markers explaining large parts of the phenotypic variance in forest trees (see above), and their variation is seldom included in breeding programs (at least to our knowledge). Nevertheless, a recent study (Jaramillo-Correa et al., 2015) has highlighted that MAS in forest trees can be a good alternative to more expensive strategies. In this work, a set of 18 SNPs associated with climate in *P. pinaster* was successfully used to predict survival in a test trial under extreme (hot and dry) conditions, which implies that variation at candidate genes can be effectively included in breeding and reforestation programmes.

The absence of markers explaining large parts of the phenotypic variance in forest trees is closely linked to the low levels of LD previously observed in these taxa, and implies that the number of markers required to pinpoint causative SNPs for a given phenotype is extremely large (Neale & Kremer, 2011). A way to circumvent this problem derives from current high-throughput genotyping technologies applied to populations of limited effective size (i.e. 'training populations'), and focuses on predicting phenotypes based on information from genome-wide markers without the prior knowledge of associations (Meuwissen, Hayes, & Goddard, 2001). Thus, with a dense enough coverage of the genome, many QTLs underlying a trait of interest can be captured in the model through their LD with genome-wide SNPs, without the need to identify individual SNP—trait associations. Simulations performed by assuming parameters similar to those found in forest trees breeding programmes (e.g. LD extent, breeding population sizes and number of QTLs) have shown that a marker density as low as two per cM can be enough to have over 50% of phenotype prediction accuracy for populations with an effective size lower than 30 individuals, even for traits with low heritability (i.e $h^2 = 0.2$) and controlled by as many as 100 QTLs (Grattapaglia & Resende, 2011). These first steps towards the application of genomic selection (GS) in forest trees enlighten a very promising path for tree breeding, as these accuracies are similar to those in programmes currently implemented in livestock (Resende, Muñoz, et al., 2012). For instance, predictions in *Pinus taeda* were highly variable, ranging from 23 to 75% for fusiform rust disease resistance, and growth (e.g. Resende, Muñoz, et al., 2012), while predictions between 50 and 70% were obtained for growth and wood traits in *Picea glauca* (Beaulieu, Doerksen, MacKay, Rainville, & Bousquet, 2014), and for wood quality in *Eucalyptus* (Resende, Resende, et al., 2012). However, given that some of these models were inconsistent across breeding populations, it is likely that, in contrast to MAS, GS might require population-specific predictive models (Beaulieu et al., 2014; Resende, Resende, et al., 2012).

9. PERSPECTIVES AND CONCLUSIONS

Population genomics is being implemented in an increasing range of forest tree species and is now helping relieve the genotyping bottleneck; large genome-wide SNP datasets are being collected that promise to reveal patterns of adaptive diversity with unprecedented detail. The challenge moving forward will be assembling the many fragmented insights provided by GWAS studies on the adaptive architecture of traits into something coherent, not unlike the assembly of many short sequence reads into a finished genome. This will require the integration of increasingly sophisticated approaches in marker-assisted and GS (Grattapaglia & Resende, 2011), including 'multi-omics' data sets (e.g. transcriptomics, proteomics, metabolomics), to generate more holistic predictions of plant phenotypes.

Another focus of the future should be on comparative population genomics of forest trees, looking at the genetic architecture of adaption in parallel across lineages inhabiting similar selection environments or at how changing environments drive genomic modifications in interacting species that develop new phenotypes (and genotypes). The need for this multispecies approach is already being felt, and frameworks such systems genomics or landscape community genomics are being proposed (Barah & Bones, 2015; Hand, Lowe, Kovach, Muhfeld, & Luikart, 2015); in forest trees these approaches could be implemented to study complex systems such as plant—insect interactions or drought adaptation, particularly for subtropical and tropical taxa, whose study is (again) lagging behind their temperate counterparts.

Connecting variability across the genome to functionally important trait variation will remain challenging, as phenotyping continues to be costly, time and space intensive, and the slow developmental pace of most trees test the patience of researchers. Relieving these bottlenecks will take both technological innovation, and the courage to leave the common garden in favour of in situ phenotyping in the field. On the technology side, high-resolution imaging and remote sensing analysis are promising sources of innovation for high-throughput phenotyping of a diverse array of ecophysiological traits, especially those associated with forest canopies (e.g. Brown et al., 2014). For example, hyperspectral and airborne sensors are, at least in principle, capable of phenotyping phenology and forest structure at the stand scale (Gleason & Im, 2012), as well as other properties of canopy reflectance associated with water stress, leaf nitrogen content, insect herbivory and other traits (Homolová, Malenovský, Clevers, García-Santos, &

Schaepman, 2013). Finer scale data such as LiDAR can provide resolution at the scale of individual trees, yielding phenotypic data on a diversity of structural and growth traits. Field-based sensors are also being developed that can simultaneously measure bud phenology and key environmental cues (e.g. heat sums) directly on individual trees growing in the field (Kleinknecht, Pers. Comm.). Together, this combination of field phenomics and genomics can spur the emergence of models for adaptive trait variation that go beyond genotype—phenotype mapping in trees growing under idealized common garden conditions. With continued innovation in sensor-based phenotyping and computational models for genomic prediction, forest tree genomics should be poised to take genotype-phenotype mapping out of the garden and into the field.

From the analytical point of view, the detection of candidate genes and their integration into predictive models will benefit from incorporating epistatic interactions between alleles. Some approaches are already being developed in this sense and showing promising results in model species data sets (e.g. Berg & Coop, 2014). Furthermore, the study of epigenetic factors implicated in adaptive responses and phenotypic plasticity must also be considered. Previous studies have already shown that epigenetic variation can account for a significant proportion of the variance in adaptation to climate in Norway spruce (Yakovlev, Fossdal, & Johnsen, 2010), while studies on the role of epigenetic factors have significantly expanded during the last decade, particularly in model species (Lisch, 2008). Lastly, forward predictive models that integrate genotypes as fitness predictors after environmental changes at different geographical scales are also being derived (e.g. Ralph & Coop, 2014), showing that local adaptation can quickly develop in species with rather limited dispersal, even when selection pressures are widespread. Such models could be thus readily applied to foretell the fate of modern natural forest tree populations and to establish mitigation measures on the onset of impeding environmental changes, once the drivers of adaptation have been correctly identified at different 'omic levels.

ACKNOWLEDGEMENTS

The authors thank I. Gamache, C. Plomion, and P. Canard for critical comments on the manuscript. This work was supported by grants from the Consejo Nacional de Ciencia y Tecnología (CB-2010-01-00153305) and the Dirección General de Asuntos del Personal Académico (IC200411 and IA201013) from the Universidad Nacional Autónoma de México to JPJ-C and the National Science Foundation (IOS-1461868 to S.R.K.).

REFERENCES

Acheré, V., Favre, J. M., Besnard, G., & Jeandroz, S. (2005). Genomic organization of molecular differentiation in Norway spruce (*Picea abies*). *Molecular Ecology, 14*, 3191–3201.

Akey, J., Jin, L., & Xion, M. (2001). Haplotypes vs single marker linkage disequilibrium tests: what do we gain? *European Journal of Human Genetics, 9*, 291–300.

Alberto, F. J., Derory, J., Boury, C., Frigerio, J.-M., Zimmermann, N. E., & Kremer, A. (2013). Imprints of natural selection along environmental gradients in phenology-related genes of *Quercus petraea*. *Genetics, 195*, 495–512.

Artlip, T. S., Callahan, A. M., Bassett, C. L., & Wisniewski, M. E. (1997). Seasonal expression of a dehydrin gene in sibling deciduous and evergreen genotypes of peach (*Prunus persica* [L.] Batsch). *Plant Molecular Biology, 33*, 61–70.

Audigeos, D., Brousseau, L., Traissac, S., Scotti-Saintagne, C., & Scotti, I. (2013). Molecular divergence in tropical tree populations occupying environmental mosaics. *Journal of Evolutionary Biology, 26*, 529–544.

Avia, K., Kärkkäinen, K., Lagercrantz, U., & Savolainen, O. (2014). Association of *FLOWERING LOCUS T/TERMINAL FLOWER1*-like gene *FTL2* expression with growth rhythm in Scots pine (*Pinus sylvestris*). *New Phytologist, 204*, 159–170.

Barah, P., & Bones, A. M. (2015). Multidimensional approaches for studying plant defence against insects: from ecology to omics and synthetic biology. *Journal of Experimental Botany, 66*, 479–493.

Bashalkhanov, S., Eckert, A. J., & Rajora, O. P. (2013). Genetic signatures of natural selection in response to air pollution in red spruce (*Picea rubens*, Pinaceae). *Molecular Ecology, 22*, 5877–5889.

Beaulieu, J., Doerksen, T., Boyle, B., Clement, S., Deslauriers, M., Beauseigle, S., et al. (2011). Association genetics of wood physical traits in the conifer white spruce and relationships with gene expression. *Genetics, 188*, 197–214.

Beaulieu, J., Doerksen, T. K., MacKay, J., Rainville, A., & Bousquet, J. (2014). Genomic selection accuracies within and between environments and small breeding groups in white spruce. *BMC Genomics, 15*, 1048.

Bedon, F., Bomal, C., Caron, S., Levasseur, C., Boyle, B., Mansfield, S. D., et al. (2010). Subgroup 4 R2R3-MYBs in conifer trees: gene family expansion and contribution to the isoprenoid- and flavonoid-oriented responses. *Journal of Experimental Botany, 61*, 3847–3864.

Benowicz, A., & El-Kassaby, Y. A. (1999). Genetic variation in mountain hemlock (*Tsuga mertensiana* Bong.): quantitative and adaptive attributes. *Forest Ecology and Management, 123*, 205–215.

Berg, J. J., & Coop, G. (2014). A population genetic signal of polygenic adaptation. *PLoS Genetics, 10*, e1004412.

Birol, I., Raymond, A., Jackman, S. D., Pleasance, S., Coope, R., Taylor, G. A., et al. (2013). Assembling the 20 Gb white spruce (*Picea glauca*) genome from whole-genome shotgun sequencing data. *Bioinformatics, 29*, 1492–1497.

Böhlenius, H., Huang, T., Charbonnel-Campaa, L., Brunner, A. M., Jansson, S., Strauss, S. H., et al. (2006). *CO/FT* regulatory module controls timing of flowering and seasonal growth cessation in trees. *Science, 312*, 1040–1043.

Bouck, A., & Vision, T. (2007). The molecular ecologist's guide to expressed sequence tags. *Molecular Ecology, 16*, 907–924.

Bradshaw, H. D., & Stettler, R. F. (1995). Molecular genetics of growth and development in *Populus*. 4. Mapping QTLs with large effects on growth, form, and phenology traits in a forest tree. *Genetics, 139*, 963–973.

Brondani, R. P. V., Williams, E. R., Brondani, C., & Grattapaglia, D. (2006). A microsatellite-based consensus linkage map for species of Eucalyptus and a novel set of 230 microsatellite markers for the genus. *BMC Plant Biology, 6*, 20.

Brown, T. B., Cheng, R., Sirault, X. R. R., Rungrat, T., Murray, K. D., Trtilek, M., et al. (2014). TraitCapture: genomic and environment modelling of plant phenomic data. *Current Opinion in Plant Biology, 18*, 73–79.

Brown, G. R., Gill, G. P., Kuntz, R. J., Langley, C. H., & Neale, D. B. (2004). Nucleotide diversity and linkage disequilibrium in loblolly pine. *Proceedings of the National Academy of Sciences of the United States of America, 101*, 15255–15260.

Burgarella, C., Navascués, M., Zabal-Aguirre, M., Berganzo, E., Riba, M., Mayol, M., et al. (2012). Recent population decline and selection shape diversity of taxol-related genes. *Molecular Ecology, 21*, 3006–3021.

Campbell, R., Pawuk, W. A., & Harris, A. S. (1989). Microgeographic genetic variation of Sitka spruce in southeastern Alaska. *Canadian Journal of Forest Research, 19*, 1004–1013.

Cappa, E. P., El-Kassaby, Y. A., Garcia, M. N., Acuña, C., Borralho, N. M. G., Grattapaglia, D., et al. (2013). Impacts of population structure and analytical models in genome-wide association studies of complex traits in forest trees: a case study in *Eucalyptus globulus*. *PloS One, 8*, e81267.

Casasoli, M., Pot, D., Plomion, C., Monteverdi, M. C., Barreneche, T., Lauteri, M., et al. (2004). Identification of QTLs affecting adaptive traits in *Castanea sativa* Mill. *Plant, Cell & Environment, 27*, 1088–1101.

Chen, J., Källman, T., Ma, X., Gyllestrand, N., Zaina, G., Morgante, M., et al. (2012). Disentangling the roles of history and local selection in shaping clinal variation of allele frequencies and gene expression in norway spruce (*Picea abies*). *Genetics, 191*, 865–881.

Chen, J., Tsuda, Y., Stocks, M., Källman, T., Xu, N., Kärkkäinen, K., et al. (2014). Clinal variation at phenology-related genes in spruce: parallel evolution in *FTL2* and *Gigantea*? *Genetics, 197*, 1025–1038.

Csilléry, K., Lalagüe, H., Vendramin, G. G., González-Martínez, S. C., Sady, B., & Oddou-Muratorio, S. (2014). Detecting short spatial scale local adaptation and epistatic selection in climate-related candidate genes in European beech (*Fagus sylvatica*) populations. *Molecular Ecology, 23*, 4696–4708.

Cullingham, C. I., Cooke, J. E., & Coltman, D. W. (2014). Cross-species outlier detection reveals different evolutionary pressures between sister species. *New Phytologist, 204*, 215–229.

De Kort, H., Vandepitte, K., Bruun, H. H., Closset-Kopp, D., Honnay, O., & Mergay, J. (2014). Landscape genomics and a common garden trial reveal adaptive differentiation to temperature across Europe in the tree species *Alnus glutinosa*. *Molecular Ecology, 19*, 4709–4721.

De Miguel, M., Cabezas, J. A., de María, N., Sánchez-Gómez, D., Guevara, M. A., Vélez, M. D., et al. (2014). Genetic control of functional traits related to photosynthesis and water use efficiency in *Pinus pinaster* Ait. drought response: integration of genome annotation, allele association and QTL detection for candidate gene identification. *BMC Genomics, 15*, 464.

De la Torre, A. R., Roberts, D. R., & Aitken, S. N. (2014). Genome-wide admixture and ecological niche modelling reveal the maintenance of species boundaries despite long history of interspecific gene flow. *Molecular Ecology, 23*, 2046–2059.

De Mita, S., Thuillet, A. C., Gay, L., Ahmadi, N., Manel, S., Ronfort, J., et al. (2013). Detecting selection along environmental gradients: analysis of eight methods and their effectiveness for outbreeding and selfing populations. *Molecular Ecology, 22*, 1383–1399.

Derba-Maceluch, M., Awano, T., Takahashi, J., Lucenius, J., Ratke, C., Kontro, I., et al. (2015). Suppression of xylan endotransglycosylase PtxtXyn10A affects cellulose microfibril angle in secondary wall in aspen wood. *New Phytologist, 205*, 666–681.

Derory, J., Scotti-Saintagne, C., Bertocchi, E., Le Dantec, L., Graignic, N., Jauffres, A., et al. (2010). Contrasting relationships between diversity of candidate genes and variation of bud burst in natural and segregating populations of European oaks. *Heredity, 104*, 438–448.

Dillon, S., McEvoy, R., Balwin, D. S., Rees, G. N., Parsons, Y., & Southerton, S. (2014). Characterisation of adaptive genetic diversity in environmentally contrasted populations of *Eucalyptus camaldulensis* Dehnh. (river red gum). *PLoS One, 8*, e103515.

Dillon, S. K., Nolan, M., Li, W., Bell, C., Wu, H.-X., & Southerton, S. G. (2009). Allelic variation in cell wall candidate genes affecting solid wood properties in natural populations and land races of *Pinus radiata*. *Genetics, 185*, 1477–1487.

Dirlewanger, E., Graziano, E., Joobeur, T., Garriga-Calderé, F., Cosson, P., Howad, W., et al. (2004). Comparative mapping and marker-assisted selection in Rosaceae fruit crops. *Proceedings of the National Academy of Sciences of the United States of America, 101*, 9891–9896.

Du, Q., Xu, B., Pan, W., Gong, C., Wang, Q., Tian, J., et al. (2013). Allelic variation in a cellulose synthase gene (*PtoCesA4*) associated with growth and wood properties in *Populus tomentosa*. *Genes, Genomes and Genetics, 3*, 2069–2084.

Eckert, A. J., Bower, A. D., Wegrzyn, J. L., Pande, B., Jermstad, K. D., Krutovsky, K. V., et al. (2009). Association genetics of coastal Douglas-fir (*Pseudotsuga menziesii* var. *menziesii*, Pinaceae). I. Cold-hardiness related traits. *Genetics, 182*, 1289–1302.

Eckert, A. J., van Heerwaarden, J., Wegrzyn, J. L., Nelson, C. D., Ross-Ibarra, J., González-Martínez, S. C., et al. (2010). Patterns of population structure and environmental associations to aridity across the range of loblolly pine (*Pinus taeda* L., Pinaceae). *Genetics, 185*, 969–982.

Eckert, A. J., Wegrzyn, J. L., Cumbie, W. P., Goldfarb, B., Huber, D. A., Tolstikov, V., et al. (2012). Association genetics of the loblolly pine (*Pinus taeda*, Pinaceae) metabolome. *New Phytologist, 193*, 890–902.

El Kayal, W., Allen, C. C. G., Ju, C. J. T., Adams, E., King-Jones, S., Zaharia, L. I., et al. (2011). Molecular events of apical bud formation in white spruce, *Picea glauca*. *Plant Cell and Environment, 34*, 480–500.

Evans, L. M., Slavov, G. T., Rodgers-Melnick, E., Martin, J., Ranjan, P., Muchero, W., et al. (2014). Population genomics of *Populus trichocarpa* identifies signatures of selection and adaptive trait associations. *Nature Genetics, 46*, 1089–1096.

Eveno, E., Collada, C., Guevara, M. A., Léger, V., Soto, A., Díaz, L., et al. (2008). Contrasting patterns of selection at *Pinus pinaster* Ait. drought stress candidate genes as revealed by genetic differentiation analyses. *Molecular Biology and Evolution, 25*, 417–437.

Falconer, D. S., & MacKay, T. F. C. (1996). *Introduction to quantitative genetics* (4th ed.). Essex, UK: Pearson Education Limited-Prentice Hall.

Flachowsky, H., Hnake, M.-V., Peil, A., Strauss, S. H., & Fladung, M. (2009). A review on transgenic approaches to accelerate breeding of woody plants. *Plant Breeding, 128*, 217–226.

Fournier-Level, A., Korte, A., Cooper, M. D., Nordborg, M., Schmitt, J., & Wilczek, A. M. (2011). A map of local adaptation in *Arabidopsis thaliana*. *Science, 334*, 86–89.

Frewen, B. E., Chen, T. H. H., Howe, G. T., Davis, J., Rohde, A., Boerjan, W., et al. (2000). Quantitative trait loci and candidate gene mapping of bud set and bud flush in *Populus*. *Genetics, 154*, 837–845.

Geraldes, A., Farzaneh, N., Grassa, C., McKown, A. D., Guy, R. D., Mansfield, S. D., et al. (2014). Landscape genomics of *Populus trichocarpa*: the role of hybridization, limited gene flow, and natural selection in shaping patterns of population structure. *Evolution, 68*, 3260–3280.

Gleason, C. J., & Im, J. (2012). Forest biomass estimation from airborne LiDAR data using machine learning approaches. *Remote Sensing of Environment, 125*, 80–91.

González-Martínez, S. C., Dillon, S., Garnier-Géré, P. H., Krutovsky, K. V., Alía, R., Burgarella, C., et al. (2011). Patterns of nucleotide diversity and association mapping.

In C. Plomion, J. Bousquet, & C. Kole (Eds.), *Genetics, genomics and breeding of conifers* (pp. 239–275). Enfield, USA: Science Publishers-CRC Press.

González-Martínez, S. C., Ersoz, E., Brown, G. R., Wheeler, N. C., & Neale, D. B. (2006). DNA sequence variation and selection of tag single nucleotide polymorphisms at candidate genes for drought-stress response in *Pinus taeda* L. *Genetics, 172*, 1915–1926.

González-Martínez, S. C., Huber, D., Ersoz, E., Davis, J. M., & Neale, D. B. (2008). Association genetics in *Pinus taeda* L. II. Water use efficiency. *Heredity, 101*, 19–26.

González-Martínez, S. C., Wheeler, N. C., Ersoz, E., Nelson, C. D., & Neale, D. B. (2007). Association genetics in *Pinus taeda* L. I. Wood property traits. *Genetics, 175*, 399–409.

Grace, L. J., Charity, J. A., Gresham, B., Kay, N., & Walter, C. (2005). Insect-resistant transgenic *Pinus radiata*. *Plant Cell Reports, 24*, 103–111.

Grattapaglia, D., & Resende, M. D. V. (2011). Genomic selection in forest tree breeding. *Tree Genetics & Genomes, 7*, 241–255.

Grivet, D., Sebastiani, F., Alía, R., Bataillon, T., Torre, S., Zabal-Aguirre, et al. (2011). Molecular footprints of local adaptation in two Mediterranean conifers. *Molecular Biology and Evolution, 28*, 101–116.

Grivet, D., Sebastiani, F., González-Martínez, S. C., & Vendramin, G. G. (2009). Patterns of polymorphism resulting from long-range colonization in the Mediterranean conifer Aleppo pine. *New Phytologist, 184*, 1016–1028.

Groover, A. T. (2005). What genes make a tree a tree? *Trends Plant Sciences, 10*, 210–214.

Guerra, F. P., Wegrzyn, J. L., Sykes, R., Davis, M. F., Stanton, B. J., & Neale, D. B. (2013). Association genetics of chemical wood properties in black poplar (*Populus nigra*). *New Phytologist, 197*, 162–176.

Guichoux, E., Garnier-Géré, P., Lagache, L., Lang, T., Boury, C., & Petit, R. J. (2013). Outlier loci highlight the direction of introgression in oaks. *Molecular Ecology, 22*, 450–462.

Günther, T., & Coop, G. (2013). Robust identification of local adaptation from allele frequencies. *Genetics, 195*, 205–220.

Gyllenstrand, N., Clapham, D., Källman, T., & Lagercrantz, U. (2007). A Norway spruce *FLOWERING LOCUS T* homolog is implicated in control of growth rhythm in conifers. *Plant Physiology, 144*, 248–257.

Hamilton, J. A., Lexer, C., & Aitken, S. N. (2013). Genomic and phenotypic architecture of a spruce hybrid zone (*Picea sitchensis* × *P. glauca*). *Molecular Ecology, 22*, 827–841.

Hamilton, M. G., Williams, D. R., Tilyard, P. A., Pinkard, E. A., Wardlaw, T. J., Glen, M., et al. (2013). A latitudinal cline in disease resistance of a host tree. *Heredity, 110*, 372–379.

Hand, B. K., Lowe, W. H., Kovach, R. P., Muhfeld, C. C., & Luikart, G. (2015). Landscape community genomics: understanding eco-evolutionary processes in complex environments. *Trends in Ecology & Evolution, 30*, 161–168.

Holliday, J. A., Ralph, S. G., White, R., Bohlmann, J., & Aitken, S. N. (2008). Global monitoring of autumn gene expression within and among phenotypically divergent populations of Sitka spruce (*Picea sitchensis*). *New Phytologist, 178*, 103–122.

Holliday, J. A., Ritland, K., & Aitken, S. (2010). Widespread, ecologically relevant genetic markers developed from association mapping of climate-related traits in Sitka spruce (*Picea sitchensis*). *New phytologist, 188*, 501–514.

Homolová, L., Malenovský, Z., Clevers, J. G. P. W., García-Santos, G., & Schaepman, M. E. (2013). Review of optical-based remote sensing for plant trait mapping. *Ecological Complexity, 15*, 1–16.

Hospital, F. (2009). Challenges for effective marker-assisted selection in plants. *Genetica, 136*, 303–310.

Houle, D., Govindaraju, D. R., & Omholt, S. (2010). Phenomics: the next challenge. *Nature Reviews Genetics, 11*, 855–866.

Howe, G. T., Aitken, S. N., Neale, D. B., Jermstad, K. D., Wheeler, N. C., & Chen, T. C. C. (2003). From genotype to phenotype: unraveling the complexities of cold adaptation in forest trees. *Canadian Journal of Botany, 81*, 1247–1266.

Ingvarsson, P. K., García, M. V., Hall, D., Luquez, V., & Jansson, S. (2006). Clinal variation in *phyB2*, a candidate gene for day-length-induced growth cessation and bud set, across a latitudinal gradient in European aspen (*Populus tremula*). *Genetics, 172*, 1845–1853.

Ingvarsson, P. K., Garcia, M. V., Luquez, V., Hall, D., & Jansson, S. (2008). Nucleotide polymorphism and phenotypic associations within and around the phytochrome B2 locus in European aspen (*Populus tremula*, Salicaceae). *Genetics, 178*, 2217–2226.

Ingvarsson, P. K., & Street, N. R. (2011). Association genetics of complex traits in plants. *New Phytologist, 189*, 909–922.

Isik, F. (2014). Genomic selection in forest tree breeding: the concept and an outlook to the future. *New Forests, 45*, 379–401.

Jaramillo-Correa, J. P., Beaulieu, J., & Bousquet, J. (2001). Contrasting evolutionary forces driving population structure at expressed sequence tag polymorphisms, allozymes and quantitative traits in white spruce. *Molecular Ecology, 10*, 2729–2740.

Jaramillo-Correa, J. P., Rodríguez-Quilón, I., Grivet, D., Lepoittevin, C., Sebastiani, F., Heuertz, M., et al. (2015). Molecular proxies for climate maladaptation in a long-lived tree (*Pinus pinaster* aiton, Pinaceae). *Genetics, 199*, 793–807.

Jennewein, S., Wildung, M. R., Chau, M., Walker, K., & Croteau, R. (2004). Random sequencing of an induced *Taxus* cell cDNA library for identification of clones involved in taxol biosynthesis. *Proceedings of the National Academy of Sciences of the United States of America, 101*, 9149–9154.

Jermstad, K. D., Bassoni, D. L., Wheeler, N. C., Anekonda, T. S., Aitken, S. N., Adams, W. T., et al. (2001). Mapping of quantitative trait loci controlling adaptive traits in coastal Douglas-fir. II. Spring and fall cold-hardiness. *Theoretical and Applied Genetics, 102*, 1152–1158.

Joost, S., Bonin, A., Bruford, M. W., Després, L., Conord, C., Erhardt, G., et al. (2007). A spatial analysis method (SAM) to detect candidate loci for selection: towards a landscape genomics approach to adaptation. *Molecular Ecology, 16*, 3955–3969.

Källman, T., De Mita, S., Larsson, H., Gyllenstrand, N., Heuertz, M., Parducci, L., et al. (2014). Patterns of nucleotide diversity at photoperiod related genes in Norway Spruce [*Picea abies* (L.) Karst.]. *PloS One, 9*, e95306.

Kang, H. M., Zaitlen, N. A., Wade, C. M., Kirby, A., Heckerman, D., Dalyk, M. J., et al. (2008). Efficient control of population structure in model organism association mapping. *Genetics, 178*, 1709–1723.

Kremer, A., & Le Corre, V. (2012). Decoupling of differentiation between traits and their underlying genes in response to divergent selection. *Heredity, 108*, 375–385.

Keller, S. R., Levsen, N., Ingvarsson, P. K., Olson, M. S., & Tiffin, P. (2011). Local selection across a latitudinal gradient shapes nucleotide diversity in balsam poplar, *Populus balsamifera* L. *Genetics, 188*, 941–952.

Keller, S. R., Levsen, N., Olson, M. S., & Tiffin, P. (2012). Local adaptation in the flowering-time gene network of balsam poplar, *Populus balsamifera*. *Molecular Biology and Evolution, 29*, 3143–3152.

Keller, G., Marchal, T., SanClemente, H., Navarro, M., Ladouce, N., Wincker, P., et al. (2009). Development and functional annotation of an 11,303-EST collection from *Eucalyptus* for studies of cold tolerance. *Tree Genetics & Genomes, 5*, 317–327.

Külheim, C., Yeoh, S. H., Maintz, J., Foley, W. J., & Moran, G. F. (2009). Comparative SNP diversity among four Eucalyptus species for genes from secondary metabolite biosynthetic pathways. *BMC Genomics, 10*, 452.

Latta, R. G. (1998). Differentiation of allelic frequencies at quantitative trait loci affecting locally adaptive traits. *The American Naturalist, 153*, 283–292.

Le Corre, V., & Kremer, A. (2003). Genetic variability at neutral markers, quantitative trait loci and trait in a subdivided population under selection. *Genetics, 164*, 2005–2019.

Leinonen, T., McCairns, R. J. S., O'Hara, R. B., & Merilä, J. (2013). $Q_{ST}-F_{ST}$ comparisons: evolutionary and ecological insights from genomic heterogeneity. *Nature Reviews Genetics, 14*, 179–190.

Leinonen, T., O'Hara, R. B., Cano, J. M., & Merilä, J. (2008). Comparative studies of quantitative trait and neutral marker divergence: a meta-analysis. *Journal of Evolutionary Biology, 21*, 1–17.

Lespinasse, D., Grivet, L., Troispoux, V., Rodier-Goud, M., Pinard, F., & Seguin, M. (2000). Identification of QTLs involved in the resistance to South American leaf blight (*Microcyclus ulei*) in the rubber tree. *Theoretical and Applied Genetics, 100*, 975–984.

Lesur, I., Le Provost, G., Bento, P., Da Silva, C., Leplé, J.-C-, Murat, F., et al. (2015). The oak gene expression atlas: insights into Fagaceae genome evolution and the discovery of genes regulated during bud dormancy release. *BMC Genomics, 16*, 112.

Levsen, N. D., Tiffin, P., & Olson, M. S. (2012). Pleistocene speciation in the genus *Populus* (salicaceae). *Systematic Biology, 61*, 401–412.

Li, P., Beaulieu, J., Corriveau, A., & Bousquet, J. (1993). Genetic variation in juvenile growth and phenology in a white spruce provenance-progeny test. *Silvae Genetica, 42*, 52–60.

Li, P., Beaulieu, J., Daoust, G., & Plourde, A. (1997). Patterns of adaptive genetic variation in eastern white pine (*Pinus strobus*) from Quebec. *Canadian Journal of Forest Research, 27*, 199–206.

Li, C., Sun, Y., Huang, H. W., & Cannon, C. H. (2014). Footprints of divergent selection in natural populations of *Castanopsis fargesii* (Fagaceae). *Heredity, 113*, 533–541.

Lisch, D. R. (2008). Epigenetic regulation of transposable elements in plants. *Annual Review of Plant Biology, 60*, 43–66.

Lowry, D. B., Hall, M. C., Salt, D. E., & Willis, J. H. (2009). Genetic and physiological basis of adaptive salt tolerance divergence between coastal and inland Mimulus guttatus. *New Phytologist, 183*, 776–788.

Luikart, G., England, P. R., Tallmon, D., Jordan, S., & Taberlet, P. (2003). The power and promise of population genomics: from genotyping to genome typing. *Nature Reviews Genetics, 4*, 981–994.

Ma, X.-F., Hall, D., Onge, K. R. S., Jansson, S., & Ingvarsson, P. K. (2010). Genetic differentiation, clinal variation and phenotypic associations with growth cessation across the *Populus tremula* photoperiodic pathway. *Genetics, 186*, 1033–1044.

Mageroy, M. H., Parent, G., Germanos, G., Giguère, I., Delvas, N., Maaroufi, H., et al. (2015). Expression of the β-glucosidase gene Pgβglu-1 underpins natural resistance of white spruce against spruce budworm. *The Plant Journal, 81*, 68–80.

Mandrou, E., Gherardi Hein, P. R., Villar, E., Vigneron, P., Plomion, C., & Gion, J.-M. (2012). A candidate gene for lignin composition in Eucalyptus: *cinnamoyl-CoA reductase (CCR)*. *Tree Genetics and Genomes, 8*, 353–364.

Marguerit, E., Bouffier, L., Chancerel, E., Costa, P., Lagane, F., Guehl, J.-M., et al. (2014). The genetics of water-use efficiency and its relation to growth in maritime pine. *Journal of Experimental Botany, 65*, 4757–4768.

McKay, J. K., & Latta, R. G. (2002). Adaptive population divergence: markers, QTL and traits. *Trends in Ecology & Evolution, 17*, 285–291.

McKown, A. D., Guy, R. D., Klápště, J., Geraldes, A., Friedman, M., Cronk, Q. C. B., et al. (2014). Geographical and environmental gradients shape phenotypic trait variation and genetic structure in *Populus trichocarpa*. *New Phytologist, 201*, 1263–1276.

McKown, A., Klápště, J., Guy, R., Geraldes, A., Porth, I., Hannemann, J., et al. (2014). Genome-wide association implicates numerous genes underlying ecological trait variation in natural populations of *Populus trichocarpa*. *New Phytologist, 203*, 535–553.

Meuwissen, T., Hayes, B., & Goddard, M. (2001). Prediction of total genetic value using genome-wide dense marker maps. *Genetics, 157*, 1819–1829.

Mimura, M., & Aitken, S. N. (2007). Adaptive gradients and isolation-by-distance with postglacial migration in *Picea sitchensis*. *Heredity, 99*, 224–232.

Morgenstern, E. K. (1969). Genetic variation in seedlings of *Picea mariana* (Mill.) BSP: II. Variation patterns. *Silvae Genetica, 18*, 161–167.

Mosca, E., Eckert, A. J., Pierro, Di, Rocchini, D., La Porta, N., Bellletti, P., et al. (2012). The geographical and environmental determinants of genetic diversity for four alpine conifers of the European Alps. *Molecular Ecology, 21*, 5530–5545.

Muranty, H., Jorge, V., Bastien, C., Lepoittevin, C., Bouffier, L., & Sánchez, L. (2014). Potential for marker-assisted selection for forest tree breeding: lessons from 20 years of MAS in crops. *Tree Genetics and Genomes, 10*, 1491–1510.

Myburg, A. A., Grattapaglia, D., Tuskan, G. A., Hellsten, U., Hayes, R. D., Grimwood, J., et al. (2014). The genome of *Eucalyptus grandis*. *Nature, 509*, 356–362.

Namroud, M.-C., Beaulieu, J., Juge, N., Laroche, J., & Bousquet, J. (2008). Scanning the genome for gene single nucleotide polymorphisms involved in adaptive population differentiation in white spruce. *Molecular Ecology, 17*, 3599–3613.

Namroud, M.-C., Guillet-Claude, C., Mackay, J., Isabel, N., & Bousquet, J. (2010). Molecular evolution of regulatory genes in spruces from different species and continents: heterogeneous patterns of linkage disequilibrium and selection but correlated recent demographic changes. *Journal of Molecular Evolution, 70*, 371–386.

Neale, D. B., & Kremer, A. (2011). Forest tree genomics: growing resources and applications. *Nature Reviews Genetics, 12*, 111–122.

Neale, D. B., & Savolainen, O. (2004). Association genetics of complex traits in conifers. *Trends in Plant Science, 9*, 325–330.

Neale, D. B., Wegrzyn, J., Stevens, K. A., Zimin, A. V., Puiu, D., Crepeau, M. W., et al. (2014). Decoding the massive genome of loblolly pine using haploid DNA and novel assembly strategies. *Genome Biology, 15*, R59.

Nei, M. (1977). F-statistics and analysis of gene diversity in subdivided populations. *Annals of Human Genetics, 41*, 225–233.

Niklas, K. J. (1997). *The evolutionary biology of plants*. Chicago: Univ. Chicago Press.

Nystedt, B., Street, N. R., Wetterbom, A., Zuccolo, A., Lin, Y. C., Scofield, D. G., et al. (2013). The Norway spruce genome sequence and conifer genome evolution. *Nature, 497*, 579–584.

Olsen, J. E., Junttila, O., Nilsen, J., Eriksson, M. E., Martinussen, I., Olsson, O., et al. (1997). Ectopic expression of oat phytochrome A in hybrid aspen changes critical daylength for growth and prevents cold acclimatization. *The Plant Journal, 12*, 1339–1350.

Olson, M. S., Levsen, N., Soolanayakanahally, R. Y., Guy, R. D., Schroeder, W. R., Keller, S. R., et al. (2013). The adaptive potential of *Populus balsamifera* L. to phenology requirements in a warmer global climate. *Molecular Ecology, 22*, 1214–1230.

Palmé, A. E., Pyhäjärvi, T., Wachowiak, W., & Savolainen, O. (2009). Selection on nuclear genes in a *Pinus* phylogeny. *Molecular Biology and Evolution, 26*, 893–905.

Paolucci, I., Gaudet, M., Jorge, V., Beritognolo, I., Terzoli, S., Kuzminsky, E., et al. (2010). Genetic linkage maps of *Populus alba* L. and comparative mapping analysis of sex determination across *Populus* species. *Tree Genetics & Genomes, 6*, 863–875.

Parchman, T. L., Gompert, Z., Mudge, J., Schilkey, F. D., Benkman, C. W., & Buerkle, C. A. (2012). Genome-wide association genetics of an adaptive trait in lodgepole pine. *Molecular Ecology, 21*, 2991–3005.

Parelle, J., Zapater, M., Scotti-Saintagne, C., Kremer, A., Jolivet, Y., Dreyer, E., et al. (2007). Quantitative trait loci of tolerance to waterlogging in a European oak (*Quercus robur* L.): physiological relevance and temporal effect patterns. *Plant, Cell & Environment, 30*, 422–434.

Pavy, N., Paule, C., Parsons, L., Crow, J. A., Morency, M-. J., Cooke, J., et al. (2005). Generation, annotation, analysis and database integration of 16 500 white spruce EST clusters. *BMC Genomics, 6*, 144.

Pelgas, B., Beauseigle, S., Achere, V., Jeandroz, S., Bousquet, J., Isabel, N., et al. (2006). Comparative genome mapping among *Picea glauca, P. mariana* × *P. rubens* and *P. abies*, and correspondence with other Pinaceae. *Theoretical and Applied Genetics, 113*, 1371–1393.

Pelgas, B., Bousquet, J., Meirmans, P. G., Ritland, K., & Isabel, N. (2011). QTL mapping in white spruce: gene maps and genomic regions underlying adaptive traits across pedigrees, years and environments. *BMC Genomics, 12*, 145.

Peña, L., & Séguin, A. (2001). Recent advances in the genetic transformation of trees. *Trends in Biotechnology, 19*, 500–506.

Petit, R. J., & Hampe, A. (2006). Some evolutionary consequences of being a tree. *Annual Review of Ecology, Evolution and Systematics, 37*, 187–214.

Porth, I., Hamberger, B., White, R., & Ritland, K. (2011). Defense mechanisms against herbivory in *Picea*: sequence evolution and expression regulation of gene family members in the phenylpropanoid pathway. *BMC Genomics, 12*, 608.

Porth, I., Klápště, J., Skyba, O., Lay, B. S. K., Geraldes, A., Muchero, W., et al. (2013). *Populus trichocarpa* cell wall chemistry and ultrastructure trait variation, genetic control and genetic correlations. *New Phytologist, 197*, 777–790.

Porth, I., Klápště, J., Skyba, O., Hannemann, J., McKown, A. D., Guy, R. D., et al. (2013). Genome-wide association mapping for wood characteristics in *Populus* identifies an array of candidate single nucleotide polymorphisms. *New Phytologist, 200*, 710–726.

Porth, I., Koch, M., Berenyi, M., Burg, A., & Burg, K. (2005). Identification of adaptation-specific differences in mRNA expression of sessile and pedunculate oak based on osmotic-stress-induced genes. *Tree Physiology, 25*, 1317–1329.

Pot, D., McMillan, L., Echt, C., Le Provost, G., Garnier-Géré, P., Cato, S., et al. (2005). Nucleotide variation in genes involved in wood formation in two pine species. *New Phytologist, 167*, 101–112.

Prunier, J., Gerardi, S., Beaulieu, J., & Bousquet, J. (2012). Parallel and lineage-specific molecular adaptation to climate in boreal black spruce. *Molecular Ecology, 21*, 4270–4286.

Prunier, J., Laroche, J., Beaulieu, J., & Bousquet, J. (2011). Scanning the genome for gene SNPs related to climate adaptation and estimating selection at the molecular level in boreal black spruce. *Molecular Ecology, 20*, 1702–1716.

Prunier, J., Pelgas, B., Gagnon, F., Desponts, M., Isabel, N., Beaulieu, J., et al. (2013). The genomic architecture and association genetics of adaptive characters using a candidate SNP approach in boreal black spruce. *BMC Genomics, 14*, 368.

Ralph, P. L., & Coop, G. (2014). The role of standing variation in geographic convergent adaptation. *bioRxiv*, 009803. http://dx.doi.org/10.1101/009803.

Rehfeldt, G. E., Ying, C. C., Spittlehouse, D. L., & Hamilton, D. A. (1999). Genetic responses to climate in *Pinus contorta*: niche breadth, climate change, and reforestation. *Ecological Monographs, 69*, 375–407.

Resende, M. F. R., Jr., Muñoz, P., Acosta, J. J., Peter, G. F., Davis, J. M., Grattapaglia, D., et al. (2012). Accelerating the domestication of trees using genomic selection: accuracy of prediction models across ages and environments. *New Phytologist, 193*, 617–624.

Resende, M. D. V., Resende, M. F. R., Jr., Sansaloni, C. P., Petroli, C. D., Missiaggia, A. A., Aguiar, A. M., et al. (2012). Genomic selection for growth and wood quality in Eucalyptus: capturing the missing heritability and accelerating breeding for complex traits in forest trees. *New Phytologist, 194*, 116–128.

Risterucci, A., Paulin, D., Ducamp, M., N'goran, J., & Lanaud, C. (2003). Identification of QTLs related to cocoa resistance to three species of *Phytophthora*. *Theoretical and Applied Genetics, 108*, 168–174.

Ritland, K., Krutovsky, K. V., Tsumura, Y., Pelgas, B., Isabel, N., & Bousquet, J. (2011). Genetic mapping in conifers. In C. Plomion, J. Bousquet, & C. Kole (Eds.), *Genetics, genomics and breeding of conifers* (pp. 196–238). Enfield, USA: Science Publishers-CRC Press.

Ruttink, T., Arend, M., Morreel, K., Storme, V., Rombauts, S., Fromm, J., et al. (2007). A molecular timetable for apical bud formation and dormancy induction in poplar. *The Plant Cell Online, 19*, 2370–2390.

Savolainen, O., Lascoux, M., & Merilä, J. (2013). Ecological genomics of local adaptation. *Nature Reviews Genetics, 14*, 807–820.

Savolainen, O., Pyhäjärvi, T., & Knurr, T. (2007). Gene flow and local adaptation in trees. *Annual Review of Ecology Evolution and Systematics, 38*, 595–619.

Scalfi, M., Troggio, M., Piovani, P., et al. (2004). A RAPD, AFLP and SSR linkage map, and QTL analysis in European beech (*Fagus sylvatica* L.). *Theoretical and Applied Genetics, 108*, 433–441.

Scotti-Saintagne, C., Mariette, S., Porth, I., Goicoechea, P. G., Barreneche, T., Bodénès, C., et al. (2004). Genome scanning for interspecific differentiation between two closely related oak species [*Quercus robur* L. and *Q. petraea* (Matt.) Liebl.]. *Genetics, 168*, 1615–1626.

Séguin, A., Lachance, D., Désjardins, A., Leplé, J.-C., & Pilate, G. (2014). Scientific research related to genetically modified trees. In T. Fenning (Ed.), *Challenges and opportunities for the World's forests in the 21st century*. Dordrecht: Springer.

Soto-Cerda, B., & Cloutier, S. (2010). Association mapping in plant genomes. In M. Çalişkan (Ed.), *Genetic diversity in plants* (pp. 29–54). Rijeka: InTech.

Srinivasan, C., Dardick, C., Callahan, A., & Scorza, R. (2012). Plum (*Prunus domestica*) trees transformed with poplar *FT1* result in altered architecture, dormancy requirement, and continuous flowering. *PLoS One, 7*, e40715.

Stölting, K. N., Nipper, R., Lindtke, D., Caseys, C., Waeber, S., Castiglioni, S., et al. (2013). Genomic scan for single nucleotide polymorphisms reveals patterns of divergence and gene flow between ecologically divergent species. *Molecular Ecology, 22*, 842–855.

Thumma, B. R., Nolan, M. F., Evans, R., & Moran, G. F. (2005). Polymorphisms in *Cinnamoyl CoA Reductase (CCR)* are associated with variation in microfibril angle in *Eucalyptus* spp. *Genetics, 171*, 1257–1265.

Tiffin, P., & Ross-Ibarra, J. (2014). Advances and limits of using population genetics to understand local adaptation. *Trends Ecology and Evolution, 29*, 673–680.

Tsumura, Y., Kado, T., Takahashi, T., Tani, N., Ujino-Ihara, T., & Iwata, I. (2007). Genome scan to detect genetic structure and adaptive genes of natural populations of *Cryptomeria japonica*. *Genetics, 176*, 2393–2403.

Tsumura, Y., Uchiyama, K., Moriguchi, Y., Ueno, S., & Ihara-Ujino, T. (2012). Genome scanning for detecting adaptive genes along environmental gradients in the Japanese conifer, *Cryptomeria japonica*. *Heredity, 109*, 349–360.

Tuskan, G. A., DiFazio, S., Jansson, S., Bohlman, J., Grigoriev, I., Hellsten, U., et al. (2006). The genome of black cottonwood, *Populus trichocharpa* (Torr. & Gray). *Science, 313*, 1596–1604.

Uchiyama, K., Iwata, H., Moriguchi, Y., Ujino-Ihara, T., Ueno, S., Tagushi, Y., et al. (2013). Demonstration of genome-wide association studies for identifying markers for wood property and male strobili traits in *Cryptomeria japonica*. *PloS One, 8*, e79866.

Wegrzyn, J. L., Eckert, A. J., Choi, M., Lee, J. M., Stanton, B. J., Sykes, R., et al. (2010). Association genetics of traits controlling lignin and cellulose biosynthesis in black cottonwood (*Populus trichocarpa*, Salicaceae) secondary xylem. *New Phytologist, 188*, 515–532.

Whiteman, A. (2014). Global trends and outlook for forest resources. In T. Fenning (Ed.), *Challenges and opportunities for the World's forests in the 21st century*. Dordrecht: Springer.
Wright, S. (1951). The genetical structure of populations. *Annals of Eugenics, 15*, 323–354.
Wright, S. I., & Gaut, B. S. (2005). Molecular population genetics and the search for adaptive evolution in plants. *Molecular Biology and Evolution, 22*, 506–519.
Xu, S. (2003). Theoretical basis of the Beavis effect. *Genetics, 165*, 2259–2268.
Yakovlev, I. A., Fossdal, C. G., & Johnsen, Ø. (2010). MicroRNAs, the epigenetic memory and climate adaptation in Norway spruce. *New Phytologist, 187*, 1154–1169.
Yu, J., Pressoir, G., Briggs, W. H., Bi, I. V., Yamasaki, M., Doebley, J. F., et al. (2006). A unified mixed-model method for association mapping that accounts for multiple levels of relatedness. *Nature Genetics, 38*, 203–208.
Zhu, M., & Zhao, S. (2007). Candidate gene identification approach: progress and challenges. *International Journal of Biological Sciences, 3*, 420–427.

SUBJECT INDEX

Note: Page numbers followed by "f" and "t" indicate figures and tables respectively.

M

N

O

P

Q

R

S

T

U

V

W

X

AUTHOR INDEX

Note: Page numbers with "f" denote figures.

B

C

D

E

H

K

L

O

P

S

T

U

V

W

X

Y

Z

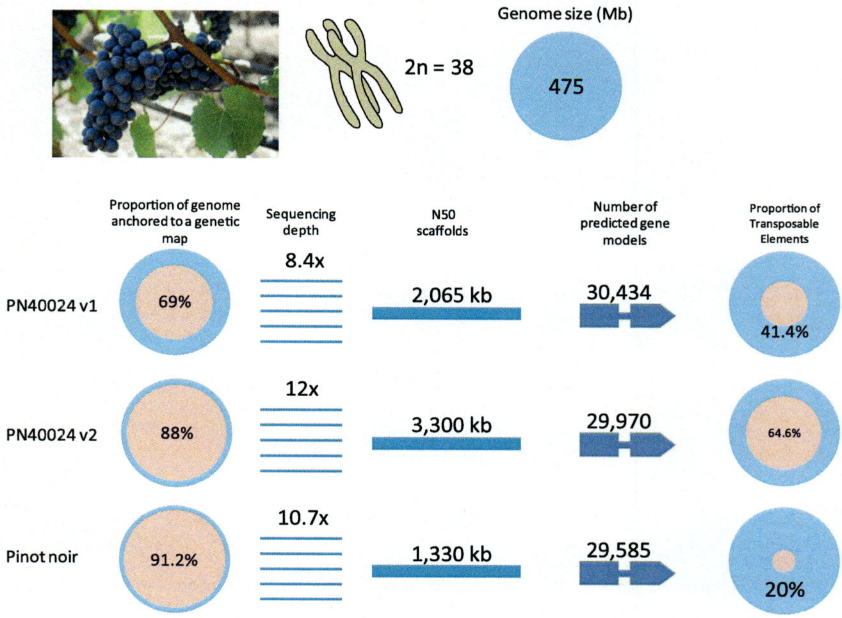

David Chagné, Figure 1 The *Vitis vinifera* genome at a glance.

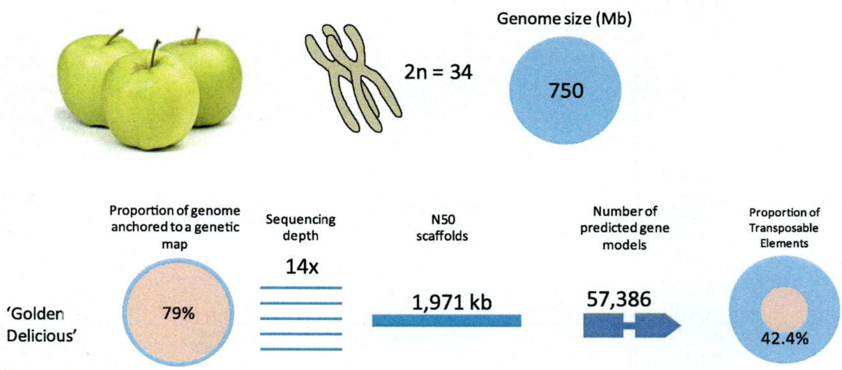

David Chagné, Figure 2 The *Malus* × *domestica* 'Golden Delicious' genome at a glance.

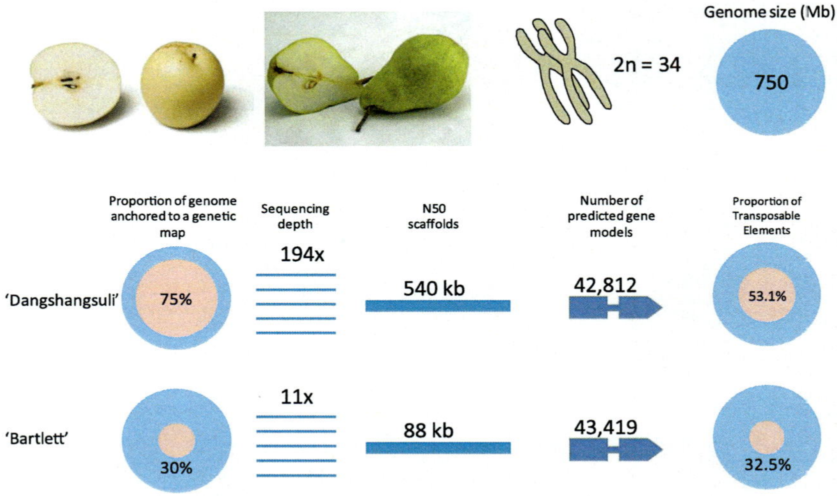

David Chagné, Figure 3 The Asian and European *Pyrus* genomes at a glance.

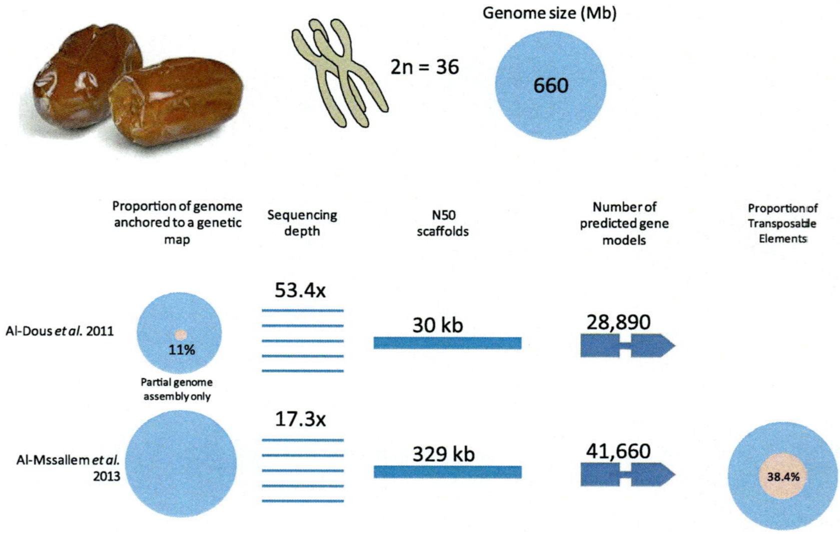

David Chagné, Figure 4 The *Phoenix dactylifera* 'Khalas' genome at a glance.

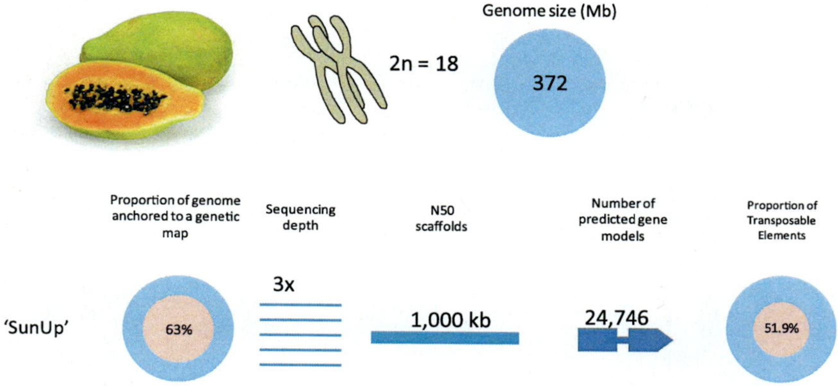

David Chagné, Figure 5 The *Carica papaya* 'SunUp' genome at a glance.

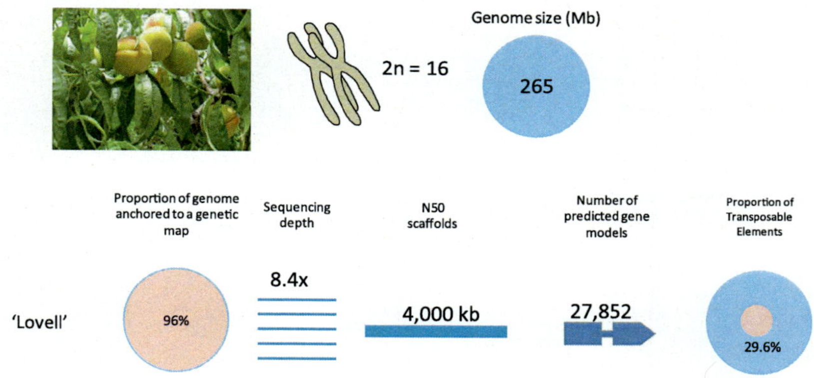

David Chagné, Figure 6 The *Prunus persica* 'Lovell' genome at a glance.

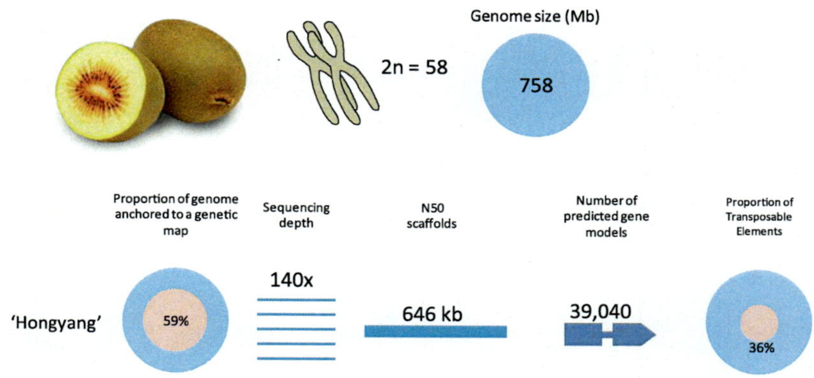

David Chagné, Figure 7 The *Actinidia chinensis* 'Hongyang' genome at a glance.

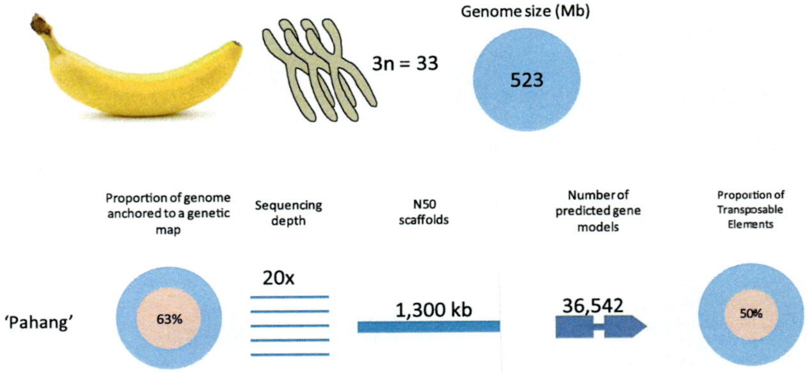

David Chagné, Figure 8 The *Musa acuminata* 'Pahang' genome at a glance.

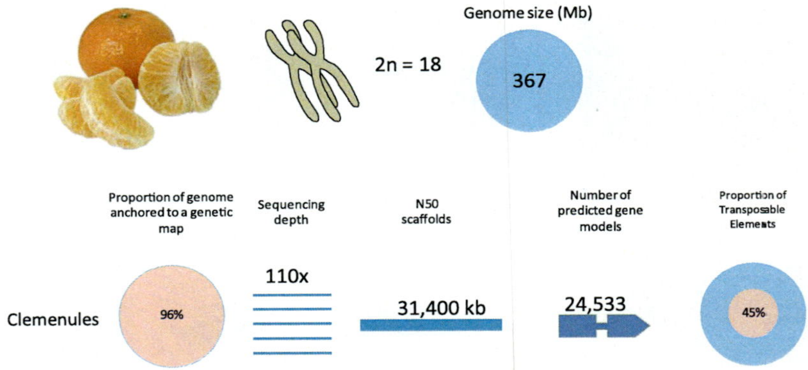

David Chagné, Figure 9 The *Citrus* × *clementina* 'Clemenules' genome at a glance.

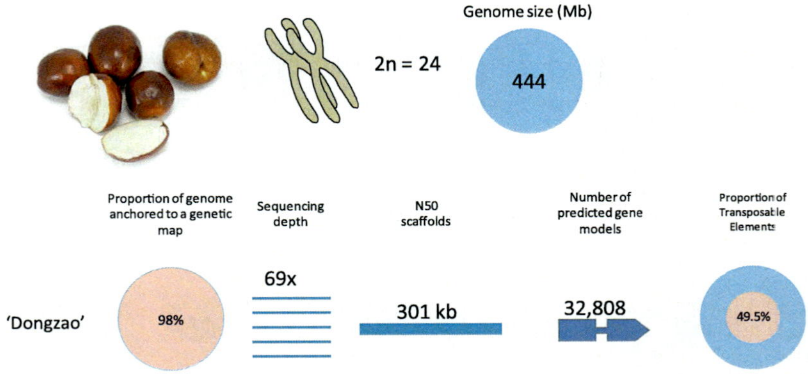

David Chagné, Figure 10 The *Ziziphus jujuba* 'Dongzao' genome at a glance.

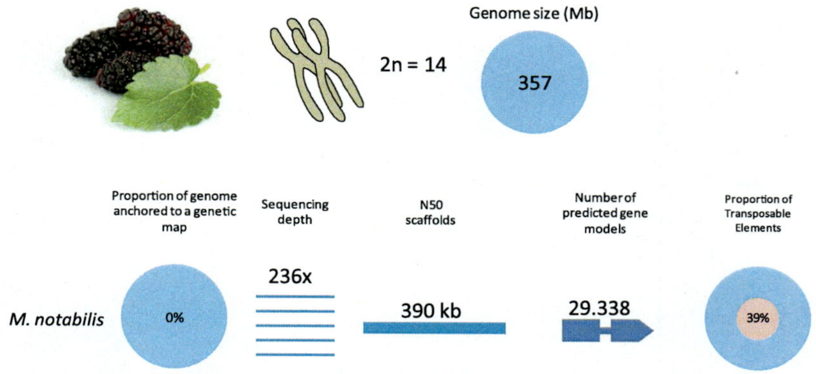

David Chagné, Figure 11 The *Morus notabilis* genome at a glance.

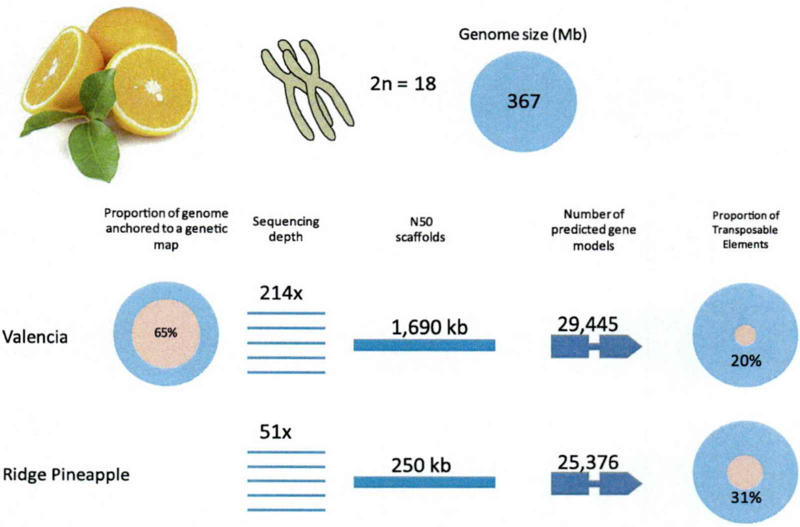

David Chagné, Figure 12 The *Citrus sinensis* Valencia and Ridge Pineapple genomes at a glance.

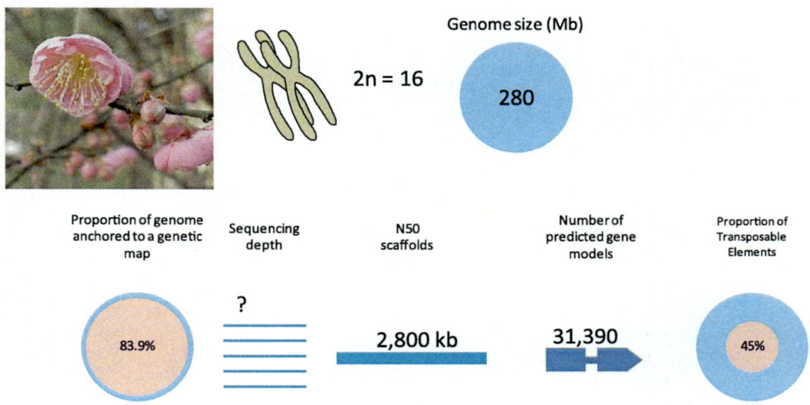

David Chagné, Figure 13 The *Prunus mume* genome at a glance.

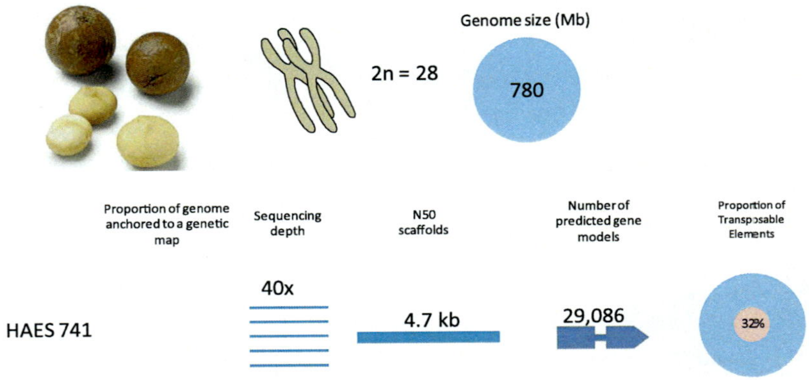

David Chagné, Figure 14 The *Macadamia integrifolia* genome at a glance.

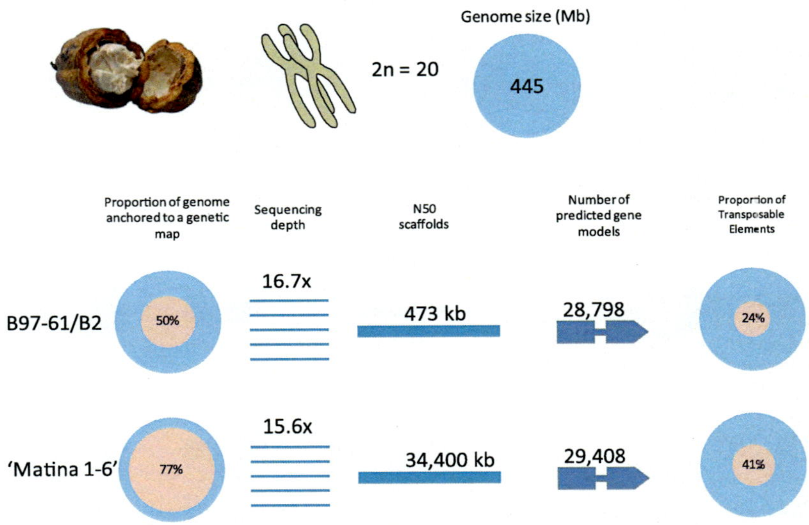

David Chagné, Figure 15 The *Theobroma cacao* genomes at a glance

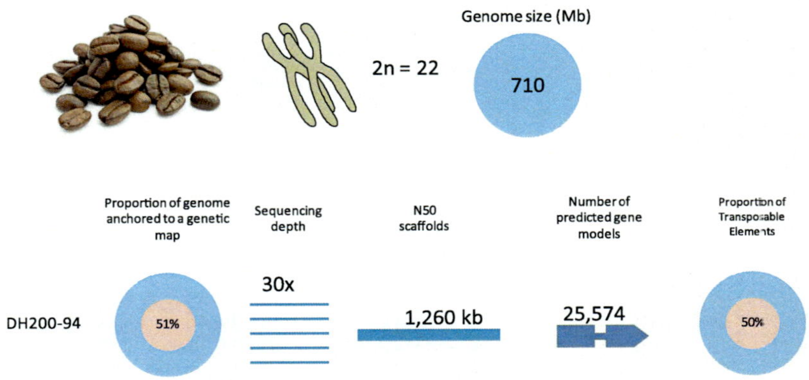

David Chagné, Figure 16 The *Coffea canephora* genome at a glance.

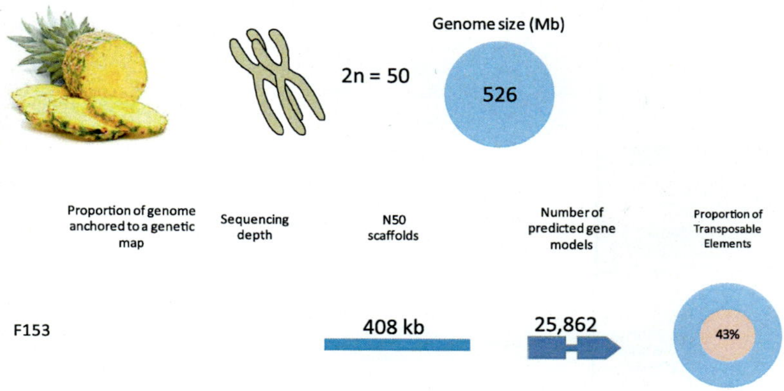

David Chagné, Figure 17 The *Ananas comosus* F153 genome at a glance.

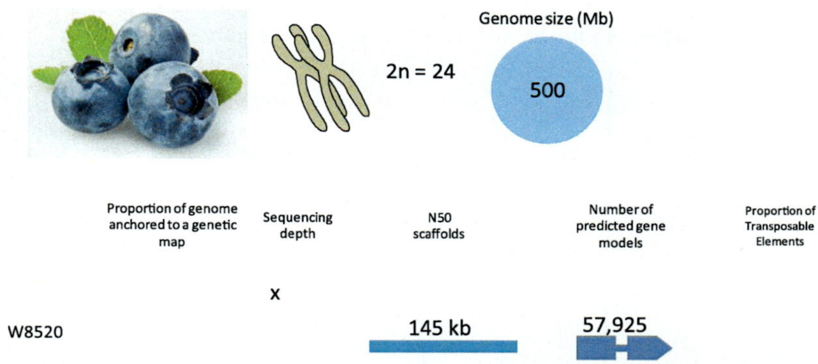

David Chagné, Figure 18 The *Vaccinium corymbosum* W8520 genome at a glance.

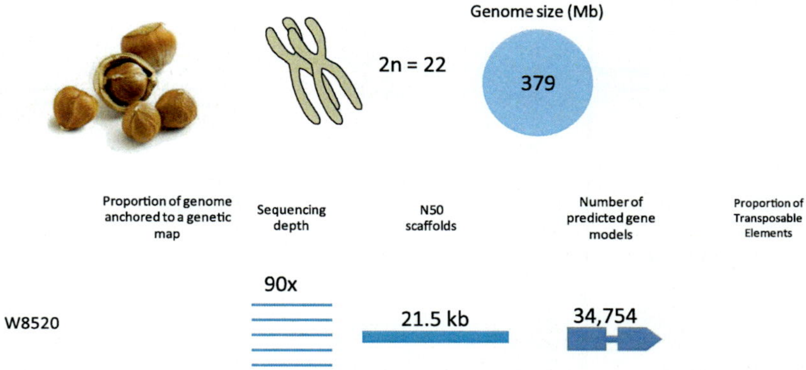

David Chagné, Figure 19 The *Corylus avellana* 'Jefferson' genome at a glance.

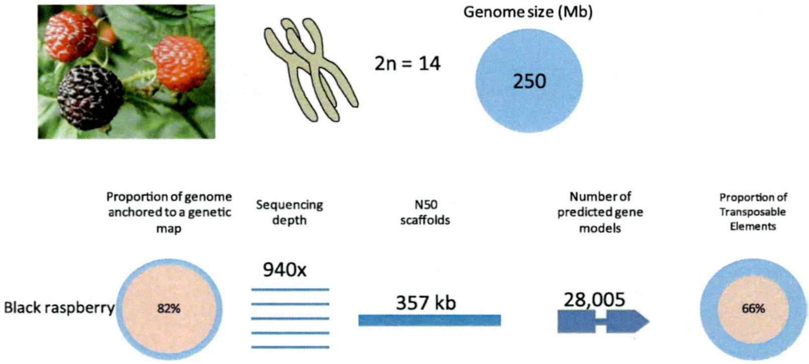

David Chagné, Figure 20 The *Rubus occidentalis* genomes at a glance.

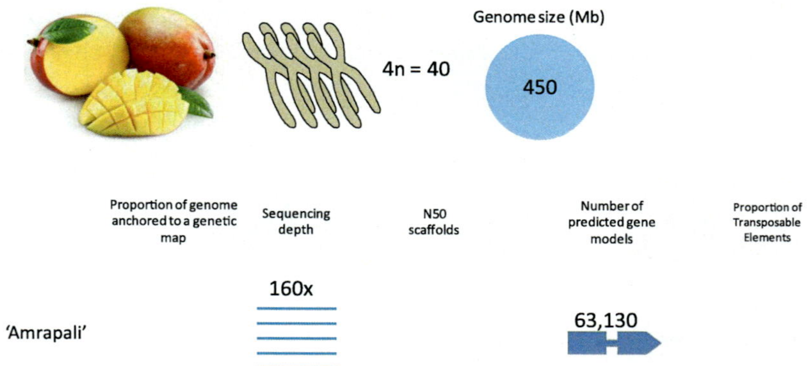

David Chagné, Figure 21 The *Mangifera indica* genome at a glance.

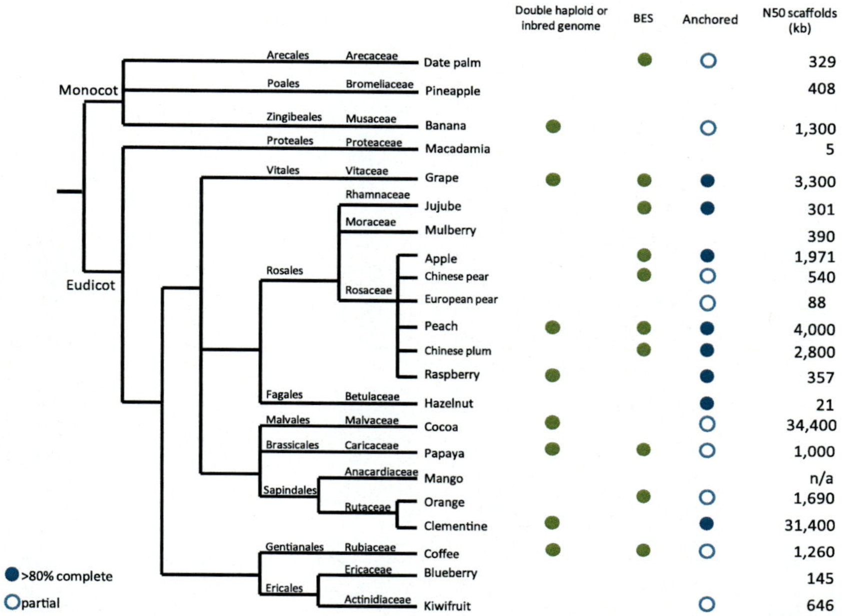

David Chagné, Figure 22 Quality metrics of fruit tree genomes sequenced to date. The species are presented using a simplified phylogeny. The use of double haploid or highly inbred accessions for sequencing, as well as whether BAC-end sequencing (BES) and genetic maps (anchored) were used, or not. The N50 for the scaffolds is indicated in kilobases.

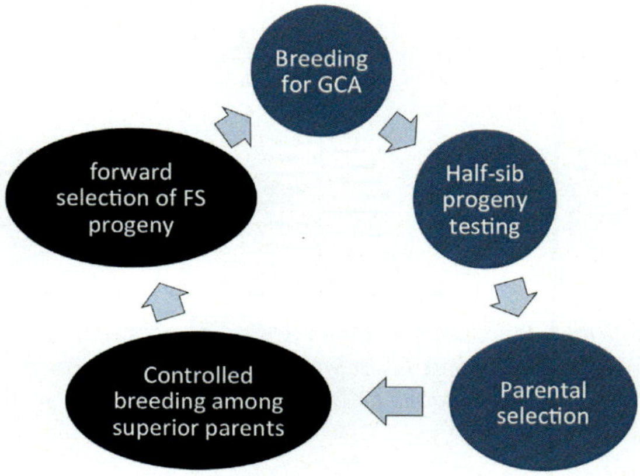

Fikret Isik *et al.*, Figure 1 A schematic representation of recurrent selection based on complementary mating design used for some conifers.

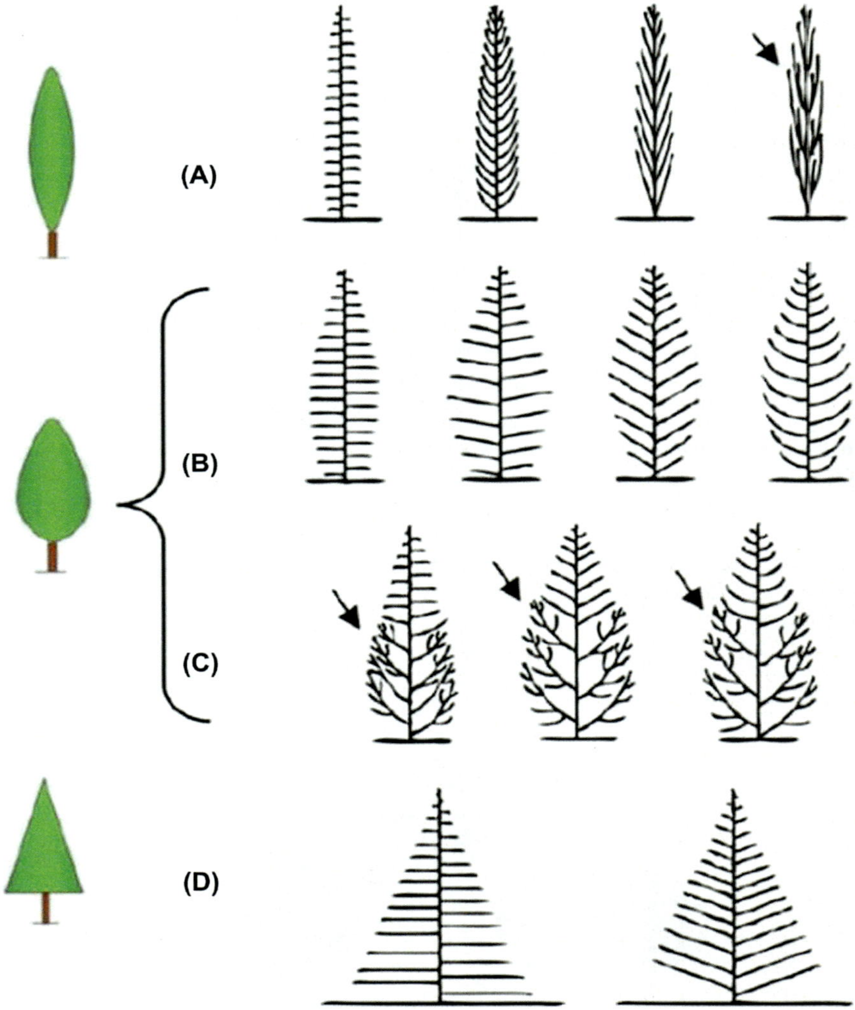

Evelyne Costes and Jean-Marc Gion, Figure 1 Different architectural types of *Cupressus sempervirens*: (A) fastigiate, (B and C) Intermediate and (D) horizontal crown shape groups. *From Barthélémy and Caraglio (2007).*

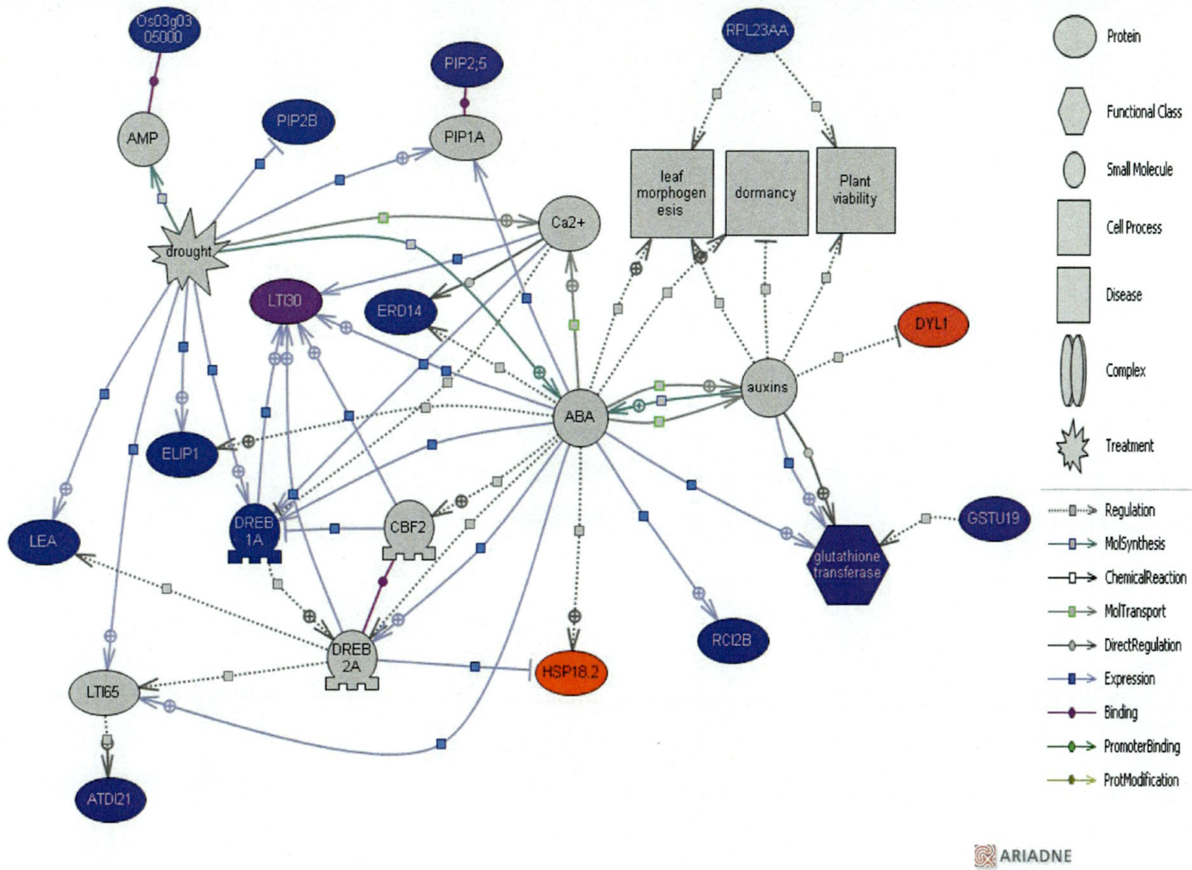

Evelyne Costes and Jean-Marc Gion, Figure 2 Functional network predicted from the genes upregulated during ecodormancy. *From Ueno et al. (2013).*

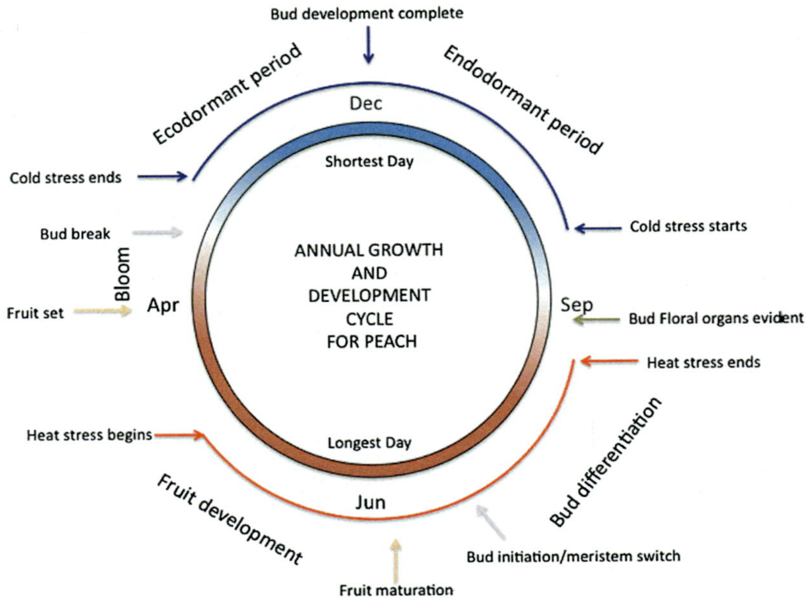

Albert G. Abbott *et al.*, Figure 1 A peach-based conceptual framework that can serve to guide investigations and interpretations of the genetic control of bud-break in fruit and forest tree species.

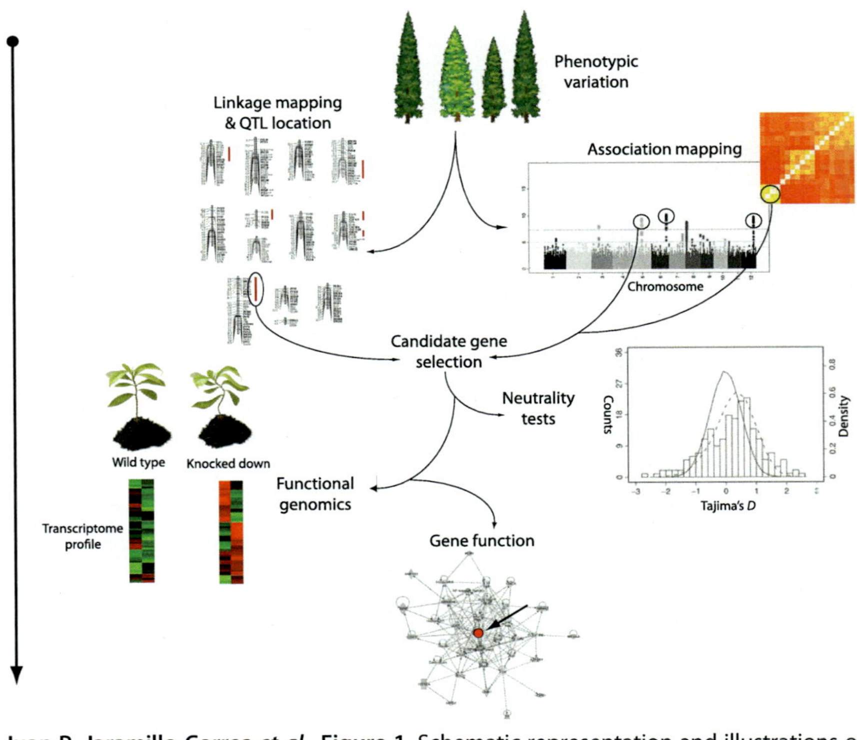

Juan P. Jaramillo-Correa *et al.*, Figure 1 Schematic representation and illustrations of some the forward-genetics approaches used to detect candidate genes involved in adaptation in forest trees that are discussed throughout this chapter.